PETROCHEMICALS

DR. ANJANA SRIVASTAVA

PETROCHEMICALS

By

Dr. Anjana Srivastava
M.Sc. Ph.D (Applied Chemistry)
Amity Institute of Applied Sciences
Amity University, NOIDA

S.K. KATARIA & SONS®
Educational Publishers
4885/109, Prakash Mahal, Dr. Subhash Bhargav Lane,
Opposite Delhi Medical Association, Daryaganj, New Delhi-110002 (INDIA)
Phone : +91-11-43551243, Mobile: +91-9871775858, 7678425209
e-mail: katariabook@gmail.com, katariabooks@yahoo.com
Website: www.skkatariaandsons.com
(e-books, subscriptions and bulk buying also available on our website)

Published by:

S.K. Kataria & Sons®

4885/109, Prakash Mahal, Dr. Subhash Bhargav Lane,
Opposite Delhi Medical Association, Daryaganj, New Delhi-110002 (INDIA)
Phone: +91-11-43551243, Mobile: +91-9871775858, 7678425209
e-mail: katariabook@gmail.com, katariabooks@yahoo.com
Website: www.skkatariaandsons.com

First Edition: 2010-2011, Reprint: 2026

Price: ₹ 395/-

Laser Typesetting by :
AISA ENTERPRISES, Delhi-110092

Printed at:
L.B. Enterprises

PREFACE

The encouragement to write the book on "Petrochemicals" came to me after persuing teaching courses on Petrochemicals at Post Graduate level at Amity University. This book is aimed to cater the need for the students who are persuing their career in Industrial chemistry. The book is aimed to provide insight to the learning student doing study about petrochemicals. I have tried my best to collect information on petrochemicals and compiled it for students, faculty and research scholar in this area.

A petrochemical is any chemical derived from fossil fuels. These include purified fossil fuels such as methane, propane, butane, gasoline, kerosene, diesel fuel, fuel oil, lubricants and asphalt.

The present book has been written in eight chapters. The first chapter accounts for the origin of petroleum which includes biotic hydrocarbons and abiogenic origin of petroleum. The global petroleum reserves with special mention of petroleum reserves in India is discussed in chapter two. Chapter three deals with the composition of crude oil. The basic constituents of crude oil has been described. Characteristics of petroleum products are summarized in chapter four. Bureau of Indian standards, New Delhi standardizes procedures and issue specifications for each petroleum products. Definition, method and significance of various test has been given. The refining of crude oil which include vacuum and atmospheric distillation, removal of impurities and de-emulsification process are discussed in chapter five.

Cracking process which include thermal and catalytic are described in chapter six. Thermal cracking include visbreaking and coking whereas catalytic process like fluid catalytic, hydrocracking catalytic alkylation, isomerization are described.

Seventh chapter covers the manufacture, properties and uses of petroleum products. These include methane, naptha, gasoline, ATP, diesel fuel, fuel oil lubricants, petroleum waxes, asphalt and petroleum coke.

Chapter eight highlights the chemicals obtained from petroleum. These include primary petroleum chemicals intermediates and their major end use products. Chemicals like acetone, acetylene, ethylene, propylene, PVC are discussed in detail.

It is hopeful that this book will be useful for students of M.Sc (Industrial chemistry) as well as students of chemical engineering.

Inspite of my best efforts, some error might have occurred in the book. Any suggestions and constructive criticism towards the contents of this book are most welcome.

AUTHOR

ACKNOWLEDGEMENT

It my great privilege and honour to present the book on **"Petrochemicals"**.

First and foremost I thank God Almighty for giving this opportunity and His grace to bring out the book in the present form.

I express my heartfelt gratitude to **Dr. Ashok K.Chauhan,** Founder President, RBEF who is a constant source of inspiration to one and all so that each individual is a better citizen tomorrow. I am also thankful to **Dr. Atul Chauhan,** Chancellor, Amity University for his support and encouragement. My special thanks to **Dr. Balvinder Shukla**, Director General Amity School of Engineering and Technology and Pro-VC Amity University, whose leadership and guidance has helped at all times.

I warmly thank **Prof. A.L.Verma,** Head, Amity Institute of Applied Science, as his kind support have been of great value. I would like to express my deep and sincere gratitude to **Dr. Sunita Rattan,** Head, Chemistry Department, whose wide knowledge and her logical way of thinking has been of great value for me. Her understanding, encouragement and personal guidance have provided a good basis to complete the book.

It is difficult to overstate my gratitude to my Ph.D. supervisor, **Dr. Ram Prasad.** His enthusiasm, inspiration and great efforts to explain things clearly and simply have motivated me to write this book. Throughout my writing period, he provided encouragement, sound advice, good teaching, good company, and lots of good ideas which have helped to make my desire come true.

I express my sincere thanks to all my colleagues for their support and co-operation during the writing and compiling of the book.

I wish to thank the publisher for the sustained interest shown by him during the entire work. A sincere expression of thanks to the entire staff of **S.K. Kataria and Sons**. I would be failing in my duties if I do not acknowledge them.

Lastly, and most importantly, I wish to thank my family – my children **Tejaswani and Nikita**, who were a constant source of inspiration and encouragement for me, my husband – **Mr. M.K. Srivastava** ,who gave me untiring help during my difficult moments and made my dream a reality and my parents as they bore me, raised me, supported me, taught me, and loved me. I dedicate this book to my family.

AUTHOR

Dedicated to

My Family

COURSE OBJECTIVES

The Indian Petrochemical Chemical Industry occupies a unique position in the Indian economy in terms of its contribution to employment and export potential. The petrochemical industry in India has been one of the fastest growing industries in the country. Since the beginning, the Indian petrochemical industry has shown an enviable rate of growth. This industry also has immense importance in the growth of economy of the country and the growth and development of manufacturing industry as well. It provides the foundation for manufacturing industries like construction, packaging, pharmaceuticals, agriculture, textiles etc

The objective of this course is to impart knowledge in the field of petrochemicals. The course will help to produce qualified professionals who can serve the Industrial sector technically and carry out quality R&D work and thus contribute to the development of the nation.

CONTENTS

CHAPTER

1

THE ORIGIN OF PETROLEUM

1.1. INTRODUCTION

INSIDE THIS CHAPTER

Petroleum is one of the most widely distributed substances in nature, but the question of its origin has never yet been satisfactorily determined, and is still a curosity for philosophers. The theories of its origin suggested by Reichenbach, Berthelot, Mendeleeff, Peckham, and others, made no attempt to account for the exceeding variety in its chemical composition, in its specific gravity, its boiling points, etc. and are all founded on some hypothetical process which differs from any with which we are acquainted; but modern geologists have agreed that, as a rule, the records of the earth's history should be read in accordance with those laws of nature which continue in force at the present day, e.g., the decomposition of fish and cetaceous animals could not now produce oil containing paraffin. Hence we can hardly believe it was possible thousands or millions of years ago, if it can be proved that any of the processes of nature with which we are familiar is calculated to produce it.

The chief characteristics of petroleum strata are enumerated as :

- The existence of adjoining beds of limestone, gypsum, etc.
- The evidence of volcanic action in close proximity to them.
- The presence of salt water in the wells.

It has been noticed that the presence of limestone close to petroleum fields in the United States and Canada, in the Caucasus, in Burma, etc., have been considered *"fossiliferous,"* or shell limestone, and have drawn the erroneous inference that the animal matter once contained in those shells originated petroleum; but no fish oil ever contained paraffin. On the other hand, the fossil shells are carbonate of lime and, as such, capable of producing petroleum under conditions such as many limestone beds have been subjected to in all ages of the earth's history. All limestone rocks were formed under water, and are mainly composed of calcareous shells, corals, encrinites, and foraminfera - the latter similar to the foraminfera of "*Atlantic ooze*" and of English chalk beds. Everywhere, under the microscope, the original connection of limestone with organic matter - its organic parentage, so to speak, and cousinship with the animal and vegetable kingdoms - is conspicuous. When pure it contains 12 per cent of carbon.

Now petroleum consists largely of carbon, its average composition being 85 per cent of carbon and 15 per cent of hydrogen, and in the limestone rocks of the United Kingdom alone there is a far larger accumulation of carbon than in all the coal measures the world contains. A range of limestone rock 100 miles in length by 10 miles in width, and 1,000 yards in depth, would contain 743,000 million tons of carbon, or sufficient to provide carbon for 875,000 million tons of petroleum. Deposits of oil-bearing shale have also limestone close at hand; *e.g.*, coral rag underlies Kimmeridge clay, as it also underlies the famous black shale in Kentucky, which is extraordinarily rich in oil.

As evidence of volcanic action in close proximity to petroleum strata, the mud volcanoes at Baku and in Burma are described, and a sulphur mine in Spain is mentioned (with which the writer is well acquainted), situated near an extinct volcano, where a perpetual gas flame in a neighbouring chapel and other symptoms indicate that petroleum is not far off. While engaged in studying the geological conditions of this mine, it was observed that **Dr. Christoff Bischoff** records in his writings that he had produced sulphur in his own laboratory by passing hot volcanic gases through chalk, which, when expressed in a chemical formula, leads at once to the postulate that, in addition to sulphur, ethylene, and all its homologues (CH), which are the oils predominating at Baku, would be produced by treating equivalents of carbonate of lime with equivalent of sulphurous acid and sulphurated hydrogen and that marsh gas and its homologues, would be produced by also by treating equivalents of carbonate of lime with equivalent of sulphurous acid. Tables are given showing the formulae for the homologues of ethylene and marsh gas resulting from the increase in regular gradation of the same constituents.

TABLE 1.1. Formulae Showing how Ethylene and its Homologues (CH) are Produced by the Action of the Volcanic Gases HS and HO on Limestone.

Carbonate of lime	+	Sulphurated hydrogen	+	Peroxide of hydrogen		Gypsum	+	Ethylene and its homologues
2CaCO	+	2HS	+	2HO	⟶	2(CaSO.HO)	+	CH
3CaCO	+	3HS	+	3HO	⟶	3(CaSO.HO)	+	CH
4CaCO	+	4HS	+	4HO	⟶	4(CaSO.HO)	+	CH
5CaCO	+	5HS	+	5HO	⟶	5(CaSO.HO)	+	CH
6CaCO	+	6HS	+	6HO	⟶	6(CaSO.HO)	+	CH
7CaCO	+	7HS	+	7HO	⟶	7(CaSO.HO)	+	CH
8CaCO	+	8HS	+	8HO	⟶	8(CaSO.HO)	+	CH
9CaCO	+	9HS	+	9HO	⟶	9(CaSO.HO)	+	CH
10CaCO	+	10HS	+	10HO	⟶	10(CaSO.HO)	+	CH
11CaCO	+	11HS	+	11HO	⟶	11(CaSO.HO)	+	CH
12CaCO	+	12HS	+	12HO	⟶	12(CaSO.HO)	+	CH
13CaCO	+	13HS	+	13HO	⟶	13(CaSO.HO)	+	CH
14CaCO	+	14HS	+	14HO	⟶	14(CaSO.HO)	+	CH
15CaCO	+	15HS	+	15HO	⟶	15(CaSO.HO)	+	CH

It is explained that these effects must have occurred, not at periods of acute volcanic eruptions, but in conditions which maybe, and have been, observed at the present time, wherever there are active solfataras or mud volcanoes at work. Descriptions of the action of solfataras by the late Sir Richard Burton and by a British consul in Iceland are quoted, and also a paragraph from Lyall's *"Principles of Geology,"* in which he remarks of the mud volcanoes at Girgenti (Sicily) that carbureted hydrogen is discharged from them, sometimes with great violence, and that they are known to have been casting out water, mixed with mud and bitumen, with the same activity as now for the last fifteen centuries. Probably at all these solfataras, if the gases traverse limestone, fresh deposits of oil-bearing strata are accumulating, and the same volcanic action has been occurring during many successive geological periods and millions of years; so that it is difficult to conceive limits to the magnitude of the stores of petroleum which may be awaiting discovery in the subterranean depths.

Gypsum may also be an indication of oil-bearing strata, for the substitution in limestone of sulphuric for carbonic acid can only be accounted for by the action of these hot sulphurous gases. Gypsum is found extensively in the petroleum districts of the United States, and it underlies the rock salt beds at Middlesboro, where, on being pierced, it has given passage to oil gas, which issues abundantly, mixed with brine, from a great depth.

Besides the space occupied by *"natural gas,"* which is very extensive, 17,000 million gallons of petroleum have been raised in America since 1860, and that quantity must have occupied more than 100,000,000 cubic yards, a space equal to a subterranean cavern 100 yards wide by 20 feet deep, and 82 miles in length, and it is suggested that beds of *"porous sandstone"* could hardly have contained so much; while vast receptacles may exist, carved by volcanic water out of former beds of rock salt adjoining the limestone, which would account for the brine that usually accompanies petroleum. It is further suggested that when no such vacant spaces were available, the hydrocarbon vapours would be absorbed into, and condensed in, contiguous clays and shales, and perhaps also in beds of coal, only partially consolidated at the time.

There is an extensive bituminous limestone formation in Persia, containing 20 per cent of bitumen, and the theory elaborated would account for bitumen and oil having been found in Canada and Tennessee embedded in limestone, which fact is cited by **Mr. Peckham** as favouring his belief that some petroleums are a *"product of the decomposition of animal remains."*

Above all, this theory accounts for the many varieties in the chemical composition of paraffin oils in accordance with ordinary operations of nature during successive geological periods.

1.1.1. Abiogenic Petroleum Origin

Abiogenic Petroleum Origin is an alternative hypothesis to the prevailing biological origin theory of petroleum origins. Most popular in Russia and Ukraine between the 1950s and 1980s, the abiogenic hypothesis has little support among contemporary petroleum geologists, who argue that abiogenic petroleum does not exist in significant amounts and that there is no indication that an application of the hypothesis is or has ever been of commercial value.

The abiogenic hypothesis argues that petroleum was formed from deep carbon deposits, perhaps dating to the formation of the Earth. The presence of methane on Saturn's moon Titan is cited as evidence supporting the formation of hydrocarbons without biology. Supporters of the abiogenic hypothesis suggest that a great deal more petroleum exists on Earth than commonly thought, and that petroleum may originate from carbon bearing fluids that migrate upward from the mantle.

Although the abiogenic hypothesis was accepted by some geologists in the former Soviet Union, most geologists now consider the biogenic formation of petroleum scientifically supported. Although evidence exists for abiogenic formation of methane and hydrocarbon gases within the Earth, studies indicate they are not produced in commercially significant quantities (*i.e.* a median abiogenic hydrocarbon content in extracted hydrocarbon gases of 0.02%).The abiogenic origin of petroleum has also recently been reviewed in detail by Glasby, who raises a number of objections, including that there is no direct evidence to date of abiogenic *petroleum* (liquid crude oil and long-chain hydrocarbon compounds) Although the biogenic theory for petroleum was first proposed by Georg Agricola in the 16th century, various abiogenic hypotheses were proposed in the nineteenth century, most notably by Alexander von Humboldt, the Russian chemist Dmitri Mendeleev and the French chemist Marcellin Berthelot. Since that time, the abiogenic hypotheses have lost ground to the view that petroleum is a fossil fuel.

Abiogenic hypotheses were revived in the last half of the twentieth century by Russian and Ukrainian Scientists, and more interest was generated by the publication of *The Deep Hot Biosphere* by Thomas Gold. Gold cited the discovery of thermophile bacteria in the Earth's crust as new support for the postulate that these bacteria could explain the existence of certain biomarkers in extracted petroleum.

1.1.1.1. History of Abiogenic Hypothesis

The abiologic theory for the origin of petroleum is usually traced to the early part of the 19th century. The hypothesis was developed well before the field of organic chemistry, much less that of biochemistry, was established so the chemical nature of the petroleum was not known. Absent intellectual framework of organic and biological chemistry, abiologic theories were inevitable. In the early 1800's, Phlogiston theory was the dominant model for explaining chemical phenomena. Furthermore, the formal study of paleontology had only started in the early 1800's. It is within this scientifically primitive but changing environment that theories on the origin of petroleum originated.

Alexander von Humboldt was the first to propose an inorganic abiogenic hypothesis for petroleum formation after he observed petroleum springs in the Bay of Cumaux (Cumana) on the northeast coast of Venezuela. In 1804 he is quoted as saying, "petroleum is the product of a distillation from great depth and issues from the primitive rocks beneath which the forces of all volcanic action lie." Abraham Gottlob Werner and the proponents of neptunism in the 1700s believed basaltic sills to be solidified oils or bitumen. While these notions have been proven unfounded, the basic idea that petroleum is associated with magmatism persisted. Other prominent proponents of what would become the abiogenic hypothesis included Mendeleev and Berthelot.

Russian geologist Nikolai Alexandrovitch Kudryavtsev proposed the modern abiotic hypothesis of petroleum in 1951. On the basis of his analysis of the Athabasca Tar Sands in Alberta, Canada, he concluded that no *"source rocks"* could form the enormous volume of hydrocarbons, and that therefore the most plausible explanation is abiotic deep petroleum. However, humic coals have since been proposed for the source rocks. Kudryavtsev's work was continued by Peter N. Kropotkin, Vladimir B. Porfir'ev, Emmanuil B. Chekaliuk, Vladilen A. Krayushkin, Georgi E. Boyko, Georgi I. Voitov, Grygori N. Dolenko, Iona V. Greenberg, Nikolai S. Beskrovny, and Victor F. Linetsky.

Astronomer Thomas Gold was the most prominent proponent of the abiogenic hypothesis in the West until his death in 2004. Currently, Jack Kenney of Gas Resources Corporation is a prominent proponent in the West.

1.1.1.2. State of Current Research

Little research is directed on establishing abiogenic petroleum or methane, althought the Carnegie Institution for Science have found that ethane and heavier hydrocarbons can be synthesized under conditions of the upper mantle. Research mostly related to astrobiology and the deep microbial biosphere and serpentinite reactions, however, continue to provide insight into the contribution of abiogenic hydrocarbons into petroleum accumulations.

- Rock porosity and migration pathways for abiogenic petroleum .
- Ocean floor hydrothermal vents as in the Lost City hydrothermal field.
- Mud volcanoes and the volatile contents of deep pelagic oozes and deep formation brines.
- Mantle peridotite serpentinization reactions and other natural Fischer-Tropsch analogs.
- Primordial hydrocarbons in meteorites, comets, asteroids and the solid bodies of the solar system
  - Primordial or ancient sources of hydrocarbons or carbon in Earth. Primordial hydrocarbons formed from hydrolysis of metal carbides of the iron peak of cosmic elemental abundance (Cr, Fe, Ni, V, Mn, Co).
- Isotopic studies of groundwater reservoirs, sedimentary cements, formation gases and the composition of the noble gases and nitrogen in many oil fields.
- The geochemistry of petroleum and the presence of trace metals related to Earth's mantle (Ni, V, Cd, As, Pb, Zn, Hg and others).

Similarly, research into the deep microbial hypothesis of hydrocarbon generation is advancing as part of the attempt to investigate the concept of panspermia and astrobiology, specifically using deep microbial life as an analog for life on Mars. Research applicable to deep microbial petroleum theories includes

- Research into how to sample deep reservoirs and rocks without contamination.
- Sampling deep rocks and measuring chemistry and biological activity.
- Possible energy sources and metabolic pathways which may be used in a deep biosphere.
- Investigations into the reworking primordial hydrocarbons by bacteria and their effects on carbon isotope fractionation.

A 2006 review article by Glasby presented arguments against the abiogenic origin of petroleum on a number of counts. A paper on *abiogenic hydrocarbon production* in a February 2008 issue of Science used data from experiments at Lost City (hydrothermal field) to report how the abiotic synthesis of hydrocarbons in nature may occur in the presence of ultramafic rocks, water, and moderate amounts of heat.

1.1.1.3. Foundations of the Hypothesis

Within the mantle, carbon may exist as hydrocarbons, chiefly methane and as elemental carbon, carbon dioxide and carbonates. The abiotic hypothesis is that the full suite of hydrocarbons found in petroleum can be generated in the mantle by abiogenic processes, and these hydrocarbons can migrate out of the mantle into the crust until they escape to the surface or are trapped by impermeable strata, forming petroleum reservoirs.

Abiogenic theories reject the supposition that certain molecules found within petroleum, known as biomarkers, are indicative of the biological origin of petroleum. They contend that some of these molecules could have come from the microbes that the petroleum encounters in its upward migration through the crust, that some of them are found in meteorites, which have presumably never contacted living material, and that some can be generated abiogenically by plausible reactions in petroleum.

Table 1.2 depicts the hypothesis which was is founded primarily upon :

TABLE 1.2.

Proponents	Item
Gold	The presence of methane on other planets, meteors, moons and comets
Gold, Kenney	Proposed mechanisms of abiotically chemically synthesizing hydrocarbons within the mantle
Kudryavtsev, Gold	Hydrocarbon-rich areas tend to be hydrocarbon-rich at many different levels
Kudryavtsev, Gold	Petroleum and methane deposits are found in large patterns related to deep-seated
	large-scale structural features of the crust rather than to the patchwork of sedimentary deposits
Gold	Interpretations of the chemical and isotopic composition of natural petroleum
Kudryavtsev, Gold	The presence of oil and methane within non-sedimentary rocks upon the Earth
Gold	The existence of methane hydrate deposits
Gold	Perceived ambiguity in some assumptions and key evidence used in the orthodox biogenic petroleum theories
Gold	Bituminous coal creation is based upon deep hydrocarbon seeps
Gold	Surface carbon budget and oxygen levels stable over geologic time scales
Kudryavtsev, Gold	Biogenic theories do not explain some hydrocarbon deposit characteristics
Szatmari	The distribution of metals in crude oils fits better with upper serpentinized mantle, primitive mantle and chondrite patterns than oceanic and continental crust, and show no correlation with sea water
Gold	The association of hydrocarbons with helium, a noble gas
Gold	Deep microbial hypothesis of hydrocarbon generation

1.1.1.4. Conventional Theories

According to generally accepted theory, petroleum is derived from ancient biomass. The theory was initially based on the isolation of molecules from petroleum that closely resemble known biomolecules

Most petroleum geologists prefer theories of oil formation, which holds that oil originated in shallow seas as vast quantities of marine plankton or plant materials died and sank into the mud. Under the resulting anaerobic conditions, the resulting organic compounds remained in reduced state. Under these conditions, anaerobic bacteria converted the lipids (fats, oils and waxes) into a waxy substance called kerogen.

As the source rock was buried deeper, overburden pressure raised temperatures into the oil window, between 80 and 180 °C. Most of the organic compounds degraded into the straight-chain **hydrocarbons** that comprise most of petroleum. This process is called generation kitchen. Once crude oil formed, it became very fluid and migrated upward through the rock strata. This process is called oil expulsion. Eventually it was either trapped in an oil reservoir or oil escaped to the surface and was biodegraded by soil bacteria.

Oil buried deeper entered the "*gas window*" of more than 160 °C and was converted into natural gas by thermal cracking. Thus, theory predicts that no oil will be found below a certain depth, only unassociated gas. At greater depths, even natural gas would be pyrolyzed.

1.1.1.5. Proposed Mechanisms of Abiogenic Petroleum

Primordial Deposits : Thomas Gold's work was focused on hydrocarbon deposits of primordial origin. Meteorites are believed to represent the major composition of material from which the Earth was formed. Some meteorites, such as carbonaceous chondrites, contain carbonaceous material. If a large amount of this material is still within the Earth, it could have been leaking upward for billions of years. The thermodynamic conditions within the mantle would allow many hydrocarbon molecules to be at equilibrium under high pressure and high temperature. Although molecules in these conditions may disassociate, resulting fragments would be reformed due to the pressure. An average equilibrium of various molecules would exist depending upon conditions and the carbon-hydrogen ratio of the material.

Creation within the Mantle : Russian researchers performed the above calculations of thermodynamic equilibrium and concluded that hydrocarbon mixes would be created within the mantle. Experiments under high temperatures and pressures produced many hydrocarbons, including n-alkanes through $C_{10}H_{22}$, from iron oxide, calcium carbonate, and water. [11] Because such materials are in the mantle and in subducted crust, there is no requirement that all hydrocarbons be produced from primordial deposits.

Hydrogen Generation : Hydrogen gas and water have been found more than 6 kilometers deep in the upper crust, including in the Siljan Ring boreholes and the Kola Superdeep Borehole. Data from the western United States suggests that aquifers from near the surface may extend to depths of 10 to 20 km. Hydrogen gas can be created by water reacting with silicates, quartz and feldspar, in temperatures in the 25° to 270°C range. These minerals are common in crustal rocks such as granite. Hydrogen may react with dissolved carbon compounds in water to form methane and higher carbon compounds.

One reaction not involving silicates which can create hydrogen is :

$$\text{Ferrous oxide} + \text{Water} \longrightarrow \text{Magnetite} + \text{Hydrogen}$$

$$3FeO + H_2O \longrightarrow Fe_3O_4 + H_2$$

The above reaction operates best at low pressures. At pressures greater than 5 GPa almost no hydrogen is created.

Serpentinite Mechanism : One proposed mechanism by which abiogenic petroleum is formed was first proposed by the Ukrainian scientist, **Prof. Emmanuil B. Chekaliuk** in 1967. He proposed that petroleum could be formed at high temperatures and pressures from inorganic carbon in the form of carbon dioxide, hydrogen and/or methane.

This mechanism is supported by several lines of evidence which are accepted by modern scientific literature. This involves synthesis of oil within the crust *via* catalysis by chemically reductive rocks. A proposed mechanism for the formation of inorganic hydrocarbons is *via* natural analogous of the Fischer-Tropsch process known as the *serpentinite mechanism* or the serpentinite process.

$$CH_4 + \frac{1}{2}O_2 \longrightarrow 2H_2 + CO$$

$$(2n+1)H_2 + nCO \longrightarrow C_nH_{2n+2} + nH_2O$$

Serpentinites are ideal rocks to host this process as they are formed from peridotites and dunites, rocks which contain greater than 80% olivine and usually a percentage of Fe-Ti spinel minerals. Most olivines also contain high nickel concentrations (up to several percent) and may also contain chromite or chromium as a contaminant in olivine, providing the needed transition metals.

However, serpentinite synthesis and spinel cracking reactions require hydrothermal alteration of pristine peridotite-dunite, which is a finite process intrinsically related to metamorphism, and further, requires significant addition of water. Serpentinite is unstable at mantle temperatures and is readily dehydrated to granulite, amphibolite, talc-schist and even eclogite. This suggests that methanogenesis in the presence of serpentinites is restricted in space and time to mid-ocean ridges and upper levels of subduction zones. However, water has been found as deep as 12 km so water-based reactions are dependent upon the local conditions. Oil being created by this process in intracratonic regions is limited by the materials and temperature.

Serpentinite synthesis: A chemical basis for the abiotic petroleum process is the serpentinization of peridotite, beginning with methanogenesis *via* hydrolysis of olivine into serpentine in the presence of carbon dioxide. Olivine, composed of Forsterite and Fayalite metamorphoses into serpentine, magnetite and silica by the following reactions, with silica from fayalite decomposition feeding into the forsterite reaction

Reaction

$$\text{Fayalite + Water} \longrightarrow \text{Magnetite + Aqueous silica + Hydrogen}$$

$$3Fe_2SiO_4 + 2H_2O \longrightarrow 2Fe_3O_4 + 3SiO_2 + 2H_2$$

Reaction

$$\text{Forsterite + Aqueous silica} \longrightarrow \text{Serpentinite}$$

$$3Mg_2SiO_4 + SiO_2 + 4H_2O \longrightarrow 2Mg_3Si_2O_5\,(OH)_4$$

When this reaction occurs in the presence of dissolved carbon dioxide (carbonic acid) at temperatures above 500 °C (formation of magnetite) takes place.

Reaction

$$\text{Olivine + Water + Carbonic acid} \longrightarrow \text{Serpentine + Magnetite + Methane}$$

$$(Fe, Mg)_2SiO_4 + nH_2O + CO_2 \longrightarrow Mg_3Si_2O_5(OH)_4 + Fe_3O_4 + CH_4$$

or, in balanced form :

$$18Mg_2SiO_4 + 6Fe_2SiO_4 + 26H_2O + CO_2 \longrightarrow 12Mg_3Si_2O_5(OH)_4 + 4Fe_3O_4 + CH_4$$

However, reaction 2(b) is just as likely, and supported by the presence of abundant talc-carbonate schists and magnesite stringer veins in many serpentinised peridotites;

Reaction

$$\text{Olivine + Water + Carbonic acid} \longrightarrow \text{Serpentine + Magnetite + Magnesite + Silica}$$

$$(Fe, Mg)_2SiO_4 + nH_2O + CO_2 \longrightarrow Mg_3Si_2O_5(OH)_4 + Fe_3O_4 + MgCO_3 + SiO_2$$

The upgrading of methane to higher n-alkane hydrocarbons is *via* dehydrogenation of methane in the presence of catalyst transition metals (*e.g.*, Fe, Ni). This can be termed spinel hydrolysis.

Spinel Polymerization Mechanism

Magnetite, chromite and ilmenite are Fe-spinel group minerals found in many rocks but rarely as a major component in non-ultramafic rocks. In these rocks, high concentrations of magmatic magnetite, chromite and ilmenite provide a reduced matrix which may allow abiotic cracking of methane to higher hydrocarbons during hydrothermal events.

Chemically reduced rocks are required to drive this reaction and high temperatures are required to allow methane to be polymerized to ethane. Above reaction also creates magnetite.

Reaction

$$\text{Methane} + \text{Magnetite} \longrightarrow \text{Ethane} + \text{Hematite}$$
$$nCH_4 + nFe3O_4 + nH_2O \longrightarrow C_2H_6 + Fe_2O_3 + HCO_3 + H^+$$

Reaction results in *n*-alkane hydrocarbons, including linear saturated hydrocarbons, alcohols, aldehydes, ketones, aromatics, and cyclic compounds.

Carbonate Decomposition

Calcium carbonate may decompose at around 500 °C through the following reaction :

Reaction

$$\text{Hydrogen} + \text{Calcium carbonate} \longrightarrow \text{Methane} + \text{Calcium oxide} + \text{Water}$$
$$4H_2 + CaCO_3 \longrightarrow CH_4 + CaO + 2H_2O$$

Note that CaO (lime) is not a mineral species found within natural rocks. Whilst this reaction is possible, it is not plausible.

1.1.1.6. Laboratory Experiments

Some research and laboratory experiments explore possible mechanisms, but there is little related geological evidence.

Carbonate Reduction

Methane has been formed in laboratory conditions *via* carbonate reduction at pressures and temperatures similar to that in the upper mantle, but a large amount of water was provided to the reaction in excess of that which is typical in mantle lithology. No *natural* rocks with a wustite-calcite composition are known to exist, which precludes this reaction from occurring in nature. Likely reactions include:

Reaction

$$\text{Ferrous oxide} + \text{Calcium carbonate} + \text{Water} \longrightarrow \text{Hematite} + \text{Methane} + \text{Calcium oxide}$$
$$8FeO + CaCO_3 + 2H_2O \longrightarrow 4Fe_2O_3 + CH_4 + CaO$$

Reaction

$$\text{Ferrous oxide} + \text{Calcium carbonate} + \text{Water} \longrightarrow \text{Magnetite} + \text{Methane} + \text{Calcium oxide}$$
$$12FeO + CaCO_3 + 2H_2O \longrightarrow 4Fe_3O_4 + CH_4 + CaO$$

Methane formation is favored under 1,200 °C at 1 GPa. At 1,500 °C hydrogen production was prevalent. Methane production is most favored at 500 °C and pressures <7 GPa; higher temperatures are expected to lead to carbon dioxide and carbon monoxide production through a reforming equilibrium with methane.

This is cited as evidence of the *plausibility* of methanogenesis under mantle conditions.

Calcite Decomposition

Carbon dioxide, can be created by calcite decomposition at 1,500 °C:

Reaction

$$\text{Calcium carbonate} \longrightarrow \text{Calcium oxide + Carbon dioxide}$$

$$CaCO_3 \longrightarrow CaO + CO_2$$

Calcite is likely molten at these temperatures, being a mixture of CaO ions and CO_2.

Ethane and Ethylene Synthesis

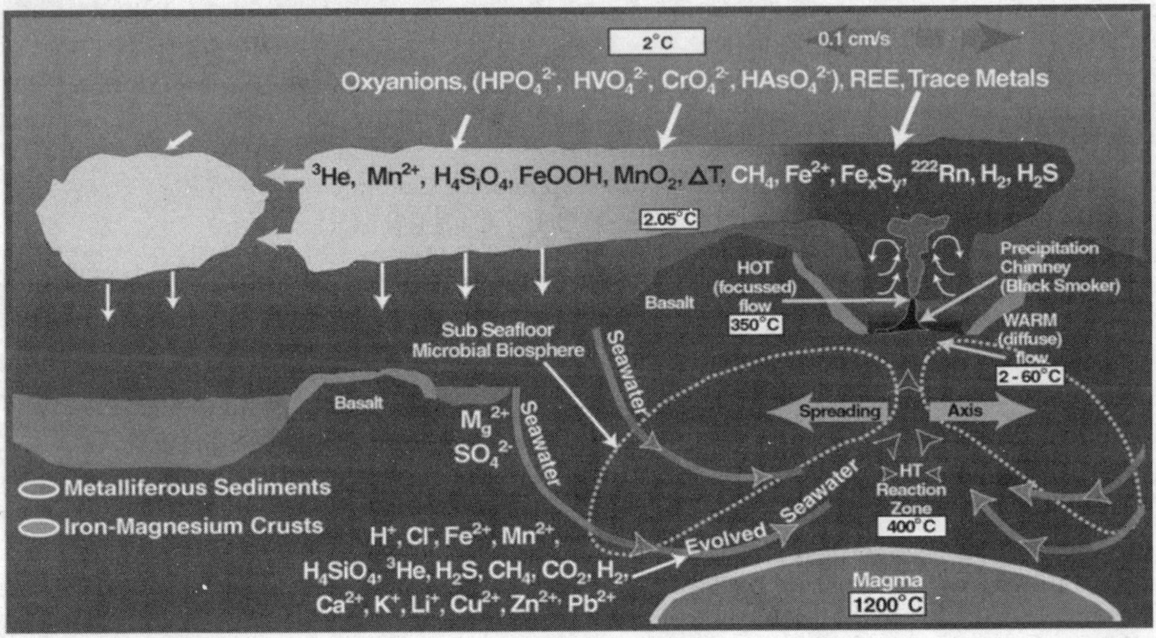

FIG. 1.1.

The synthesis of ethane and ethylene has been done at 800 °C, using electric discharges in laboratory experiments. This experiment was in a hot gas, rather than hot mantle fluids. The calculated reactions are :

$$\text{Carbon dioxide + Methane} \longrightarrow \text{Carbon monoxide + Ethane + Water}$$

$$CO_2 + 2CH_4 \longrightarrow CO + C_2H_6 + H_2O$$

and

$$\text{Carbon dioxide + Ethane} \longrightarrow \text{Carbon monoxide + Ethylene + Water}$$

$$CO_2 + C_2H_6 \longrightarrow CO + C_2H_4 + H_2O$$

1.1.1.7. Evidence of Abiogenic Mechanisms

- Scaled particle theory for a simplified perturbed hard-chain, statistical mechanical model predicts that methane compressed to 30 or 40 kbar at 1000 °C (conditions in the mantle) yields hydrocarbons having properties similar to petroleum.
- Experiments in diamond anvil high pressure cells have confirmed this theory

1.2. BIOTIC (MICROBIAL) HYDROCARBONS

The "*deep biotic petroleum hypothesis*", similar to the abiogenic petroleum origin hypothesis, holds that not all petroleum deposits within the Earth's rocks can be explained purely according to the orthodox view of petroleum geology. Thomas Gold used the term *the deep hot biosphere* to describe the microbes which live underground. This hypothesis is different from biogenic oil in that the role of deep-dwelling microbes is a biological source for oil which is not of a sedimentary origin and is not sourced from surface carbon. Deep microbial life is only contaminant of primordial hydrocarbons. Parts of microbes yield molecules as biomarkers.

Deep biotic oil is considered to be formed as a byproduct of the life cycle of deep microbes. Shallow biotic oil is considered to be formed as a byproduct of the life cycles of shallow microbes.

1.2.1. Microbial Biomarkers

Extremophile organisms living within the crust (deep heat-loving bacteria thermophiles) are considered a plausible source of biomarkers which are not sourced from kerogen.

Hopanoids, called the "*most abundant natural products on Earth*", were believed to be indicators of oil derived from ferns and lichens but are now known to be created by many bacteria, including archaea.

Sterane was thought to have come from processes involving surface deposits but is now known to be produced by several prokaryotes including methanotrophic proteobacteria.

The case for shallow bacterial life creating petroleum is apparent from circumstantial evidence at "tar seeps" in sandstone outcrops where live oil is encountered down-dip (*e.g.* Midway-Sunset field, San Joaquin Valley, California). Bacteria are considered to have "*degraded*" higher gravity oil to bitumens.

Extrapolation of bacterial degradation to still higher gravity oils and finally to methane leads to the suggestion that all petroleum up to tar and most of the carbon in coal are derivatives of methane, which is progressively stripped of its hydrogen by bacteria and archaea. The resultant partial methane molecules, CH_3, CH_2, CH, may be called "*an-hydrides*". Anhydride hypothesis, a New Theory of Petroleum and Coal Generation, is offered by **C. Warren Hunt** (1999).

Due to the difficulty in culturing and sampling thermophilic bacteria little was known of their chemistry. As more is learned of bacterial chemistry, more biomarker chemicals can be attributed to bacterial sources. Although extremophile micro-organisms exist deep underground and some metabolize carbon, some of these biomarkers are so far only known from surface plants and remain the most reliable chemical evidence of biogenic genesis of petroleum.

This evidence is consistent with the biogenic hypothesis, although it might be true that these hydrocarbons have merely been in contact with ancient plant residues. There also is evidence that low temperature relatives of hyperthermophiles are widespread, so it is also possible for biological deposits to have been altered by low temperature bacteria which are similar to deeper heat-loving relatives.

It must also be acknowledged that, if extremophilic bacteria prove to be the source of some parts of known oils, that this remains a *biological* process. Thorough rebuttal of biogenic origins based on biomarkers has been offered by *Kenney, et al.* (2001).

1.2.2. Isotopic Evidence

Methane is ubiquitous in crustal fluid and gas. Research continues to attempt to characterise crustal sources of methane as biogenic or abiogenic using carbon isotope fractionation of observed gases (**Lollar & Sherwood** 2006). There are few clear examples of abiogenic methane-ethane-butane, as the same processes favor enrichment of light isotopes in all chemical reactions, whether organic or inorganic. ä^{13}C of methane overlaps that of inorganic carbonate and graphite in the crust, which are heavily depleted in ^{12}C, and attain this by isotopic fractionation during metamorphic reactions.

One argument for abiogenic oil cites the high carbon depletion of methane as stemming from the observed carbon isotope depletion with depth in the crust. However, diamonds, which are definitively of mantle origin, are not as depleted as methane, which implies that methane carbon isotope fractionation is not controlled by mantle values. Commercially extractable concentrations of helium (greater than 0.3%) are present in natural gas from the Panhandle-Hugoton fields in the USA, as well as from some Algerian and Russian gas fields.

Helium trapped within most petroleum occurrences, such as the occurrence in Texas, is of a distinctly crustal character with an *Ra* ratio of less than 0.0001 that of the atmosphere. The Chimaera gas seep, near Antalya (SW Turkey), new and thorough molecular and isotopic analyses including methane (~87% v/v; D13C1 from -7.9 to -12.3 ‰; D13D1 from –119 to –124‰), light alkanes (C2+C3+C4+C5 = 0.5%; C6+: 0.07%; D13C2 from -24.2 to -26.5 ‰; D13C3 from -25.5 to -27 ‰), hydrogen (7.5 to 11 %), carbon dioxide (0.01-0.07%; D13CCO2: -15 ‰), helium (~80 ppmv; R/Ra: 0.41) and nitrogen (2-4.9%; D15N from -2 to -2.8 ‰) converge to indicate that the seep releases a mixture of organic thermogenic gas, related to mature Type III kerogen occurring in Paleozoic and Mesozoic organic rich sedimentary rocks, and abiogenic gas produced by low temperature serpentinization in the Tekirova ophiolitic unit.

1.2.3. Biomarker Chemicals

Certain chemicals found in naturally occurring petroleum contain chemical and structural similarities to compounds found within many living organisms. These include terpenoids, terpenes, pristane, phytane, cholestane, chlorins and porphyrins, which are large, chelating molecules in the same family as heme and chlorophyll. Materials which suggest certain biological processes include tetracyclic diterpane and oleanane.

The presence of these chemicals in crude oil is assumed to be as a result of the inclusion of biological material in the oil. This is predicated upon the theory that these chemicals are released by kerogen during the production of hydrocarbon oils.

However, since the advent of abiogenic hypothesis, the veracity of these assumptions has been called into question and new lines of evidence used to provide alternative explanations. Some are provided by many scientists from around the world including Russia.

1.2.4. Trace Metals

Nickel (Ni), vanadium (V), lead (Pb), arsenic (As), cadmium (Cd), mercury (Hg) and others metals frequently occur in oils. Some heavy crude oils, such as Venezuelan heavy crude have up to 45% vanadium pentoxide content in their ash, high enough that it is a commercial source for vanadium. These metals are common in Earth's mantle, thus their compounds in oils are often called as *abiomarkers*.

Analysis of 22 trace elements in 77 oils correlate significantly better with chondrite, serpentinized fertile mantle peridotite, and the primitive mantle than with oceanic or continental crust, and shows no correlation with seawater.

1.2.5. Reduced Carbon

Sir Robert Robinson studied the chemical makeup of natural petroleum oils in great detail, and concluded that they were mostly far too hydrogen-rich to be a likely product of the decay of plant debris. However, several processes which generate hydrogen could supply kerogen hydrogenation which is compatible with conventional petroleum generation theories. Olefins, the unsaturated hydrocarbons, would have been expected to predominate by far in any material that was derived in that way. He also wrote: *"Petroleum is considered to be a primordial hydrocarbon mixture into which bio-products have been added."*

This has however been demonstrated later to be a misunderstanding by Robinson, related to the fact that only short duration experiments were available to him. Olefins are thermally very unstable (that is why natural petroleum normally does not contain such compounds) and in laboratory experiments that last more than a few hours, the olefins are no longer present.

The presence of low-oxygen and hydroxyl-poor hydrocarbons in natural living media is supported by the presence of natural waxes (n=30+), oils (n=20+) and lipids in both plant matter and animal matter, for instance fats in phytoplankton, zooplankton and so on. These oils and waxes, however, occur in quantities too small to significantly affect the overall hydrogen/carbon ratio of biological materials. However, after the discovery of highly aliphatic biopolymers in algae, and that oil generating kerogen essentially represent concentrates of such materials, no theoretical problem exists anymore. Furthermore, the millions of source rock samples that has been analysed for petroleum yield by the petroleum industry have eliminated any pre-existing doubt about the enormous quantities of petroleum generated in sedimentary basin by thermal cracking of kerogen.

1.3. DIRECT OBSERVATIONS

The following are the direct tests of the abiogenic hypothesis of petroleum or impartial evidence generated by observations of the Earth which can be used to argue the hypothesis for or against, and are presented as such.

- The Lost City Hydrothermal Vent Field was determined to have abiogenic hydrocarbon production. Proskurowski *et al.* wrote, "Radiocarbon evidence rules out seawater bicarbonate as the carbon source for FTT reactions, suggesting that a mantle-derived inorganic carbon source is leached from the host rocks. Our findings illustrate that the abiotic synthesis of hydrocarbons in nature may occur in the presence of ultramafic rocks, water, and moderate amounts of heat."
- The Siljan Ring meteorite crater, Sweden, was proposed by Thomas Gold as the most likely place to test the hypothesis because it was one of the few places in the world where the granite basement was cracked sufficiently (by meteorite impact) to allow oil to seep up from the mantle; furthermore it is infilled with a relatively thin veneer of sediment, which was sufficient to trap any abiogenic oil but was modelled as untenable for a biogenic origin of any oil (it had not developed the 'oil window' and structural traps typical of biogenic plays). Drilling of the Siljan Ring with the Gravberg-1 7,500 m

borehole penetrated the lowest reservoirs. Hydrocarbons were found, though in an economically unviable form of sludge. It was proposed that the eight barrels of magnetite paste produced were contaminated from the diesel fuel based drilling fluid used to do the drilling, but the diesel was demonstrated to be not of the kind of oil found in the shaft. This well also sampled over 13,000 feet (4,000 m) of methane bearing inclusions. [2] To be safe, a second hole was drilled a few miles away with no diesel fuel based drilling fluid and this produced (according to Gold) 15 tons of oil. [3] However, no documentation of any such discovery has ever been provided. The *"discovery"* is not mentioned in any well log, including the well completion log. No other personell at the drilling location ever reported anything apart from some additional magnetite rich mud contaminated with minor organics from the Gravberg well. When Gold published the paper stating this discovery, several prominent geologists tried to stop the publication because of it's *"unscientific"* content.

- Direct observation of bacterial mats and fracture-fill carbonate and humin of bacterial origin in deep boreholes in Iran, Australia Sweden and Canada.

1.4. EXAMPLE PROPOSED A BIOGENIC METHANE DEPOSITS

Panhandle-Hugoton field (Anadarko Basin) in Texas-Oklahoma, USA is the most important gas field with commercial helium content. The White Tiger oil field in Vietnam has been proposed as an example of abiogenic oil because it is 4,000 m of fractured basement granite, at a depth of 5,000 m. However, others argue that it contains biogenic oil which leaked into the basement horst from conventional source rocks within the Cuu Long basin. A major component of mantle-derived carbon is indicated in commercial gas reservoirs in the Pannonian and Vienna basins of Hungary and Austria. Natural gas pools interpreted as being mantle-derived are the Shengli Field and Songliao Basin, northeastern China. The Chimaera gas seep, near Antalya (SW Turkey), has continuously been active for millennia and it is known to be the source of the first Olympic fire in the Hellenistic period. Chimaera represents the biggest emission of abiogenic methane on land discovered so far; deep and pressurized gas accumulations necessary to sustain the gas flow for millennia, likely charged by an active inorganic source, may be present.

1.5. THE GEOLOGICAL ARGUMENT FOR ABIOGENIC OIL

Given the known occurrence of methane and the probable catalysis of methane into higher atomic weight hydrocarbon molecules, the abiogenic hypothesis considers the following to be key observations in support :

- The serpentinite synthesis, graphite synthesis and spinel catalysation models prove the process is viable.
- The likelihood that abiogenic oil seeping up from the mantle is trapped beneath sediments which effectively seal mantle-tapping faults.
- Mass-balance calculations for supergiant oilfields which argue that the calculated source rock could not have supplied the reservoir with the known accumulation of oil, implying deep recharge (**Kudryavtsev**, 1951).

1.5.1. Incidental Evidence

The proponents of abiogenic oil use several arguments which draw on a variety of natural phenomena in order to support the hypothesis :

- The modelling of some researchers which shows the Earth was accreted at relatively low temperature, thereby perhaps preserving primordial carbon deposits within the mantle, to drive abiogenic hydrocarbon production.
- The presence of methane within the gases and fluids of mid-ocean ridge spreading centre hydrothermal fields.

1.5.2. The Geological Argument Against

Oil deposits are not directly associated with tectonic structures.

Key arguments against chemical reactions, such as the serpentinite mechanism, as being the major source of hydrocarbon deposits within the crust are :

- The lack of available pore space within rocks as depth increases. This is contradicted by numerous studies which have documented the existence of hydrologic systems operating over a range of scales and at all depths in the continental crust.
- The presence of no commercial hydrocarbon deposits within the crystalline shield areas of the major cratons especially around key deep seated structures which are predicted to host oil by the abiogenic hypothesis Limited evidence that major serpentinite belts underlie continental sedimentary basins which host oil.
- Lack of conclusive proof that carbon isotope fractionation observed in crustal methane sources is entirely of abiogenic origin.
- Mass balance problems of supplying enough carbon dioxide to serpentinite within the metamorphic event before the peridotite is fully reacted to serpentinite.
- Drilling of the Siljan Ring failed to find commercial quantities of gas, thus providing a counter example to Kudryavtsev's Rule and failing to locate the predicted abiogenic gas . Helium in the Siljan Gravberg-1 well was depleted in ^{3}He and not consistent with a mantle origin.
- The distribution of sedimentary basins is caused by plate tectonics, with sedimentary basins forming on either side of a volcanic arc, which explains the distribution of oil within these sedimentary basins.
- Kudryavtsev's Rule has been explained for oil and gas (not coal): Gas deposits which are below oil deposits can be created from that oil or its source rocks. Because natural gas is less dense than oil, as kerogen and hydrocarbons are generating gas the gas fills the top of the available space. Oil is forced down, and can reach the spill point where oil leaks around the edge(s) of the formation and flows upward. If the original formation becomes completely filled with gas then all the oil will have leaked above the original location. Ubiquitous presence of diamondoids in natural hydrocarbons such as oil, gas and condensates are composed of carbon from biological sources, unlike the carbon found in normal diamonds. Arguments against the incidental evidence.
- Gas ruptures during earthquakes are more likely to be sourced from biogenic methane generated in unconsolidated sediment from existing organic matter, released by earthquake liquefaction of the reservoir during tremors.
- The presence of methane hydrate is arguably produced by bacterial action upon organic detritus falling from the littoral zone and trapped in the depth due to pressure and temperature.

- The likelihood of vast concentrations of methane in the mantle is very slim, given mantle xenoliths have negligible methane in their fluid inclusions; conventional plate tectonics explains deep focus quakes better, and the extreme confining pressures invalidate the hypothesis of gas pockets causing quakes.
- Further evidence is the presence of diamond within kimberlites and lamproites which sample the mantle depths proposed as being the source region of mantle methane.

REVIEW QUESTIONS

1. Discuss the abiogenic origin of petroleum.
2. What are the conventional theories for origin of petroleum ?
3. List the proposed mechanisms of abiogenic petroleum.
4. What were the evidences of abiogenic mechanism for origin of petroleum ?
5. What do you mean by biotic hydrocarbons ?
6. Explain serpentinite synthesis for origin of petroleum.

❑❑❑

CHAPTER

2

GLOBAL PETROLEUM RESOURCES : A VIEW TO THE FUTURE

INSIDE THIS CHAPTER

2.1. INTRODUCTION

Geologists are often asked: How much oil and gas are left in the world? There is no simple answer to this question. Petroleum is vital to many industries, and is of importance to the maintenance of industrialized civilization itself, and thus is critical concern to many nations. Oil accounts for a large percentage of the world's energy consumption, ranging from a low of 32% for Europe and Asia, up to a high of 53% for the Middle East. Other geographic regions' consumption patterns are as follows: South and Central America (44%), Africa (41%), and North America (40%). The world at large consumes 30 billion barrels (4.8 km^3) of oil per year, and the top oil consumers largely consist of developed nations. In fact, 24% of the oil consumed in 2004 went to the United States alone, though by 2007 this had dropped to 21% of world oil consumed.

In the US, in the states of Arizona, California, Hawai, Nevada, Oregon and Washington, the Western States Petroleum Association (WSPA) is responsible for producing, distributing, refining, transporting and marketing petroleum. This is non-profit trade association that was founded in 1907, and is the oldest petroleum trade association in the United States. Petroleum, in one form or another, is not a recent discovery. More than four thousand years ago, according to Herodotus and confirmed by Diodorus Siculus, asphalt was employed in the construction of the walls and towers of Babylon; there were oil pits near Ardericca (near Babylon), and a pitch spring on Zacynthus. Great quantities of it were found on the banks of the river Issues, one of the tributaries of the Euphrates. Ancient Persian tablets indicate the medicinal and lighting uses of petroleum in the upper levels of their society.

Oil was exploited in the Roman province of Dacia, now in Romania, where it was called *picula*.

The earliest known oil wells were drilled in China in 347 CE or earlier. They had depths of up to about 800 feet (240 m) and were drilled using bits attached to bamboo poles. The oil was burned to evaporate brine and produce salt. By the 10th century, extensive bamboo pipelines connected oil wells with salt springs. The ancient records of China and Japan are said to contain many allusions to the use of natural gas for lighting and heating. Petroleum was known as *burning water* in Japan in the 7th century. In his book *Dream Pool Essays* written in 1088, the polymathic scientist and statesman Shen Kuo of the Song Dynasty coined the word *Shíyóu*, literally "*rock oil*" for petroleum, which remains the term used in contemporary Chinese.

The first streets of Baghdad were paved with tar, derived from petroleum that became accessible from natural fields in the region. In the 9th century, oil fields were exploited in the area around modern Baku, Azerbaijan, to produce naphtha. These fields were described by the Arab geographer Abu al-Hasan 'Alî al-Mas'ûdî in the 10th century, and by Marco Polo in the 13th century, who described the output of those wells as hundreds of shiploads. Petroleum was distilled by the Persian **alchemist Muhammad ibn Zakarîya Râzi** (Rhazes) in the 9th century, producing chemicals such as kerosene in the alembic (*al-ambiq*), and which was mainly used for kerosene lamps. Arab and Persian chemists also distilled crude oil in order to produce flammable products for military purposes. Through Islamic Spain, distillation became available in Western Europe by the 12th century. It has also been present in Romania since the 13th century, being recorded as pãcurã.

The earliest mention of petroleum in the Americas occurs in **Sir Walter Raleigh's** account of the Trinidad Pitch Lake in 1595; whilst thirty-seven years later, the account of a visit of a Franciscan, Joseph de la Roche d'Allion, to the oil springs of New York was published in Sagard's *Histoire du Canada*. A Russian traveller, Peter Kalm, in his work on America published in 1748 showed on a map the oil springs of Pennsylvania. In 1710 or 1711 (sources vary) the Russian-born Swiss physician and Greek teacher Eyrini d'Eyrinis (also spelled as Eirini d'Eirinis) discovered asphaltum at Val-de-Travers, (Neuchâtel). He established a bitumen mine *de la Presta* there in 1719 that operated until 1986. Oil sands were mined from 1745 in Merkwiller-Pechelbronn, Alsace under the direction of **Louis Pierre Ancillon de la Sablonnière**, by special appointment of Louis XV. The Pechelbronn oil field was active until 1970, and was the birth place of companies like Antar and Schlumberger. The first modern refinery was built there in 1857.

The modern history of petroleum began in 1846 with the discovery of the process of refining kerosene from coal by Nova Scotian Abraham Pineo Gesner. Ignacy **£ukasiewicz** improved Gesner's method to develop a means of refining kerosene from the more readily available "rock oil" ("**petroleum**") seeps in 1852 and the first rock oil mine was built in Bóbrka, near Krosno in Galicia(Poland/Ukraine) in the following year. In 1854, Benjamin Silliman, a science professor at Yale University in New Haven, was the first to fractionate petroleum by distillation. These discoveries rapidly spread around the world, and Meerzoeff built the first Russian refinery in the mature oil fields at Baku in 1861. At that time Baku produced about 90% of the world's oil.

The first commercial oil well in Romania was drilled in 1857, and the world's first oil refinery opened at Ploiesti, Romania being the first country in the world with a crude oil output officially recorded in international statistics, namely 275 tonnes

- The first oil well in North America was in Oil Springs, Ontario, Canada in 1858, dug by James Miller Williams. The US petroleum industry began with Edwin Drake's drilling of a 69-foot (21 m) oil well in 1859, on Oil Creek near Titusville, Pennsylvania, for the Seneca Oil Company (originally yielding 25 barrels per day (4.0 m^3/d), by the end of the year output was at the rate of 15 barrels per day (2.4 m^3/d)). See also the Pennsylvanian oil rush The industry grew through the 1800s, driven by the demand for kerosene and oil lamps. It became a major national concern in the early part of the 20th century; the introduction of the internal combustion engine provided a demand that has largely sustained the industry to this day. Early "local" finds like those in Pennsylvania and Ontario were quickly outpaced by demand, leading to *"oil booms"* in Texas, Oklahoma.

By 1910, significant oil fields were discovered in Canada (specifically, in the province of Alberta), the Dutch East Indies (1885, in Sumatra), Persia (1908, in Masjed Soleiman), (1863, in Zorritos District) Peru, Venezuela, and Mexico, and were being developed at an industrial level.

During World War II, oil facilities were a major strategic asset and were extensively bombed.

Even until the mid-1950s, coal was still the world's foremost fuel, but oil quickly took over. Following the 1973 energy crisis and the 1979 energy crisis, there was significant drop in oil supply levels. This brought to light the concern that oil is a limited resource that will eventually run out, at least as an economically viable energy source. At the time, the most common and popular predictions were quite dire. However, a period of increased production and reduced demand caused an oil glut in the 1980s.

Today, about 90% of vehicular fuel needs are met by oil. Petroleum also makes up 40% of total energy consumption in the United States, but is responsible for only 2% of electricity generation. Petroleum's worth as a portable, dense energy source powering the vast majority of vehicles and as the base of many industrial chemicals makes it one of the world's most important commodities. Access to it was a major factor in several military conflicts of the twentieth century, including World War II The top three oil producing countries are Saudi Arabia, Russia, and the United States. About 80% of the world's readily accessible reserves are located in the Middle East, with 62.5% coming from the five Arabian countries: UAE, Iraq, Qatar and Kuwait. However, with high oil prices, (above $100/barrel) Venezuela has larger reserves than Saudi Arabia due to crude reserves derived from bitumen. Figure 2.1 shows the oil producing countries.

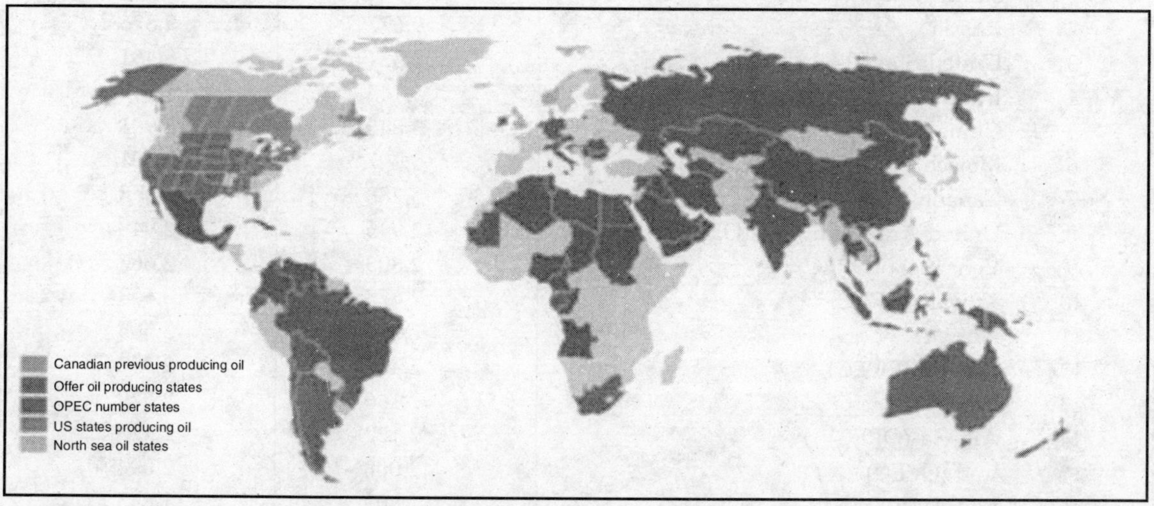

Fig. 2.1. Shows The Oil Producing Countries.

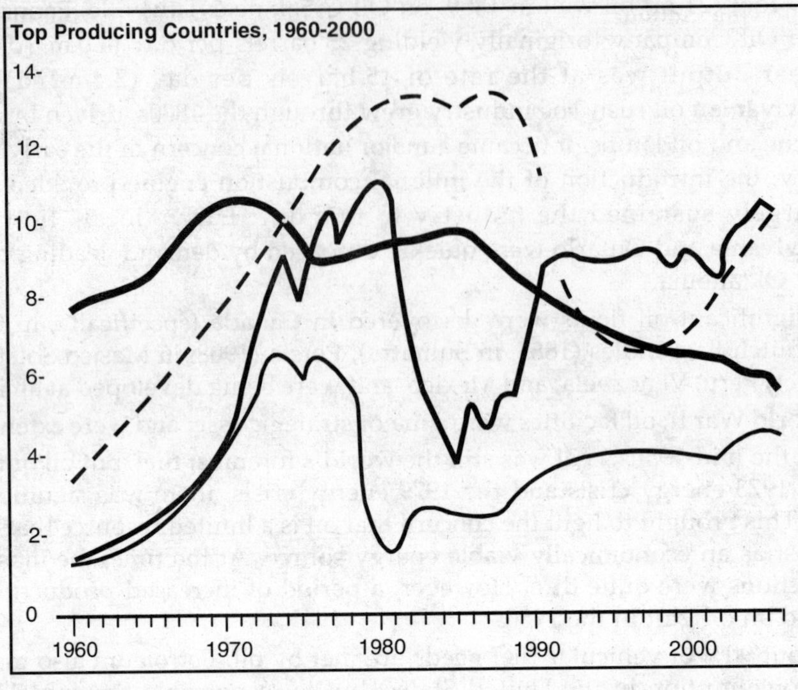

FIG. 2.2. GRAPH OF TOP OIL PRODUCING COUNTRIES 1960-2006, INCLUDING SOVIET UNION.

In petroleum industry parlance, *production* refers to the quantity of crude extracted from reserves, not the literal creation of the product. The crude oil producing countries are given below in the **Table 2.1.**

TABLE 2.1. Showing the Crude Oil Producing Countries.

	Producing Nation	10^3bbl/d (2006)	10^3bbl/d (2007)
1	Saudi Arabia (OPEC)	10,665	10,234
2	Russia[1]	9,677	9,876
3	United States[1]	8,331	8,481
4	Iran (OPEC)	4,148	4,043
5	China	3,845	3,901
6	Mexico[1]	3,707	3,501
7	Canada[2]	3,288	3,358
8	United Arab Emirates (OPEC)	2,945	2,948
9	Venezuela (OPEC)[1]	2,803	2,667
10	Kuwait (OPEC)	2,675	2,613
11	Norway[1]	2,786	2,565
12	Nigeria (OPEC)	2,443	2,352
13	Brazil	2,166	2,279
14	Algeria (OPEC)	2,122	2,173
15	Iraq (OPEC)[3]	2,008	2,094
16	Libya (OPEC)	1,809	1,845
17	Angola (OPEC)	1,435	1,769

	Producing Nation	10^3bbl/d (2006)	10^3bbl/d (2007)
18	United Kingdom	1,689	1,690
19	Kazakhstan	1,388	1,445
20	Qatar (OPEC)	1,141	1,136
21	Indonesia	1,102	1,044
22	India	854	881
23	Azerbaijan	648	850
24	Argentina	802	791
25	Oman	743	714
26	Malaysia	729	703
27	Egypt	667	664
28	Australia	552	595
29	Colombia	544	543
30	Ecuador (OPEC)	536	512
31	Sudan	380	466
32	Syria	449	446
33	Equatorial Guinea	386	400
34	Yemen	377	361
35	Vietnam	362	352
36	Thailand	334	349
37	Denmark	344	314
38	Congo	247	250
39	Gabon	237	244

Source: U.S. Energy Information Administration

[1]Peak production of conventional oil already passed in this state

[2]Although Canadian conventional oil production is declining, total oil production is increasing as oil sands production grows. If oil sands are included, it has the world's second largest oil reserves after Saudi Arabia.

[3]Though still a member, Iraq has not been included in production figures since 1998

Geology, economics, politics, technology and social science all play a role in assessing our resources and determining the economic viability of those resources.

2.2. UNITED STATE GEOLOGICAL SURVEY ESTIMATE OF PETROLEUM RESOURCES (USGS)

In 2000, the U.S. Geological Survey completed an assessment of the world's conventional petroleum resources, exclusive of the United States. This assessment is different from those before it: Overall the 2000 assessment of potential petroleum resources is higher than previous assessments, largely because it is the first USGS world assessment to include field growth estimates

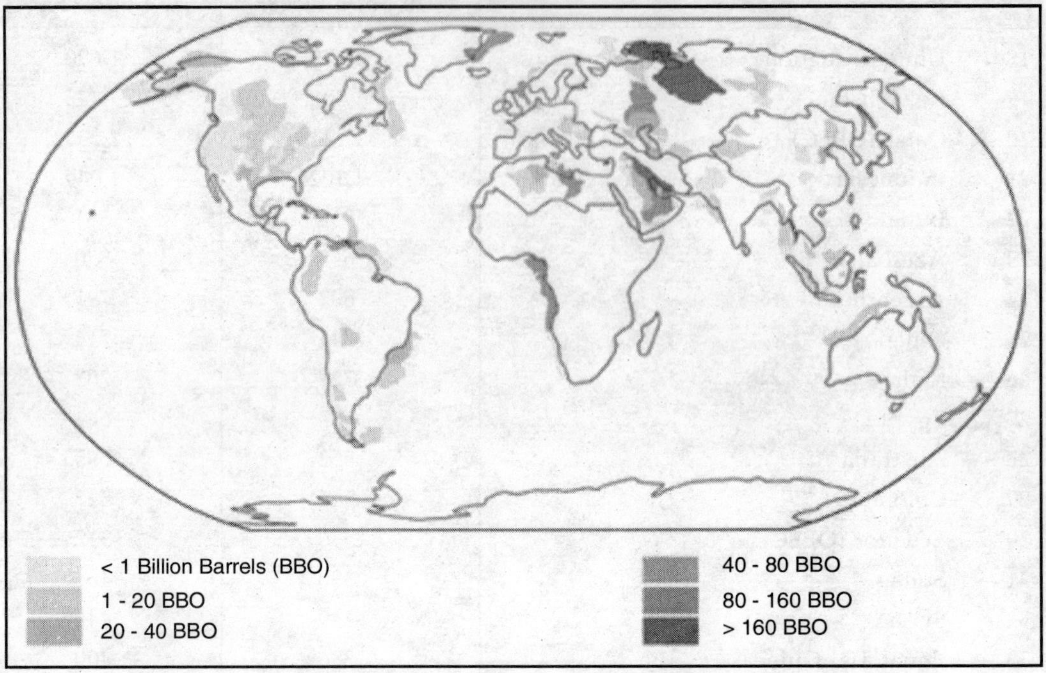

FIG. 2.3. THE WORLD'S OIL: CONVENTIONAL OIL ENDOWMENT (CUMULATIVE PRODUCTION PLUS MEAN ESTIMATES OF REMAINING OIL RESERVES AND UNDISCOVERED OIL RESOURCES) BY PROVINCE IN BILLION BARRELS OF OIL (BBO) FOR 128 OIL PROVINCES. WORLD DATA ARE FROM THE U.S. GEOLOGICAL SURVEY (USGS) 2000 ASSESSMENT OF WORLD PETROLEUM RESOURCES; U.S. DATA ARE FROM USGS (1995 ASSESSMENT) AND FROM THE MINERALS MANAGEMENT SERVICE (1996 ASSESSMENT). USGS IMAGE.

Based on a thorough investigation of the petroleum geology of each province, the assessment couples geologic analysis with a probabilistic methodology to estimate remaining potential. Including the assessment numbers for the United States from USGS and the Minerals Management Service (MMS), the world's endowment of recoverable oil — which consists of cumulative production, remaining reserves, reserve growth and undiscovered resources — is estimated at about 3 trillion barrels of oil. Of this, about 24 percent has been produced and an additional 29 percent has been discovered and booked as reserves. The natural gas endowment is estimated at 15.4 quadrillion cubic feet (2.5 trillion barrels of oil equivalent), of which only about 11 percent has been produced and an additional 31 percent has been discovered and booked as reserves.

The USGS assessment does not cover all sedimentary basins of the world. Relatively small volumes of oil or gas have been found in an additional 279 provinces, and significant accumulations may occur in these or other basins that were not assessed. The estimates are therefore conservative.

Areas that contain the greatest volumes of undiscovered conventional oil include the Middle East, western Siberia, the Caspian region, and the Niger and Congo deltas. Significant undiscovered oil resource potential was suggested in a number of areas that do not have important production history, such as northeast Greenland and offshore Suriname. Areas that contain the greatest volumes of undiscovered conventional gas include the West Siberian Basin, the shelves of the Barents and Kara Seas, the Middle East and offshore Norway. Significant additional undiscovered gas resources may occur in a number of areas where large discoveries have been made but remain undeveloped. Examples include East Siberia and the Northwest Shelf of Australia. About half of the assessed undiscovered petroleum potential of the world is offshore, especially outside the established provinces of the United States, former Soviet Union, Middle East and North Africa. Arctic basins, which hold about 25 percent of the undiscovered petroleum resources, make up the next great frontier.

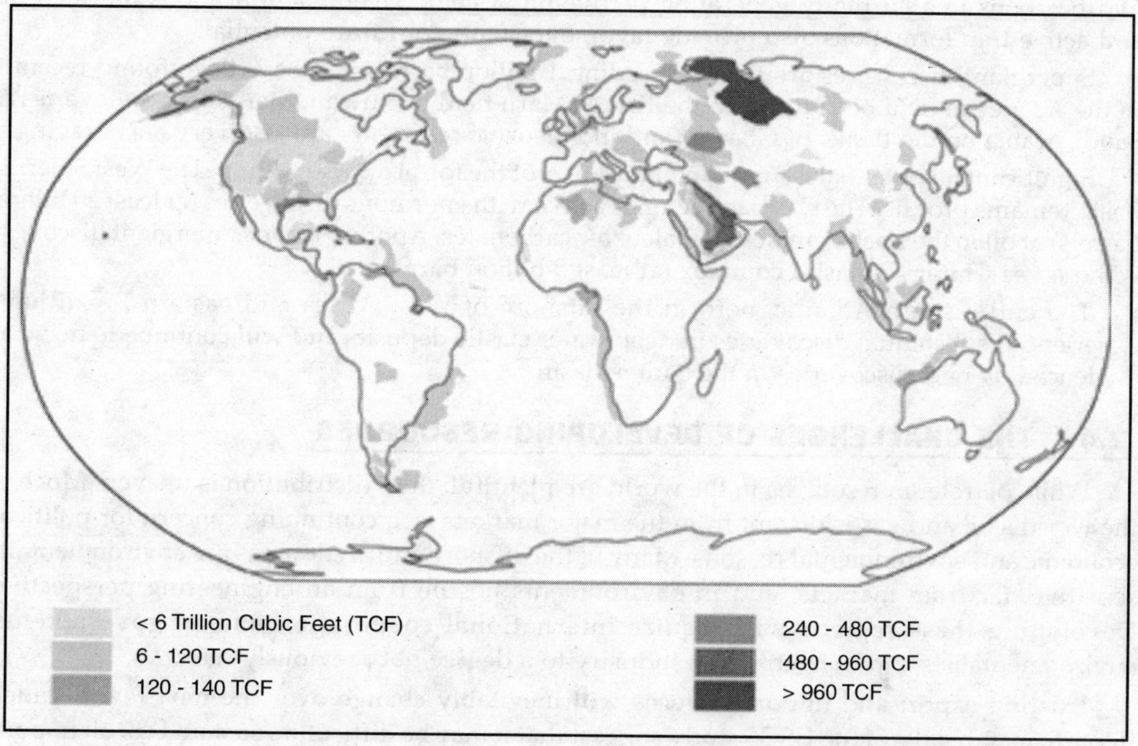

Fig. 2.4 The world's gas: Gas endowment (cumulative production plus mean estimates remaining of oil reserves and undiscovered natural gas resources) by province, in trillions of cubic feet of gas (TCF) for 128 gas provinces. Note that 6 TCF equal 1 BBO in energy equivalent units. World data are from the U.S. Geological Survey (USGS) 2000 assessment; U.S. data are from USGS (1995 assessment) and the Minerals Management Service (1996 assessment). USGS image.

The assessment suggests that some recent claims of an imminent oil shortage cannot be supported. Furthermore, large volumes of natural gas can replace oil in most market sectors. The rate of production of resources depends on many factors, including investments in exploration and development, political conditions and the growth or decline in demand from the global economy.

It is impossible to forecast rates of production purely from geologic information. Given this fact, it does appear likely that the global production from conventional oil accumulations identified in the USGS 2000 assessment will be in decline before the middle of the century. The relative abundance of natural gas will encourage its substitution for oil in some market sectors. Advancements in new technologies — such as gas-to-liquids and various hydrogen based systems that extract hydrogen from methane — will facilitate this substitution.

2.3. THE MIDDLE EAST

The USGS assessment of undiscovered oil resources for the Middle East was higher than previous assessments. This new estimate is based on detailed petroleum migration modelling that integrated geochemical data with recently acquired structure and seismic data. The Jurassic has both rich source rocks and carbonate reservoirs. Cretaceous and Tertiary reservoirs trap hydrocarbons in a currently generating petroleum system. Several significant evaporite seals and active trap formations also provide favorable factors for future potential.

Supergiant discoveries are those exceeding 1 billion barrels and have been found recently in the Azadegan field of Iran and in the Kra Al Maru field of Kuwait. Many new discoveries in Saudi Arabia on the flanks of Ghawar and in Paleozoic reservoirs are also very encouraging.

Significant oil potential is found in many areas of the former Soviet Union. The West Siberian Basin remains prolific. The Caspian region is active with enormous discoveries (at least 20 billion barrels) of oil in the Kashagan field in Paleozoic carbonates. Another recent supergiant discovery is the Azeri-Chirag-Gunashli complex (at least 5 billion barrels).

The entire south Atlantic, both in the offshore of West Africa and eastern Brazil have experienced substantial discoveries in deep water clastic deposits and will continue to do so as evidenced by new discoveries in the Santos Basin.

2.4. THE CHALLENGES OF DEVELOPING RESOURCES

While petroleum resources in the world are plentiful, their distribution is uneven. Much of the world's oil and gas is distant from the major markets — a continuing concern for political, economic and environmental reasons. Many of the offshore and arctic areas are environmentally sensitive, far from markets, and in environments hostile from an engineering perspective. Developing these regions will require international cooperation among governments, environmentalists and the petroleum industry to a degree not previously seen.

Existing export and import patterns will inevitably change over the next few decades. In particular, results of the USGS study suggest that it may be difficult to sustain U.S. oil imports from Mexico and natural gas imports from both Canada and Mexico. *For example*, the USGS estimates for technically recoverable, conventional natural gas resources in the Western Canada Sedimentary Basin are an order of magnitude lower than previous Canadian assessments of the region. Recent Canadian assessments have been lowered. The differences among assessments are related to the minimum size at which a field is considered economically viable. Smaller fields are more numerous than larger fields. The smallest field size that USGS considered viable on a world scale was 3 billion cubic feet of gas for a Canadian gas pool.

Although many energy scenarios call for increased natural gas consumption in the United States, we may need to increase imports of liquefied natural gas or make major domestic natural gas discoveries to meet this new demand. Currently, the United States imports about 15 percent of its natural gas from Canada.

Many organizations and individuals regard the USGS 2000 assessment as the most credible and scientific world petroleum assessment available. However, assessments are dynamic and each represents a snapshot in time. It is worth considering the evolution of some of the non-geologic factors that necessitated a revision from previously published USGS assessment numbers. Because these parameters are still evolving, we can also speculate how they may impact future assessments.

2.5. THE RESOURCE PYRAMID

The amount of oil or gas in the world can be viewed as a pyramid with a small amount of high quality resource that is cheap to extract, and with increasing amounts of lower quality resource that cost more to extract. The upper part of the pyramid is well defined, as these resources are mostly known and are generally considered "*conventional*." The lower part of the pyramid is less well understood and the amount of petroleum in accumulations that are now largely uneconomic — such as hydrates or basin-centered gas — is highly speculative. An assessment draws a slice through the pyramid defining the resource of oil or gas that is estimated to become economic within the foreseeable future.

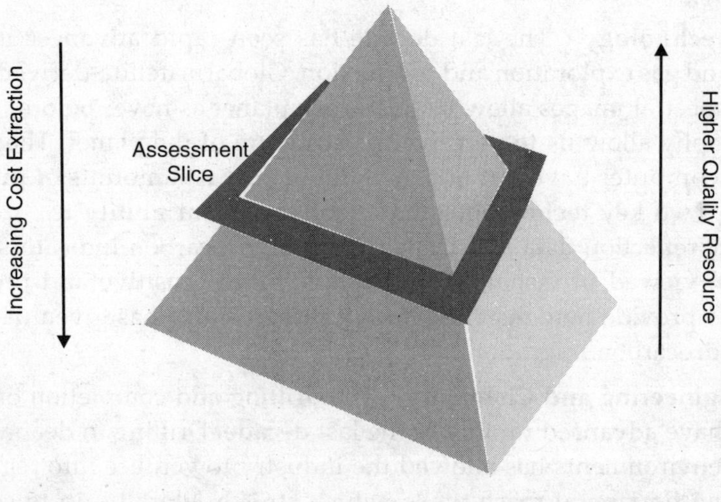

FIG. 2.5. THE RESOURCE PYRAMID : THE PYRAMID CHARTS THE ECONOMIC FEASIBILITY OF AN AREA'S PETROLEUM RESERVES. IT SHOWS A SMALL VOLUME OF PRIME RESOURCE AT TOP AND, AT BOTTOM, LARGER VOLUMES OF LOWER QUALITY RESOURCES THAT ARE MORE EXPENSIVE TO EXTRACT. THE AMOUNT AND NATURE OF OCCURRENCE OF OIL AND GAS NEAR THE BASE OF THE PYRAMID IS NOT WELL UNDERSTOOD. ASSESSMENTS SLICE THROUGH THE PYRAMID, DEFINING ONLY RESOURCES THAT HAVE ECONOMIC POTENTIAL WITHIN THE FORESEEABLE FUTURE. USGS IMAGE.

Over time, the relative positions within the pyramid of the various accumulations of the world's oil and gas vary. The geological abundance of petroleum (prior to extraction) remains the same, but our perception of it changes for a variety of reasons. Hydrocarbon accumulations that were once thought to be only of scientific interest are transformed into "*unconventional resources*" and eventually become "*conventional*" as they rise higher in the pyramid. For example, the recovery costs for oil from Alberta's oil sands have fallen dramatically over the last 20 years and are now about $8 per barrel. By 2005, fully 10 percent of North America's oil production will come from Alberta's oil sands.

The Alberta oil sands contain 1.6 trillion barrels of oil, of which 311 billion barrels are recoverable with current technology. Similarly, a significant amount of U.S. natural gas production now comes from sources once considered unconventional but now viewed as conventional. Resources may also fall in the pyramid. In 1860, Titusville was the oil resource at the top of the pyramid. In 1901, Spindletop was. Although they still produce, western Pennsylvania and eastern Texas are no longer near the top of the pyramid. At the beginning of the 20th century, the Middle East as an oil resource was viewed at the bottom of the pyramid.

2.5.1. Elements Influencing the Pyramid

The human and social factors affecting data and information incorporated into a resource assessment can be viewed schematically as a sphere of interacting factors that circle outside the pyramid and act on it. In large part, these factors determine the relative position of various petroleum accumulations within the pyramid and ultimately determine what a viable petroleum resource is. Human and social factors are dynamic and appear to change at an ever-increasing pace. USGS gains insights into potential future viable resources related to these factors by talking with petroleum industry leaders.

Elements affecting the pyramid are :

Geoscience Technology : The last decade has seen rapid advances in the science and technology of oil and gas exploration and production. Global, satellite-derived data and detailed topographic and spectral images allow us to see our planet as never before. Deep seismic data and earth tomography allow us to perceive the structure of the planet. The digital revolution and the personal computer have put our own data and vast amounts of public domain data at our fingertips. Two key technologies have enhanced our ability to "*see*" the subsurface realm: 3-D seismic reflection data and DHIs (Direct Hydrocarbon Indicators determined from seismic attributes) viewed in visualization centers. More sensitive and precise logging and geochemistry tools provide new insights. Recent research also has given us more clues about fluid flow and hydrocarbon migration.

Petroleum Engineering and Chemistry : The drilling and completion of wildcat tests and production wells have advanced rapidly in the last decade. Drilling in deep water in the Arctic and other hostile environments has allowed the industry to venture into regions once thought "*out of reach.*" Long horizontal reach wells, which stretch laterally up to several kilometers, extend the reach of drilling and enhance recovery, have become common. Well stimulation and improved technologies, which increase the rate of petroleum flow from a reservoir by injecting hydraulic fluids to fracture the rocks, have greatly increased recovery factors and improved recovery economics for many accumulations.

Extraction of heavy oil, tight gas and coal bed methane is now economically viable in North America, but still considered "*unconventional*" in many parts of the world. The increasing ability to process huge data sets promises advanced models of fluid flow and geochemistry that will continue to improve the ability to identify prospects, and will help to boost recovery factors. A small percentage rise in recovery factors worldwide would increase recoverable oil and gas resources considerably.

The International Oil and Gas Business : The complex interrelations of national oil companies, large international integrated energy companies, and independent operators all contribute to our perceptions of world oil and gas resources. Sanctions, subsidies, taxes, regulations and environmental policies all impact the potential availability of resources and our relative understanding of them. Despite all of our geological expertise, much oil and gas is still found by drilling, and if large geographic areas are "*off limits*" for long periods, there is no chance of serendipitous discoveries.

Gas is assuming a more prominent role in the petroleum industry. Much needs to be learned about the viability of unconventional gas deposits such as coalbed methane, tight gas, fractured shale gas, gas hydrates and dissolved saltwater gas. We have many questions to answer: Can we effectively find and produce these dispersed, low concentration hydrocarbons? Will markets be found for geographically stranded gas? If a coalbed methane accumulation is found in Siberia, is it a reserve, a resource or something that has no potential to be developed within the foreseeable future?

The Political Domain : Thirty years ago it was common to speak of the *"500 sedimentary basins"* of the world. At that time, the Cold War divided the world, and most western geoscientists knew little of eastern block basins. Now we speak of more than 1,500 sedimentary basins worldwide. Political agendas and policies have a major impact on whether or not a particular hydrocarbon accumulation is considered a resource. Regulations, environmental policies, subsidies and economic agendas in one region may have a global impact.

Global Economics : In the competitive petroleum business of finding and producing oil and gas, performance is measured by the value created. The energy business floats on an economic sea of high and low tides and periodic storms. The health of the global economy, the price of oil, and cycles of demand and over supply raise and lower all boats. In such dynamic, commodity driven environments, the definition of a resource changes with time and market conditions.

The Human Element : A critical component in the resource equation is the human mind. The mind can create models of the dynamic subsurface realm and intelligently estimate the resource base. But we face two problems. In this time of prodigious quantities of data and information, the number of knowledgeable and experienced oil finders is decreasing because fewer are being trained. The annual survey of universities conducted by the American Geological Institute shows that the number of geoscience degrees granted in 2001 was a third of the number granted 20 years ago.

The second problem is that the average citizen has little knowledge of the sub-surface realm and even less about the fluid dynamics of oil and gas deposits. Thus, most policymakers and users of energy are not equipped to enter the debate about the nature of the world's oil and gas resources. As a consequence, the debate often has migrated to the end members - each one with a set of personal, corporate, political or environmental agendas. In large part, it is a misunderstanding of the nature of resources that allows those with agendas to argue vehemently over the size of remaining resources.

The resource pyramid will continue to evolve. An understanding of the geological, technological, economic, political and social forces that drive that change is critical to understanding resources. Raw oil and gas resource numbers, used out of context, are the shot and powder of manipulators. The economic petroleum geoscientist must clearly be at the table when corporate managers, politicians, lawyers, economists and activist groups discuss world energy resources.

2.6. OIL IN THE CASPIAN

At the turn of the 20th century, the area around what is now Baku, Azerbaijan, on the Absheron peninsula, was the world's largest oil exporter. *"Middle Eastern oil started in the Caspian,"* says Igor Effimoff, a Caspian expert who is now chief operating officer of Teton Petroleum Company and who worked in the Caspian for Pennzoil Caspian Corp. and Larmag Energy N.V.

FIG. 2.6. A MAP OF THE CASPIAN REGION AND ITS PIPELINES, SUPPLIED COURTESY OF IGOR EFFIMOFF, TETON PETROLEUM CO.

Nonetheless, much of the oil frontier in the offshore of the Caspian region remains underexplored and untapped. "For some years, the bulk of the Caspian area lay relatively fallow because the Soviet Union had diverted its resources to West Siberia, where huge reserves of oil and gas had been discovered at shallow depths," *Effimoff says*. "Not everything has been found. There are many opportunities remaining in the Caspian." With the breakup of the Soviet Union, which had controlled most of the region, five independent countries now border the Caspian Sea. All of them are interested in developing their resources. According to the U.S. Department of Energy's Energy Information Administration (EIA), the Caspian region's proven, probable and possible oil reserves (50 percent of probable) could be as high as 233 billion barrels. EIA says the Caspian region, which it defines as the Caspian Sea, all of Azerbaijan, Kazakhstan and Turkmenistan, and the portions of Russia and Iran near the Sea , is comparable to the North Sea in its hydrocarbon potential. It is also possesses a huge natural gas resource. EIA estimates that the region's proven, probable and possible natural gas reserves are as high as 293 trillion cubic feet. Azerbaijan is developing what could be the largest new field discovered in the last 25 years, Shah Deniz.

The Caspian countries are accessing foreign investors and the latest technologies not only technology in the marine environment but also drilling deeper, more complicated wells, onshore and offshore.

At the same time, the United States is interested in developing oil resources as alternatives to Middle East oil, and is putting the Caspian region in the limelight. Despite its fortuitous geology and access to new tools, complicated politics will mean the region probably won't begin to realize its full potential for several years, says Andrew Neff, Eurasian Energy Analyst with EIA.

"The uncertainty surrounding the ownership of the Sea's resources are probably the biggest obstacle to be overcome in order for the Caspian to realize its full hydrocarbon potential. The Soviet Union and Iran signed bilateral treaties to share the Caspian Sea; but under new politics, debate continues among the five countries over how to divide the Sea. Turkmenistan and, more strongly, Iran oppose a plan suggested by Russia, Kazakhstan and Azerbaijan.

The second obstacle, is a lack of export routes to transport the oil and gas to consumers.

"Investments in export infrastructure will need to be made first, which brings us back to the lack of an agreement on dividing the Sea."

Also in recent years, some western companies drilling in the Caspian Sea encountered dry holes or uncommercial quantities. Some of these failed sites were near highly productive areas. "The disappointments have been mainly driven by variations in geology that were not recognized before the wells were drilled stating that the region's geology is extremely complicated. Caspian region remains high because of the huge potential it promises. Already, consortia of international energy companies have invested billions in large projects for producing three oil fields and the Shah Deniz gas field, along with pipelines to transport the resources to market. These projects access portions of the Caspian Sea where boundaries are not disputed.

2.7. OIL IN IRAQ

Of all the uncertainties in assessing world oil resources, one of the greatest is the future of Iraq. The oil reserves in Iraq are among the largest in the world, second only to Saudi Arabia. More than 100 billion barrels of oil are known to exist within the country, and the U.S. Geological Survey (USGS) estimates an additional 45 billion barrels remain to be discovered. Yet war and sanctions have driven wild fluctuations in Iraq's oil production over the past 25 years, and the crystal ball for future production is hazy at best.

"Iraq contains whole petroleum systems: world-class source rocks, overlain by excellent reservoirs and terrific evaporite seals," sas said by Thomas Ahlbrandt, a research geologist with USGS.

Many of the source rocks began forming in the Jurassic. Shallow seas periodically inundated the northeastern section of the Arabian plate, where Iraq currently sits. Those seas teemed with life, spurred by the warmth of the equatorial latitudes. Thick layers of organic-rich shales and marine carbonates accumulated in expansive basins and compressed into source rocks that remained little disturbed for millions of years.

In the Tertiary, subduction of the Arabian plate beneath the Eurasian plate drove the source rocks down into a *"thermal generating window"*. High heat and pressure converted the shale and carbonate-rich rocks into oil and gas. Much of the oil migrated up into anticlines sealed by thick layers of salt and anhydrite left behind from the Jurassic seas as they dried out. Despite its immense reserves, Iraq's oil production is low. In 2001, Iraq produced 2.45 million barrels of oil per day, accounting for only 12 percent of the total oil produced in the Persian Gulf. The United States produced more than three times that much, even though its reserves are much smaller.

"They have the oil resources but it is a question of whether they have the wells or facilities to increase production."

Low production reflects years of sanctions and deteriorating drilling infrastructure. On Aug. 6, 1990, four days after Iraq invaded Kuwait, the United Nations banned all oil exports from the country. Oil production dropped to 0.3 million barrels of oil per day, less than a tenth of what it was before the Persian Gulf and Iran/Iraq wars began. For five years, production barely exceeded what the country itself consumed. Wells and supporting facilities fell into disrepair. In 1996, Iraq began exporting oil again under the U.N. Oil-for-Food program. The program set strict limits on the amount of oil that Iraq could export. All revenue from the sales had to go toward food, medicine or other humanitarian supplies for the people of Iraq. Oil production steadily increased for the first few years of the program.

However, production has leveled off in the past 3 years, even though the U.N. has removed the ceiling for the amount of oil that Iraq can produce under the Oil-for-Food program. Sanctions continue to discourage international investment needed to boost production capacity. Russia, France and China have contracts to develop oil fields in the region, but sanctions have slowed fulfillment of those contracts.

Iraq claims production will increase quickly if the U.N. lifts sanctions. "Iraq anticipates producing up to 3 million barrels per day within a year or two, and up to 5 million a day within five years, assuming they are not limited by demand, OPEC or sanctions." Yet the future of production is unclear, especially given the possibility of a U.S.-led war in Iraq.

The United States does import some oil from Iraq, although the numbers are low. In 2001, imports from Iraq accounted for only 5.5 percent of the total U.S. imports. In contrast, imports from the Persian Gulf as a whole added up to 20 percent - a significant contribution since the United States relies on imports for more than half of its oil consumption each year.

2.8. SCENARIO OF PETROLEUM RESERVES IN INDIA

According to *Oil and Gas Journal (OGJ)*, India had 5.6 billion barrels of proven oil reserves as of January 2009, the second-largest amount in the Asia-Pacific region after China. India's crude oil reserves tend to be light and sweet, with specific gravity varying from 38° API in the offshore Mumbai High field to 32° API at other onshore basins.

India produced roughly 880 thousand bbl/d of total oil in 2008, of which approximately 650 thousand bbl/d was crude oil, with the rest of production resulting from other liquids and refinery gain. India has over 3,600 operating oil wells, according to *OGJ*. Although oil production in India has slightly trended upwards in recent years, it has failed to keep pace with demand and is expected to decline slightly in 2009.

India's oil consumption has continued to be robust in recent years. In 2007, India consumed approximately 2.8 million bbl/d, making it the fifth largest consumer of oil in the world. Demand grew to nearly 3 million bbl/d in 2008. It has been anticipated that consumption growth rates flattening in 2009 largely due to slowing economic growth rates and the recent global financial crisis. The Fig. 2.7 below shows the production and consumption data of India from 1990-2009.

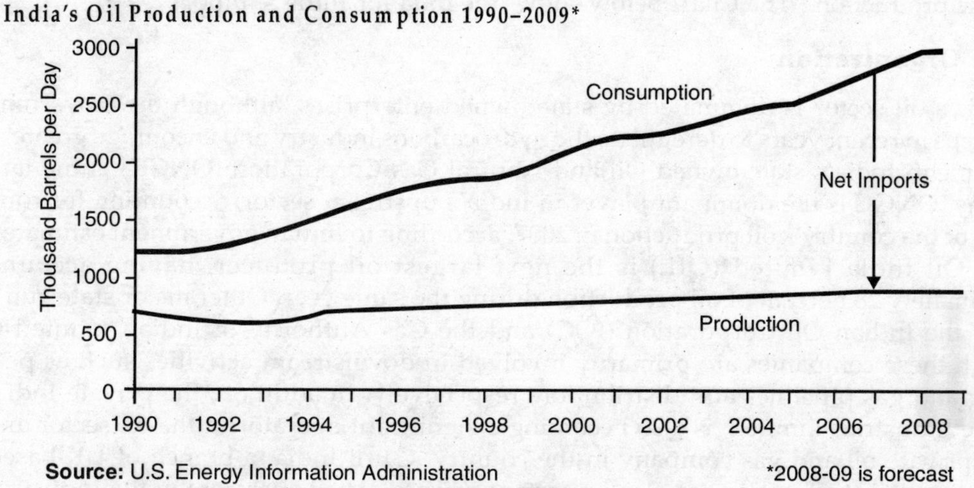

Fig. 2.7. Shows the production and consumption data of India from 1990-2009.

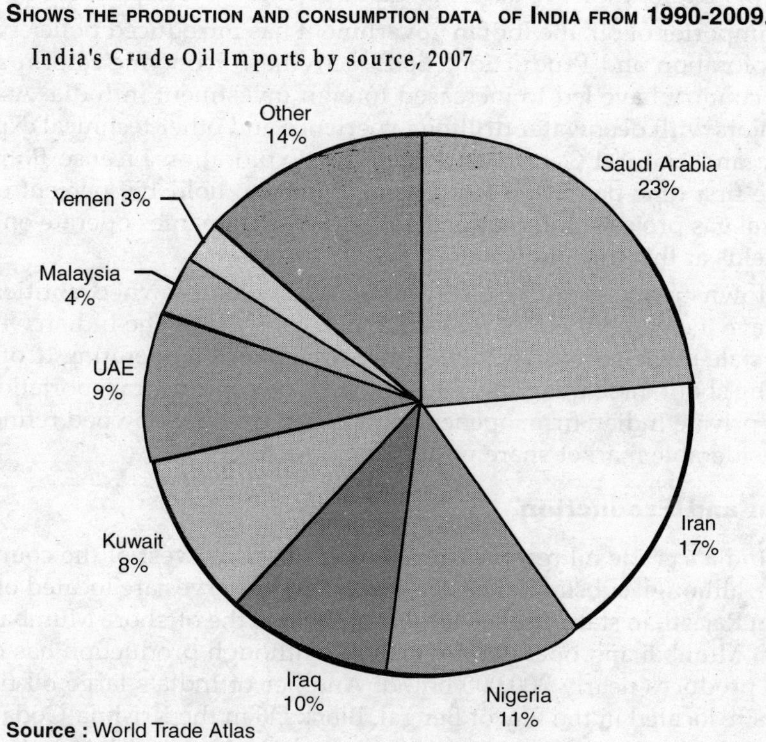

Fig. 2.8. India's Oil Import.

The combination of rising oil consumption and relatively flat production has left India increasingly dependent on imports to meet its petroleum demand. In 2006, India was the seventh largest net importer of oil in the world. With 2007 net imports of 1.8 million bbl/d, India is currently dependent on imports for 68 percent of its oil consumption. The EIA expects India to become the fourth largest net importer of oil in the world by 2025, behind the United States, China, and Japan.

The Government of India's largest crude oil import partner is Saudi Arabia, followed by Iran. Nearly three-fourths of India's crude oil imports come from the Middle East. The Indian government expects this geographical dependence to rise in light of limited prospects for domestic production. The Chart below show the data for India's import.

Sector Organization

India's oil sector is dominated by state-owned enterprises, although the Government has taken steps in recent years to deregulate the hydrocarbons industry and encourage greater foreign involvement. India's state-owned Oil and Natural Gas Corporation (ONGC) is the largest oil company. ONGC is the dominant player in India's upstream sector, accounting for roughly 71 percent of the country's oil production in 2007, according to Indian government estimates. State-owned Oil India Limited (OIL) is the next largest oil producer, having accounted for approximately 28 percent of oil production during the same year. Other major state-run players include the Indian Oil Corporation (IOC) and the Gas Authority of Indian Limited (GAIL), although these companies are primarily involved in downstream activities such as petroleum refining and gas pipelines and distribution, respectively. In addition, the private Indian firm, Reliance Industries Limited, is also becoming a significant operator in the oil sector and is the largest private oil and gas company in the country. Cairn India, a branch of UK-based Cairn Energy, and BG Exploration are also important private sector operators in the industry.

As a net importer of oil, the Indian government has introduced policies aimed at increasing domestic Exploration and Production (E&P) activities. Economic reform and other efforts to open up the country have led to increased foreign investment in India. As part of an effort to attract oil majors with deepwater drilling experience and other technical expertise, the Ministry of Petroleum and Natural Gas created the New Exploration License Policy (NELP) in 2000, which for the first time permitted foreign companies to hold 100 percent equity ownership in oil and natural gas projects. International oil and gas companies operate only a relatively small number of fields at this time, however.

India's downstream sector is also dominated by state-owned entities, although private companies have increased their market share in recent years. The Indian Oil Corporation (IOC) is the largest state-owned company in the downstream sector, operating 10 of India's 18 refineries and controlling about three-quarters of the domestic oil pipeline transportation network. Reliance Industries, a private Indian firm, opened India's first privately-owned refinery in 1999, and has gained a considerable market share in India's oil sector.

Exploration and Production

Most of India's crude oil reserves are located offshore, west of the country, and onshore in the northeast, although substantial newly-discovered reserves are located offshore in the Bay of Bengal and in Rajasthan state. India's largest oil field is the offshore Mumbai High field, located north-west of Mumbai and operated by ONGC. Although production has declined at Mumbai High, it still produces nearly 300,000 bbl/d. Another of India's large oil fields is the Krishna-Godavari basin, located in the Bay of Bengal. Block D6 in the Krishna-Godavari basin, operated by Reliance Industries, could account for as much as 40 percent of India's current domestic hydrocarbon output when production peaks. Oil production from this block began in September 2008. Onshore, GAIL is planning to invest $18 million in exploratory drilling in Assam State, an important location for oil reserves. ONGC also plans to invest as much as $2.4 billion in exploratory work in the northeast, particularly in Mizoram, Tripura, and Nagaland states.

In light of declining production at the majority of India's fields, companies are investing in enhanced oil recovery methods. ONGC plans to invest nearly $1.5 billion in such projects, and a multitude of these schemes have been approved for many of the company's fields.

To help meet growing oil demand and support the country's energy security, India has promoted various E&P projects in an effort to boost domestic oil production. However, new E&P projects are expected to be difficult due to their deepwater location or terrain type. In order to address these challenges, Indian companies are recruiting foreign firms with greater experience and more sophisticated technology. *For example*, ONGC recently assigned a participating interest to Rocksource ASA, a Norwegian company with technological expertise in deepwater drilling, and to Petrobras for the development of an eastern offshore deepwater block. The participation of private foreign firms over the last five years has helped develop previously unexploited deepwater areas and allow India to tap more of its domestic oil resources.

The primary mechanism through which the Indian Government has promoted new E&P projects has been the NELP framework. The latest round of auctions, NELP VII, has been less successful than the government expected and with interest stemming largely from domestic firms rather than foreign investors. India plans to launch the NELP VIII bidding round in mid-March 2009.

In recent years, Indian NOCs have increasingly looked to acquire equity stakes in E&P projects overseas. The most active company abroad is ONGC Videsh Ltd (OVL), the overseas investment arm of ONGC. ONGC Videsh conducts oil and natural gas operations in 13 countries, including Vietnam, Myanmar, Russia (Sakhalin Island), Iran, Iraq, Sudan, Brazil, and Columbia. One of ONGC Videsh's most high profile investments is its share in the Greater Nile Petroleum Operating Company (GNPOC), which has engaged in E&P work in Sudan since 1997. ONGC Videsh acquired a 25 percent equity stake in the company in 2003, with the balance held by the China National Petroleum Company (CNPC, 40 percent), Petronas (30 percent), and the Sudan National Oil Company (Sudapet, 5 percent). The GNPOC acreage in Sudan holds proved crude oil reserves of more than one billion barrels with current production levels at roughly 300,000 bbl/d from 10 fields. In addition to the upstream activities, the GNPOC companies operate a 935-mile crude oil pipeline that pumps oil to Port Sudan for export . ONGC Videsh also holds a 20 percent stake in the ExxonMobil-led consortium that operates the Sakhalin-I project in Russia. According to company estimates, the oil fields associated with Sakhalin-I hold recoverable crude oil reserves of 2.3 billion barrels. For more information on the Sakhalin-I project, please see the Russian CAB.

In addition to ONGC, other Indian companies are also actively involved in E&P projects abroad. OIL, for example, is working on projects in Libya, Gabon, Nigeria, and Sudan, and IOC, GAIL, and Reliance are also pursuing overseas oil E&P assets.

Downstream/Refining

According to *OGJ*, India had 2.26 million bbl/d of crude oil refining capacity at 18 facilities as of January 2008. The country has the eighth largest refinery capacity in the world. This year, privately-owned Reliance Industries surpassed state-owned India Oil Corporation (IOC) in terms of refining capacity in the country due to additions to its recently upgraded facility. Reliance's only refinery, the Jamnagar facility, is India's largest, with an initial capacity of 660,000 bbl/d. Reliance recently enlarged the Jamnagar site to add an additional capacity of 580,000 bbl/d, making it the largest refining complex in the world with a refining capacity of 1.24 million bbl/d.

Other key upcoming refinery projects include Essar Oil's Vadinar refinery expansion of 110,000 bbl/d in 2010 and 320,000 bbl/d in 2013, IOC's Paradeep refinery expansion of 300,000 bbl/d in 2014, and 300,000 bbl/d capacity additions to refineries in Bina and Bhatinda by 2013. India is slated to add 1.6 million bbl/d of refining capacity through 2015 based on current proposed projects.

Due to expectations of high demand for petroleum products in the region, further investment in the Indian refining sector is likely. As part of the country's 11th Five Year plan from 2007 to 2012, the Government would like to promote India as a competitive refining destination, and industry experts expect the country to be a significant exporter of refined products to Asia in the near future. The remaining challenge for the country will be to obtain a secure supply of crude oil to feed its refineries.

Refined Fuel Subsidies

Beginning in 2002, the Indian government introduced some measures aimed at deregulation in the downstream oil sector. Private refiners may now directly market some of their own petroleum products to their customers. Additionally, the government phased out the Administered Price Mechanism (APM) on oil products in 2002, replacing it with the new Market Determined Price Mechanism (MDPM). However, while the MDPM is notionally benchmarked to international oil prices, the Indian government continues to heavily subsidize domestic prices of oil products such as diesel, LPG, and kerosene for consumers. As such, demand for petroleum products in India has been substantially influenced by the government's pricing scheme. With diesel prices significantly lower than other fuels, such as gasoline, demand for diesel rose substantially, by as much as 25 percent between 2006 and the first half of 2008, according to industry analysts.

The combination of high international crude oil prices and low domestic fuel rates before the fourth quarter of 2008 led private refiners in the country, particularly Reliance and its Jamnagar facility, to have an overt export focus in order to avoid suffering significant financial losses in the domestic market. Although partial compensation in the form of oil bonds has been extended to ease the financial burden imposed by capped fuel prices to the three state-owned retailers IOC, Hindustan Petroleum (HPCL), and Bharat Petroleum (BPCL), the private refiners have not received the same government support. Upstream firms such as ONGC also subsidized about one-third of state refiners' revenue losses through discounts on crude sales. The Government pushed through small domestic fuel price increases, as the price of petroleum products rose by 15 percent in the first half of 2008, according to industry reports, but rates remained well below competitive international levels. The rate hikes have had little effect on rising demand for diesel fuel and have not entirely closed the losses incurred by Oil Marketing Companies (OMCs). Because of a reversal in crude oil prices from all-time highs in July 2008 and a desire to stimulate the domestic economy, the Indian government cut retail prices in December 2008 and January 2009 by approximately 12 percent for diesel and over 20 percent for gasoline.

The various Petroleum companies in India are :

Indian Oil Corporation : Indian Oil Corporation is an Indian public-sector petroleum company. It is India's largest commercial enterprise, ranking 105th on the Fortune Global 500 listing (2009). It began operation in 1959 as Indian Oil Company Ltd. The Indian Oil Corporation was formed in 1964, with the merger of Indian Refineries Ltd. Indian Oil and its subsidiaries account for a 47% share in the petroleum products market, 40% share in refining capacity and

67% downstream sector pipelines capacity in India. The Indian Oil Group of Companies owns and operates 10 of India's 19 refineries with a combined refining capacity of 60.2 million metric tons per year.On 30th June 2009 . Indian Oil will complete 50 years of its existence and a series of events are being planned to celebrate its Golden Jubilee Year

Overview Indian Oil operates the largest and the widest network of fuel stations in the country, numbering about 17606 (15557 regular ROs & 2049 Kissan Sewa Kendra). It has also started Auto LPG Dispensing Stations (ALDS). It reaches Indane cooking gas to over 47.5 million households through a network of 4,990 Indian distributors. In addition, Indian Oil's Research and Development Center (R&D) at Faridabad supports, develops and provides the necessary technology solutions to the operating divisions of the corporation and its customers within the country and abroad. Subsequently, Indian Oil Technologies Limited - a wholly owned subsidiary, was set up in 2003, with a vision to market the technologies developed at Indian Oil's Research and Development Center. It has been modeled on the R&D marketing arms of Royal Dutch Shell and British Petroleum. Indian Oil's product range covers petrol, diesel, LPG, auto LPG, aviation turbine fuel, lubricants, naphtha, bitumen, paraffin, kerosene etc. extra Premium petrol, Xtra Mile diesel, Servo lubricants, Indane LPG, Autogas LPG, Indian Oil Aviation are some of its prominent brands.

Recently Indian Oil has also introduced a new business line of supplying LNG(Liquefied natural gas) by the cryogenic transportation. The branding called *"LNG at Doorstep"*. Lng headquarters are located in scope complex, Lodhi Road Delhi. The brief out line of the Indian refineries are

Refineries

- **Digboi Refinery :** The **Digboi Refinery** was set up at Digboi in 1901 by Assam Oil Company Ltd.. The Indian Oil Corporation Ltd (IOC) took over the refinery and marketing management of Assam Oil Company Ltd. with effect from 1981 and created a separate division. This division has both refinery and marketing operations. The refinery at Digboi had an installed capacity 0.50 MMTPA (million metric tonnes per annum). The refining capacity of the refinery was increased to 0.65 MMTPA by modernization of refinery in July, 1996. A new delayed Coking Unit of 1,70,000 TPA capacity was commissioned in 1999. A new Solvent Dewaxing Unit for maximizing production of micro crystalline wax was installed and commissioned in 2003. The refinery has also installed Hydrotreater to improve the quality of diesel, in Upper Assam, is India's oldest refinery and was commissioned in 1901.
- **Guwahati Refinery : Guwahati Refinery** is an oil refinery near Guwahati, India owned by the Indian Oil Corporation. It was the first in the public sector and was set up in collaboration with Romania at a cost of Rs.17.29 crores and commissioned on 1 January 1962 with a design capacity of 0.75 MMTPA. The present capacity of this Refinery is 1.00 MMTPA. A hdrotreater unit for improving the quality of diesel has been installed and was commissioned in 2002. The refinery has also installed in 2003 Indmax Unit, a novel technology developed by its R&D Centre for upgrading heavy ends LPG, motor spirit and diesel oil.
- **Barauni Refinery : Barauni Refinery** in the Bihar state of India was built in collaboration with the Soviet Union at a cost of Rs.49.4 crores and went on stream in July, 1964. The initial capacity of 2 MMTPA was expanded to 3 MMTPA by 1969. The present capacity of this refinery is 6.00 MMTPA. A Catalytic Reformer Unit (CRU) was also added to the refinery in 1997 for production of unleaded motor spirit. Gujarat Refinery, at Koyali in Gujarat in Western India, is IndianOil's largest refinery. The refinery was commissioned

in 1965. It also houses the first hydrocracking unit of the country. Its present capacity is 13.70 MMTPA.

- **Haldia Refinery :** The **Haldia Refinery** for processing 2.5 MMTPA of Middle East crude was commissioned in January, 1975 with two sectors - one for producing fuel products and the other for Lube base stocks. The refinery is in Haldia near Kolkata (West Bengal).The fuel sector was built with French collaboration and the Lube sector with Romanian collaboration. The refining capacity of the Refinery was increased to 2.75 MMTPA in 1989 through debottle necking measures. The refining capacity was further expanded to 3.75 MMTPA with the commissioning of new crude distillation unit of 1.0 MMTPA in March, 1997. The present refining capacity of this Refinery is 6.00 MMTPA
- **Mathura Refinery :** The **Mathura Refinery**, owned by Indian Oil Corporation, is located in Mathura, Uttar Pradesh. The refinery processes low sulphur crude from Bombay High, imported low sulphur crude from Nigeria, and high sulphur crude from the Middle East. The refinery, which cost Rs.253.92 crores to build, was commissioned in January, 1982.[1] Construction began on the refinery in October 1972. The foundation stone was laid by Indira Gandhi, the former Prime Minister of India. The FCCU and Sulphur Recovery Units were commissioned in January, 1983. The refining capacity of this refinery was expanded to 7.5 MMTPA in 1989 by debottlenecking and revamping. A DHDS Unit was commissioned in 1989 for production of HSD with low sulphur content of 0.25% wt. (max.). The present refining capacity of this refinery is 8.00 MMTPA. In January 2009, the plant shut down for a period of time due to a strike.[2]The refinery was in the news for allegedly causing the white marble of the Taj Mahal to yellow. It is located about 50 kilometers away from the Taj Mahal.[3] It is currently asking the Indian government to allow an expansion, raising the capacity to 11 million tonnes.[4] The refinery also wants to create a new garbage disposal site, which has garnered new outrage from environmental activists because the site will be located even closer to the Taj Mahal. The India Government hired a panel to examine the effects of the refinery on the Taj Mahal. The panel found that the air has high levels of suspended particulate matter, caused by factory emissions, dust, construction, and exhaust from automobiles. These are causing the Taj Mahal to change color.[6]Mathura Refinery is known to be the number one amongst all 14 refineries in India.
- **Panipat Refinery : Panipat Refinery** is an oil refinery that was set up in 1998. It is located in Baholi village, Panipat, Haryana. Panipat is the seventh refinery belonging to Indian Oil Corporation Limited. The cost of the refinery's construction was Rs 9024 crores. It has a capacity of 12 MMTPA.

The refinery's highlights include :

- Zero discharge of effluent gases.
- The presence of four ambient air monitoring stations that were in place well before the refinery was in use.
- It is an eco-friendly refinery, as indicated by a green belt outside it.
- The establishment of a totally electronic based communication system within the refinery.
- It has the lowest manpower of all refineries in the region with similar capacities.

Panipat Refinery holds several quality certifications from DNV (Det Norske Veritas), including:

1. **ISO 9002** : **ISO 9000** is a *family* of standards for quality management systems. ISO 9000 is maintained by ISO, the International Organization for Standardization and is

administered by accreditation and certification bodies. The rules are updated, as the requirements motivate changes over time. Some of the requirements in ISO 9001:2008 (which is one of the standards in the ISO 9000 family) include

- a set of procedures that cover all key processes in the business;
- monitoring processes to ensure they are effective;
- keeping adequate records;
- checking output for defects, with appropriate and corrective action where necessary;
- regularly reviewing individual processes and the quality system itself for effectiveness; and
- facilitating continual improvement

A company or organization that has been independently audited and certified to be in conformance with ISO 9001 may publicly state that it is *"ISO 9001 certified"* or *"ISO 9001 registered"*. Certification to an ISO 9001 standard does not guarantee any quality of end products and services; rather, it certifies that formalized business processes are being applied.

2. **ISO 14001 :** The **ISO** 14000 is a standard for environmental management systems that is applicable to any business, regardless of size, location or income. The aim of the standard is to reduce the environmental footprint of a business and to decrease the pollution and waste a business produces. The most recent version of ISO 14001 was released in 2004 by the International Organization for Standardization (ISO) which has representation from committees all over the world. The ISO 14000 environmental management standards exist to help organizations minimize how their operations negatively affect the environment. In structure it is similar to ISO 9000 quality management and both can be implemented side by side. In order for an organization to be awarded an ISO 14001 certificate they must be externally audited by an audit body that has been accredited by an accreditation body. In the UK, this is UKAS. Certification auditors need to be accredited by the International Registrar of Certification Auditors. The certification body has to be accredited by the Registrar Accreditation Board in the USA, or the National Accreditation Board in Ireland.

- **ISO 14001** Environmental Management Systems—Requirements with guidance for use.
- **ISO 14004** Environmental Management Systems—General guidelines on principles, systems and support techniques.
- **ISO 14015** Environmental assessment of sites and organizations.
- **ISO 14020** series (14020 to 14025) Environmental labels and declarations.
- **ISO 14031** Environmental Performance Evaluation—Guidelines.
- **ISO 14040** series (14040 to 14049), Life Cycle Assessment, LCA, discussed pre-production planning and environment goal setting.
- **ISO 14050** terms and definitions.
- **ISO 14062** discusses making improvements to environmental impact goals.
- **ISO 14063** Environmental Communication—Guidelines and examples
- **ISO 19011** which specifies one audit protocol for both 14000 and 9000 series standards together. This replaces **ISO 14011** meta-evaluation—how to tell if your intended regulatory tools worked. 19011 is now the only recommended way to determine this.

3. **OHSAS 18001**

Group companies and joint ventures

- **Indian Oil Technologies Ltd :** Indian Oil Technologies Ltd.is the marketing arm for the entire range of technologies developed at the R&D Centre of Indian Oil Corporation Limited. This website gives an overview of the various technologies, products, quality services and solutions, developed and tested to meet or exceed customer requirements. These technologies are tailor made and meant to provide competitive advantage. Indian Oil Technologies is built on the strength of its credibility proven in its sister divisions.

  Indian Oil Technologies Ltd. headquarters are located at Indian Oil R&D Centre, Faridabad. The :

  - Indian Oil (Mauritius) Ltd.
  - Lanka IOC PLC - Group company for Sri Lanka retail and storage operations which is listed on Colombo's stock exchange. It was locked into a bitter subsidy payment dispute with Sri Lanka's Government which has since been resolved.
  - IOC Middle East FZE
  - Chennai Petroleum Corporation Ltd.
  - Bongaigoan Refinery and Petrochemicals Ltd.
  - Green Gas Ltd. - joint venture with Gas Authority of India for city-wide gas distribution networks.
  - Indo Cat Pvt. Ltd., with Intercat, USA, for manufacturing 15,000 tonnes per annum of FCC (fluidised catalytic cracking) catalysts & additives in India, for catering to rising global demand.
  - Numerous exploration and production ventures with Oil India Ltd., Oil and Natural Gas Corporation.

Indian Oil is the highest ranked Indian company in the prestigious Fortune Global 500 listing, the 116th position(in 2008) based on fiscal 2007 performance. It is also the 18th largest petroleum company in the world and the number one petroleum trading company among the National Oil Companies in the Asia-Pacific region. IOCL was featured on the 2008 Forbes Global 2000 at position 303.

Oil Industry Development Board : India has begun the development of a strategic crude oil reserve sized at 37.4 million barrels, enough for two weeks of consumption.[1] Petroleum stocks have been transferred from the Indian Oil Corporation (IndianOil) to the Oil Industry Development Board (OIDB). The OIDB then created the Indian Strategic Petroleum Reserves Ltd (ISPRL) to serve as the controlling Government agency for the strategic reserve. The primary energy consumption in India in 1998 was 285 kg. OE as against 5800 kg OE in North America. This is excluding the non-commercial energy sources like wood and animal waste. Although there is a conscious shift towards the use of oil as an alternate fuel, coal still constitutes over 56% of the total energy consumption.

Oil and Natural Gas Corporation

Oil and Natural Gas Corporation Limited (ONGC) (incorporated on June 23, 1993) is an Indian public sector petroleum company. It is a Fortune Global 500 company ranked 335th, and contributes 77% of India's crude oil production and 81% of India's natural gas production. It is the highest profit making corporation in India. It was set up as a commission on August 14, 1956. Indian Government holds 74.14% equity stake in this company. ONGC is one of Asia's largest and most active companies involved in exploration and production of oil. It is involved in exploring for and exploiting hydrocarbons in 26 sedimentary basins of India. It produces about 30% of India's crude oil requirement. It owns and operates more than 11,000 kilometers of pipelines in India. Until recently (March 2007) it was the largest company in terms of market cap in India. In August 1956, the Oil and Natural Gas Commission was formed. Raised from mere Directorate status to Commission, it had enhanced powers. In 1959, these powers were further enhanced by converting the commission into a statutory body by an act of Indian Parliament

Since its foundation stone was laid, ONGC is transforming India's view towards Oil and Natural Gas by emulating the country's limited upstream capabilities into a large viable playing field. ONGC, since 1959, has made its presence noted in most parts of India and in overseas territories. ONGC found new resources in Assam and also established the new oil province in Cambay basin (Gujarat). In 1970 with the discovery of Bombay High (now known as Mumbai High), ONGC went offshore. With this discovery and subsequent discovery of huge oil fields in the Western offshore, a total of 5 billion tonnes of hydrocarbon present in the country was discovered. The most important contribution of ONGC, however, is its self-reliance and development of core competence in exploration and production activities at a globally competitive level.

Post 1990, the liberalized economic policy was brought into effect, subsequently partial disinvestments of government equity in Public Sector Undertakings were sought. As a result, ONGC was re-organized as a limited company and after conversion of business of the erstwhile Oil & Natural Gas Commission to that of Oil and Natural Gas Corporation Ltd in 1993, 2 percent of shares through competitive bidding were disinvested. Further expansion of equity was done by 2 percent share offering to ONGC employees. Another big leap was taken in March 1999, when ONGC, Indian Oil Corporation (IOC) and Gas Authority of India Ltd.(GAIL) agreed to have cross holding in each other's stock. Consequently the Government sold off 10 per cent of its share holding in ONGC to IOC and 2.5 per cent to GAIL. With this, the Government holding in ONGC came down to 84.11 per cent. In 2002-03 ONGC took over Mangalore Refinery and Petrochemicals Limited (MRPL) from Birla Group and announced its entrance into retailing business. ONGC also went to global fields through its subsidiary, ONGC Videsh Ltd. (OVL). ONGC has made major investments in Vietnam, Sakhalin and Sudan and earned its first hydrocarbon revenue from its investment in Vietnam.

In 2009, ONGC discovered a massive oil field, with up to 1 billion barrel reserves of heavy crude, in the Persian Gulf off the coast of Iran. Additionally, ONGC also signed a deal with Iran to invest US$3 billion to extract 1.1 billion cubic feet of natural gas from the Farzad B gas field.[1]

Oil India

Oil India Ltd., (OIL) is a large natural gas and petroleum company in India under the administrative control of the Ministry of Petroleum and Natural Gas of the Government of India. OIL is engaged in the business of exploration, development and production of crude oil and natural gas, transportation of crude oil and production of liquid petroleum gas. OIL is the pioneer in exploration and production of hydrocarbons in India, and traces its roots back to Oil India Private Ltd., formed in 1959 with The Burmah Oil Company Ltd. holding two-thirds of equity and Government of India holding one-third. Oil India Private evolved into Oil India Ltd., which was an equal partnership between Burmah Oil and Government of India. In 1983 the company became a public sector undertaking of the Government of India. The Company presently produces over 3.2 MMTPA (million tons per annum) of crude oil, over 5 MMSCMD of Natural Gas and over 50,000 Tones of LPG annually. Most of this emanates from its traditionally rich oil and gas fields concentrated in the Northeastern part of India and contribute to over 65% of total Oil&Gas produced in the region. The search for newer avenues has seen OIL spreading out its operations in onshore / offshore Orissa and Andaman, deserts of Rajasthan, plains of Uttar Pradesh, riverbeds of Brahmaputra and offshore Saurashtra. In Rajasthan, OIL discovered gas in 1988, heavy oil / bitumen in 1991 and started production of gas in 1996. The company has accumulated over a hundred years of experience in the field of oil and gas production, since the discovery of Digboi oilfield in 1889. It is possibly the only company to do so. From well completion to wellbore servicing, installation, operation and maintenance of modern surface handling facilities, the company has the skill and expertise to manage the entire range of operations required for on shore oil and gas production. The company has over 100,000 square kilometers of license areas for oil and gas exploration. It has emerged as a consistently profitable international company with exploration blocks as far as Libya and sub-Saharan Africa. In recent years, OIL has stepped up E & P activities significantly including Gas monetization in the North-East India. OIL has set up the NEF (North East Frontier) project to intensify its exploration activities in the frontier areas in North East, which are logistically very difficult and geologically complex. Presently, seismic surveys are being carried out in Manbhum, Pasighat and other Trust Belt areas. The Company operates a crude oil pipeline in the North East for transportation of crude oil produced by both OIL and ONGCL in the region to feed Numaligarh, Guwahati, Bongaigaon and Barauni refineries and a branch line to feed Digboi refinery. A 1157 kilometers long fully automated telemetric pipeline with 212 kilometers of looping having a total capacity to transport over 6.0 MMTPA remains the lifeline of the Company. Commissioned in 1962, the double skinned crude oil pipeline traverses 78 river crossings including the mighty Brahmaputra River meandering through paddy fields, forests and swamps. There are 9 pumping stations, 17 Repeater stations and a terminal at Barauni. The engines that drive the giant pumps along the pipeline have crossed over two hundred thousand hours of service and established a world record of machine run - hours. The Company is currently in the process of constructing a 660 KM long Product Pipeline from Numaligarh to Siliguri. The Pipeline is expected to be completed by mid 2007. OIL also sells its produced gas to different customers in Assam viz. BVFCL, ASEB, NEEPCO, IOC (AOD), and APL and to RSEB in Rajasthan. The company also produces Liquefied Gas (LPG) in its plant at Duliajan, Assam.

Strategic Petroleum Reserve

In support of the country's energy security, Indian officials have declared that the country intends to develop a Strategic Petroleum Reserve (SPR). The decision has been made to set up a strategic reserve of 5 million tons (36.6 million barrels) of crude oil in underground structures in Mangalore, Visakhapatnam, and Padur. The project is expected to come online in 2012. The location of the storage facilities was selected to be along the coast so that the reserves could be easily transported to refineries during a supply disruption. The SPR project is being managed by the Indian Strategic Petroleum Reserves Limited (ISPRL), which is part of Oil Industry Development Board (OIDB), a state-controlled organization. Despite these plans, India does not have any strategic crude oil stocks at this time.

The crude oil pipelines and the product pipe lines in India are shown below in Fig. 2.9

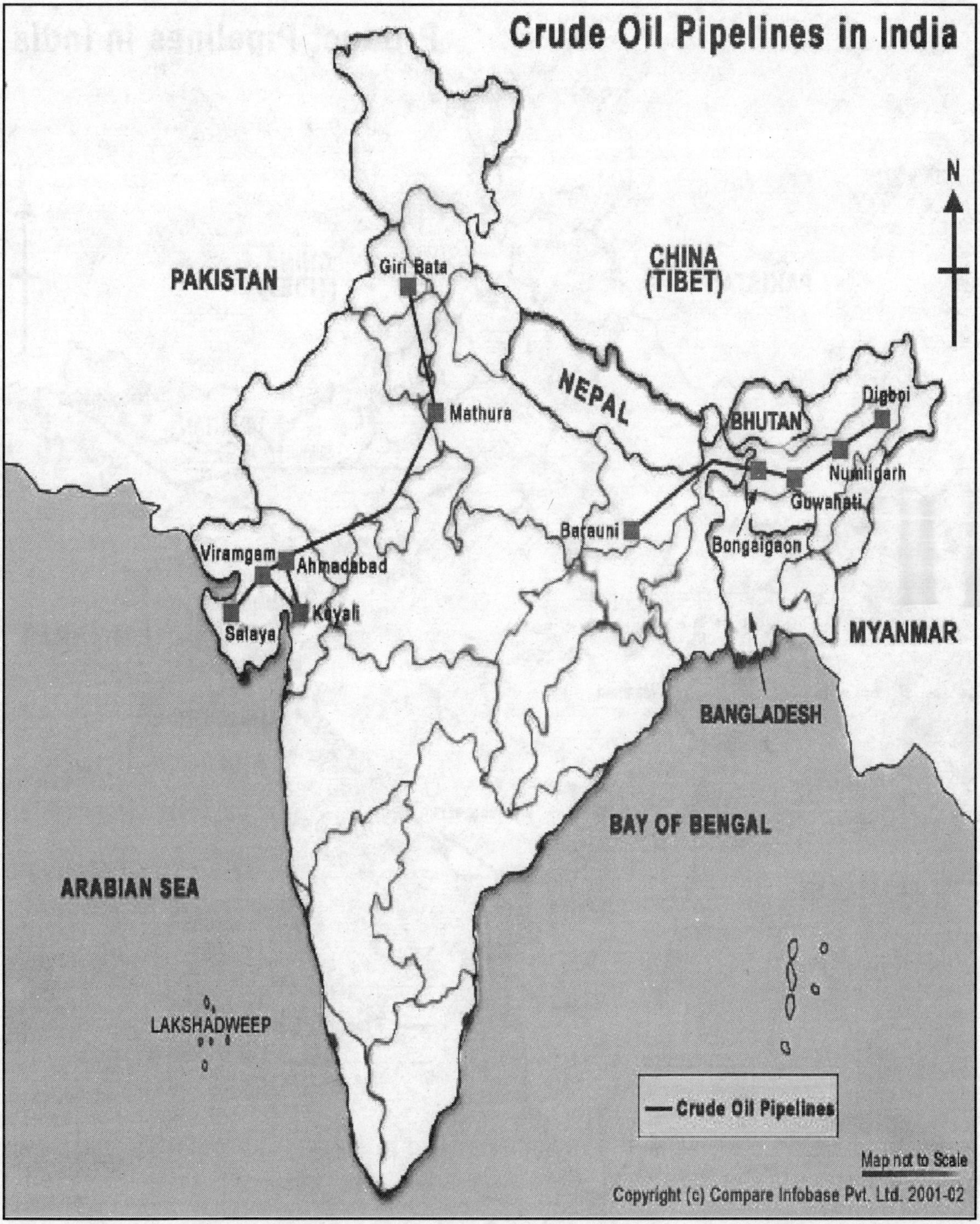

FIG. 2.9. THE CRUDE OIL PIPELINES AND THE PRODUCT PIPE LINES IN INDIA

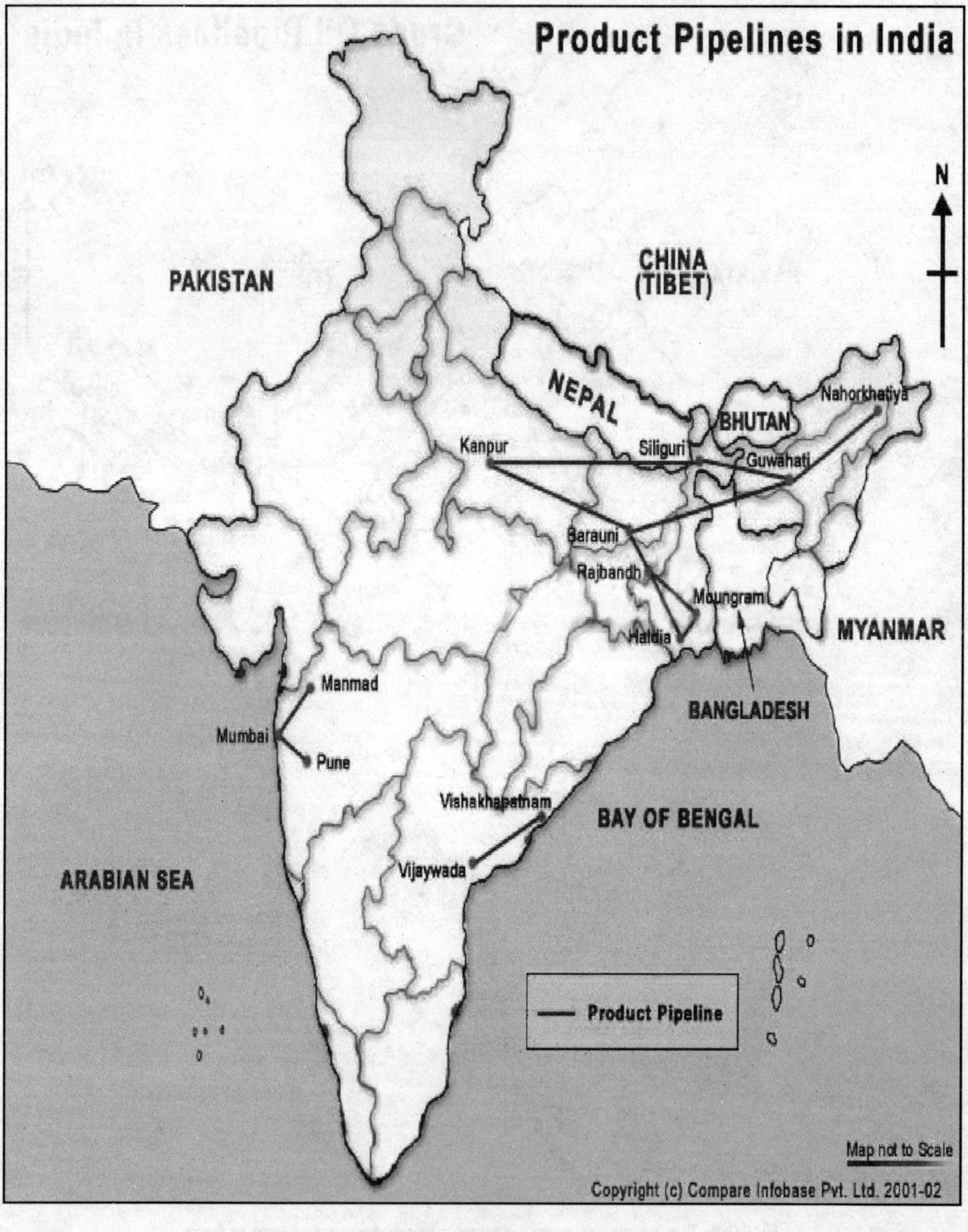
Product Pipelines in India
N
PAKISTAN
CHINA
(TIBET)
NEPAL
BHUTAN
Nahorkhatiya
Kanpur
Siliguri
Guwahati
Barauni
Rajbandh
Moungram
Haldia
MYANMAR
Manmad
Mumbai
Pune
BANGLADESH
Vishakhapatnam
BAY OF BENGAL
Vijaywada
ARABIAN SEA
Product Pipeline
Map not to Scale
Copyright (c) Compare Infobase Pvt. Ltd. 2001-02

FIG. 2.10.

REVIEW QUESTIONS

1. Write short note on resource pyramid for oil or gas.
2. What are the elements which influence the pyramid ?
3. Discuss briefly the scenario of petroleum reserves in India.
4. List the various reference in India.
5. What is ISO 140001 ? Explain.

❑❑❑

CHAPTER

3

COMPOSITION OF CRUDE OIL

INSIDE THIS CHAPTER

3.1. INTRODUCTION

Crude oil is the term for *"unprocessed"* oil, the stuff that comes out of the ground. It is also known as **petroleum**. Crude oil is a **fossil fuel**, meaning that it was made naturally from decaying plants and animals living in ancient seas millions of years ago — most places crude oil were once sea beds. **Crude oils vary in colour**, from clear to tar-black, **and in viscosity**, from water to almost solid. Crude oil varies greatly in appearance depending on its composition. It is usually black or dark brown (although it may be yellowish or even greenish). In some places it is found at the surface in the form of seepages, in others it occurs trapped at greater or lesser depths in rocky formations. It is usually a dark coloured or black liquid with a characteristic odour imparted by small quantities of compounds such as those containing sulphur or nitrogen.

In the reservoir it is usually found in association with natural gas, which being lighter forms a gas cap over the petroleum, and saline water which, being heavier than most forms of crude oil, generally sinks beneath it. Crude oil may also be found in semi-solid form mixed with sand and water, as in the Athabasca oil sands in Canada, where it is usually referred to as crude bitumen. In Canada, bitumen is considered a sticky, tar-like form of crude oil which is so thick and heavy that it must be heated or diluted before it will flow. Venezuela also has large amounts of oil in the Orinoco oil sands, although the hydrocarbons trapped in them are more fluid and are usually called extra heavy oil. These oil sands resources are called unconventional oil to distinguish them from oil which can be extracted using traditional oil well methods. Between them, Canada and Venezuela contain an estimated 3.6 trillion barrels (570×10^9 m^3) of bitumen and extra heavy oil, about twice the volume of the world's reserves of conventional oil.

Petroleum is used mostly, by volume, for producing fuel oil and gasoline (petrol), both important *"primary energy"* sources. 84 per cent by volume of the hydrocarbons present in petroleum is converted into energy-rich fuels (petroleum-based fuels), including gasoline, diesel, jet, heating, and other fuel oils, and liquefied petroleum gas. The lighter grades of crude oil produce the best yields of these products, but as the world's reserves of light and medium oil are depleted, oil refineries are increasingly having to process heavy oil and bitumen, and use more complex and expensive methods to produce the products required. Because heavier crude oils have too much carbon and not enough hydrogen, these processes generally involve removing carbon from or adding hydrogen to the molecules, and using fluid catalytic cracking to convert the longer, more complex molecules in the oil to the shorter, simpler ones in the fuels.

3.2. CHARACTERISTICS OF CRUDE OILS

Crude oils range widely in their physical and chemical properties. The properties are given in Table 3.1 below :

TABLE 3.1. Characteristic of Crude Oil.

Properties	Ankaleshwar	North/ Gujarat	Bombay High	Basrah	Kirkok
1. Sp. Gravity @ 15°C	0.7887	0.8868	0.8366	0.852	0.840
2. API	48	28	38	34.5	35.5
3. Colour	Brownish black	Black	Brownish black	Black	—
4. Total S, volt %	0.05 volt %	0.07 vol %	0.13	1.9	1.9
5. H_2S, ppm	—	—	—	Nil	200
6. RVP @ 37.8°C KPa	66.86	15.23	21.37	54.45	55–65.5
7. Pour Point, °C	18	27	30	–15	—
8. Flash point °C	< 27	—	< 27	15	—
9. Kinematic viscosity @ 37.8°C est	2.02	19.68 @ 50°C	3.816	5.9	—
10. Water content vol %	—	1.5	0.04	Nil	0.2
11. Salt contents wt. %	3–5	30–35	8.0	7.0	5.0
12. Ash content, wt. %	—	—	0.0047	0.006	—
13. KUOP	12	11.8	11.7	—	—

UOP – Universal Oil Production Co. **RVP** – Read Vapour Pressure **Source** – Petroleum Refining R. Prasad.

3.3. COMPOSITION

Crude oils contain **hydrocarbons**. Hydrocarbons are molecules that contain hydrogen and carbon and come in various lengths and structures, from straight chains to branching chains to rings. The proportion of hydrocarbons in the mixture is highly variable and ranges from as much as 97% by weight in the lighter oils to as little as 50% in the heavier oils and bitumens.

3.3.1. Hydrocarbon

There exist a great variety of these hydrocarbons, depending upon the number of carbon and hydrogen atoms in each molecule. The relative percentage of each varies from oil to oil, determining the properties of each oil. The composition of carbon and hydrogen by weight is given below in **Table 3.2.**

TABLE 3.2. Composition of Carbon and Hydrogen in Crude Oil.

Composition by Weight	
Element	Percent range
Carbon	83 to 87%
Hydrogen	10 to 14%

There are two things that are characteristics of hydrocarbons are :

- Hydrocarbons contain a lot of **energy**. Many of the things derived from crude oil like gasoline, diesel fuel, paraffin wax.
- Hydrocarbons can take on many different forms. The smallest hydrocarbon is **methane** (CH_4), which is a gas and is lighter than air. Longer chains with 5 or more carbons are liquids. Very long chains are solids like wax or tar. By chemically cross-linking hydrocarbon chains synthetic rubber to nylon to the plastic can be obtained. Hydrocarbon chains are very versatile.

Saturated Hydrocarbons

The hydrocarbon consists of carbon with all its valencies combined with separate hydrogen atoms. This type of compound is known as a *'Saturated'* compound, since there are no possibilities of further combination of the molecule with anything else. A series of these saturated hydrocarbons exists in which chains of carbon atoms are linked together with hydrogen occupying the remaining carbon valencies. At each end of the chain will be CH_3 groups.

The whole series of such hydrocarbons is known as the *'paraffin'* series, and when the carbon atoms form a straight chain, the series is known as the *'normal paraffin'* series.

Paraffins

- General formula: $\mathbf{C_nH_{2n+2}}$ (n is a whole number, usually from 1 to 20)
- straight- or branched-chain molecules

These can be gasses or liquids at room temperature depending upon the molecule examples: methane, ethane, propane, butane, isobutane, pentane, hexane.

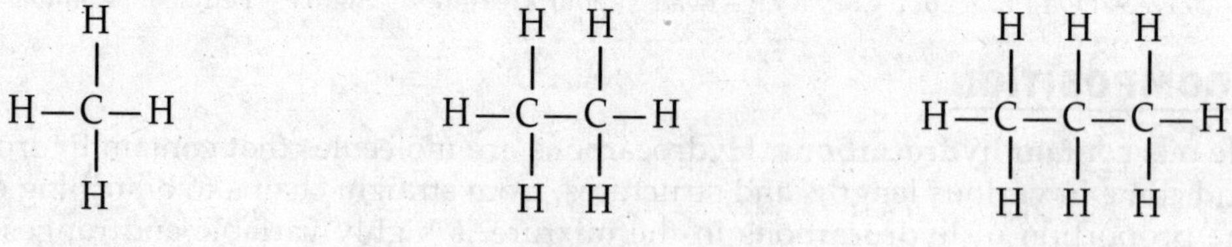

Methane CH_4 Ethane C_2H_6 Propane C_3H_8

From pentane (C_5H_{12}) onwards, Latin or Greek numerals are used to reveal the number of C-atoms per molecule.

$$CH_3{-}CH_2{-}CH_2{-}CH_3$$

n-Pentane

$$CH_3{-}CH_2{-}CH_2{-}CH_2{-}CH_2{-}CH_3$$

n-Hexane

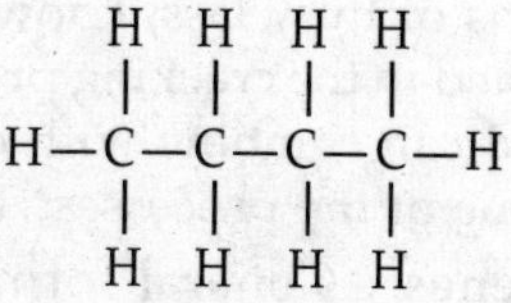

Normal Butane
nC_4H_{10}

From Butane (C_4H_{10}) onwards, alkanes may exist in two or more isomers differing in structure.

Example : Butane may also occur as :

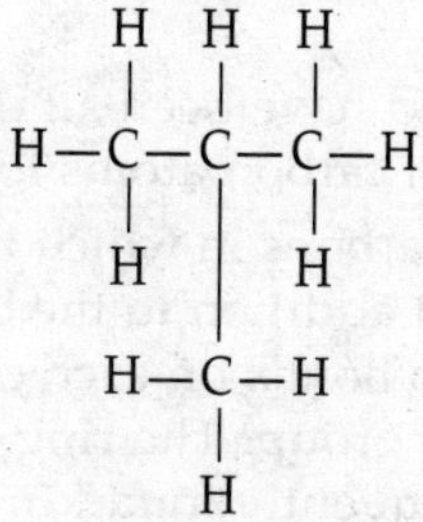

Isobutane
C_4H_{10}

The number of isomers increases for each next hydrocarbon in the series as given in Table 3.3.

TABLE 3.3. Number of Isomers for Different Hydrocarbons.

Molecule Name	Number of Isomers
C_4H_{10} (Butane)	2
C_5H_{12} (Pentane)	3
C_6H_{14} (Hexane)	5
C_7H_{16} (Heptane)	9
C_8H_{18} (Octane)	18
C_9H_2O(nonane)	35
$C_{10}H_{22}$ (decane)	75
$C_{15}H_{32}$ (Pentadecane)	4347
$C_{20}H_{42}$ (eicosane)	366319
$C_{40}H_{82}$ (tetra contane)	62491178805831

Unsaturated Hydrocarbons

In addition to the hydrocarbons in which every valency or bond is satisfied by a separate atom, there are hydrocarbons in which carbon is linked to carbon by two or more of the valencies. In this case the number of hydrogen atoms that can be linked to the carbon atoms is reduced. *For example*, in the hydrocarbon C_2H_4 or $H_2C = CH_2$ the carbon atoms are linked together by two bonds, and only two additional hydrogen atoms can then be linked to each carbon atom. It is noticed that each carbon atom still has four valencies, but they are not all used up in separate combination with other atoms. Such a hydrocarbon is called an *'unsaturated'* hydrocarbon.

Hydrocarbons of this class, known as 'olefins', are not found in natural crude petroleum, but they are formed in the cracking process. Being unsaturated they can readily take part in chemical reactions and can combine with each other (polymerize), both of which facts form significant bases for some of the processes. The unsaturated hydrocarbons are :

- **Alkenes :** General formula : C_nH_{2n} (n is a whole number, usually from 1 to 20) linear or branched chain molecules ontaining one carbon-carbon double-bond. They can be liquid or gas. **Examples:** ethylene, butene, isobutene.
- **Dienes** and **Alkynes :** General formula : C_nH_{2n-2} (n is a whole number, usually from 1 to 20) linear or branched chain molecules containing two carbon-carbon double-bonds. They can be liquid or gas. **Examples :** acetylene, butadienes.

Ring Hydrocarbons

The hydrocarbons so far discussed have been all of the '*chain*' variety. There are however, other series of hydrocarbons in which carbon atoms form closed rings.

(*i*) These compounds are hydrocarbons in which the carbon atoms forming a ring are fully saturated. This means that, in addition to the bonds connecting the carbon atoms with each other, the remaining two bonds on each carbon atom are linked to separate atoms of hydrogen or an equivalent group. The ring may contain various numbers of carbon atoms, though those most frequently found in petroleum contain 5–7.

(*ii*) The second important series of ring hydrocarbons is the series whose simplest member is benzene. This is a six-carbon ring hydrocarbon in which each carbon atom in the ring has only one hydrogen atom attached.

Naphthenes or Cyclo-paraffins

- General formula : C_nH_{2n} (n is a whole number usually from 1 to 20)
- Ringed structures with one or more rings
- Rings contain only single bonds between the carbon atoms
- Typically liquids at room temperature

Examples: cyclohexane, methyl cyclopentane

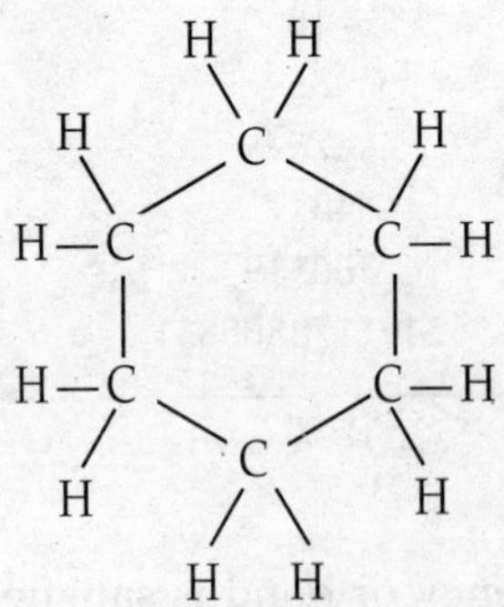

Cyclohexane
C_8H_{12}

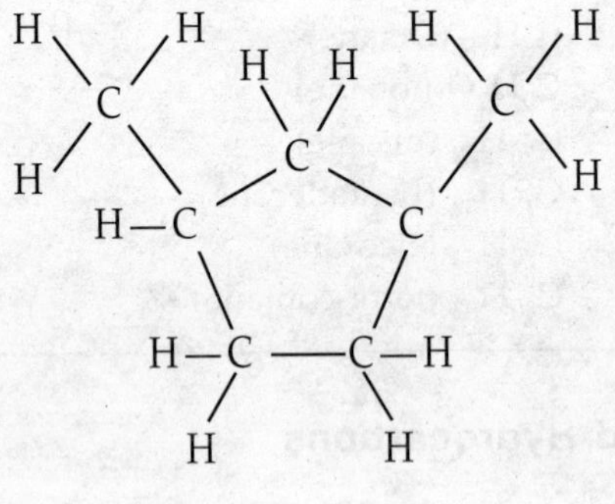

Dimethyl Cyclopentane
C_7H_{14}

The Aromatics

- General formula : C_6H_5 - Y (Y is a longer, straight molecule that connects to the benzene ring).

- Ringed structures with one or more rings.
- Rings contain six carbon atoms, with alternating double and single bonds between the carbons.
- Generally liquids

Examples : Benzene, napthalene

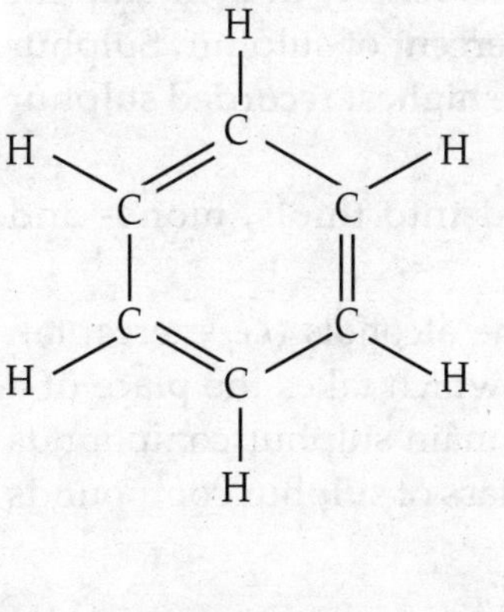

Benzene
C_6H_6

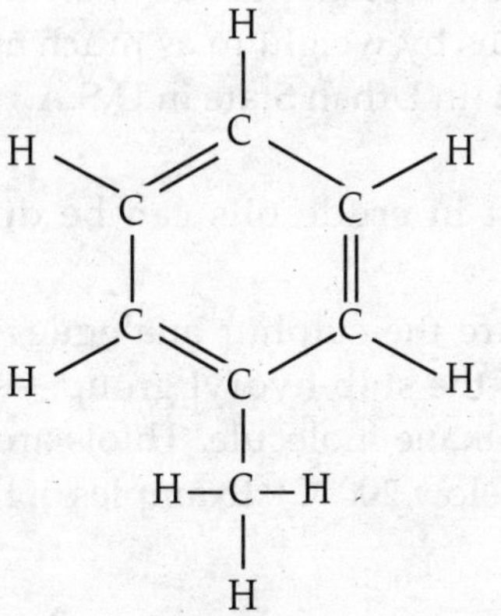

Toluene
C_7H_8

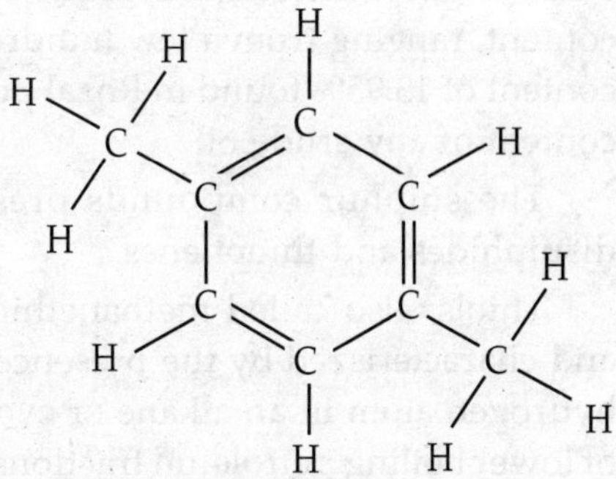

Benzene
C_6H_6

The average composition of hydrocabons in crude oil is summarized below in the **Table 3.4.**

TABLE 3.4. The Average Composition of Hydrocabons.

Composition by Weight		
Hydrocarbon	Average	Range
Paraffins	30%	15 to 60%
Naphthenes	49%	30 to 60%
Aromatics	15%	3 to 30%

3.3.2. Other Types of Compounds in Petroleum

The organic compounds in crude oil contain nitrogen, oxygen and sulfur, and trace amounts of metals such as iron, nickel, copper and vanadium. The exact molecular composition varies widely from formation to formation but the proportion of chemical elements vary over fairly narrow limits as shown in Table 3.5.

TABLE 3.5. Composition of Various Elements in Crude Oil.

Compostion by Weight	
Element	Percent range
Nitrogen	0.1 to 2%
Oxygen	0.1 to 1.5%
Sulfur	0.5 to 6%
Metals	less than 1000 ppm

Sulphur Compounds

Sulphur is the most abundant atomic constituent of crude oils, other than carbon and hydrogen. One of the origins of sulphur compounds in the crude oils is the natural sulphur content of tissues of the plant and animal remains deposited in ancient geological beds. The other origin is the biogenic reduction of sulphate. Crude oils vary considerably in their sulphur content, ranging from a few hundredths by weight to as much as 8 percent of sulphur. Sulphur content of 13.95% found in Rozal point (in Uthah State in U.S.A.) is the highest recorded sulphur content of any crude oil.

The sulphur compounds present in crude oils can be divided into thiols, mono- and disulphides and thiophenes.

Thiols, also called methanethiol are the sulphur analogues of the alcohols (*i.e.* mercaptan and characteriszed by the presence of the sulp-hydryl group, —SH, which takes the place of a hydrogen atom in an alkane or cycloalkane molecule. Thiols are the main sulphur compounds of lower boiling petroleum fractions (below 200°C). Examples of this class of sulphur compounds are given below :

CH_3SH	CH_3CH_2SH	$(CH_3)_2CHSH$
Methanethiol (methyl mercaptan)	**Ethanethiol (ethyl mercaptan)**	**Isopropanethiol (isoprepyl mercaptan)**

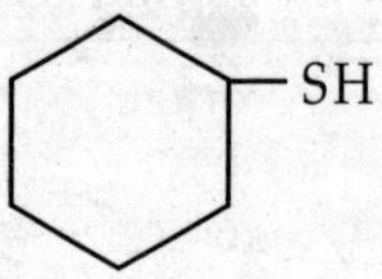

Cyclohexanethiol (cyclohexyl mercaptan)

The presence of thiols in the petroleum fractions causes corrosion problems, catalyst poisoning and unpleasant laboratory and plant odours. Thiols are not particularly poisonous in low concentration, however, and they are often used as odourants in LPG. The lower thiols are the most malodourous. One part in 50 million of ethanethiol can be detected in air by the human nose. The odour is strong at 0.6 ppm and distinct at 0.03-0.07 ppm. In high concentrations, the odour changes to something like that of chloroform.

Thiols are stronger acids than alcohols, and it is used to remove low molecular weight thiols from light gasolines with caustic soda solution. Hydroprocessing techniques are employed to desulphurise other oil fractions and here combined sulphur is eliminated as hydrogen sulphide from all type of compound containing this heteroatom.

Monosulphides (thiaalkanes and thiacycloalkanes) are the compounds formally derived by replacement of both hydrogens of HSH by hydrocarbon groups. These appear in very low abundance in crude oils. Examples are given below :

$CH_3SCH_2CH_3$

2-thiabutane
(methyl ethyl sulphate)

$CH_3SCH_2CH_3$

2-thiapentane
(methyl propyl sulphide)

$(CH_3CH_2)S$

3-thiapentane
(diethyl sulphide)

$CH_3CH(CH_3)SCH(CH_3)CH_3$

2, 4-dimethyl-3-thiapentane
(diisopropyl sulphide)

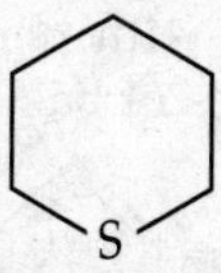

Thiacyclohexane (cyclohexyl sulphide)

As in the case of thiols, the sulphides are generally very malodourous.

A few low molecular weight disulphides (dithiaalkanes) have also been shown to be present in crudes. The occurrence of disulphides in crude oils may be owing to the secondary reaction of thiols with an oxidant such as air or free sulphur. Organic disulphides are isologs of the organic perioxides, ROOR, but are much more stable. Examples are given below :

CH_3SSCH_3

2,3-dithiabutane
(methyl disulphide)

$CH_3CH_2SSCH(CH_3)CH_3$

2-dimethyl-3, 4-dithiahexane
(ethyl isopropyl disulphide)

$CH_3CH_2SSCH_2CH_3$

3,4-diethiahexane
(diethyl disulphide)

Some examples of thiophenes found in crudes are given below:

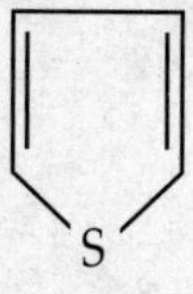

Thiophene

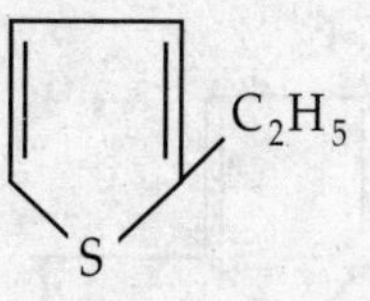

2-ethylthiophene

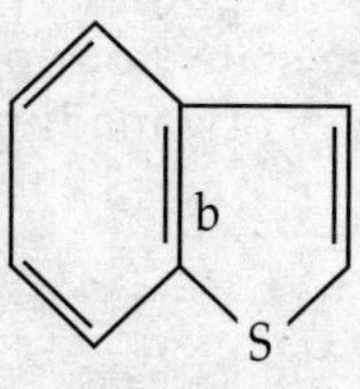

benzo[b]thiophene

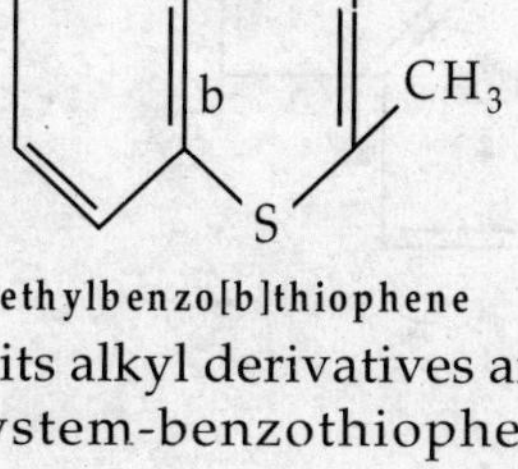

2-methylbenzo[b]thiophene

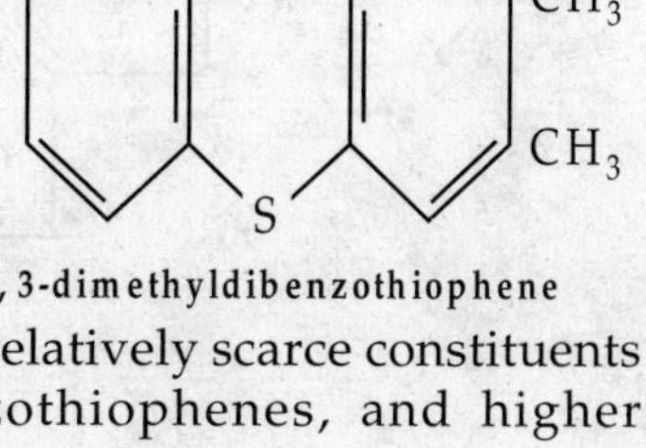

2, 3-dimethyldibenzothiophene

Thiophene and its alkyl derivatives are normally relatively scarce constituents of crude oils, but condensed system-benzothiophenes, dibenzothiophenes, and higher polycyclics (e.g. benzonaphthothiophenes) are important components of all high sulphur crudes. Thiophene derivatives are particularly abundant in crude oils containing high proportions of aromatics, resins and asphaltenes.

Nitrogen Compounds

Nitrogen occurs in crude oils to a lesser extent than sulphur, generally less than half of the sulphur content and frequently amounting to only about one-tenth. The nitrogen content is usually greater in heavier grades of crude oil. The nitrogen compounds are divided into : (*i*) Basic nitrogen compounds and (*ii*) Non-basic nitrogen compounds.

Basic nitrogen compounds (titratable with perchloric acid) are predominant in the lower boiling fractions of the crude oil. Examples of basic nitrogen compounds are given below :

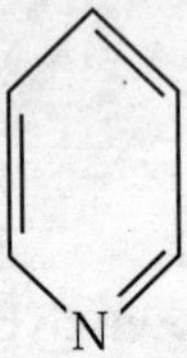

Pyridine

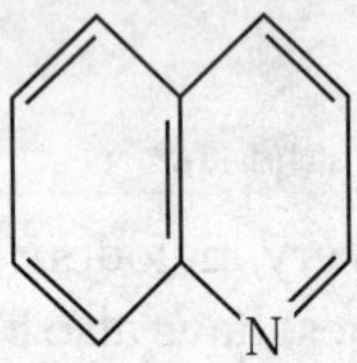

Quinoline

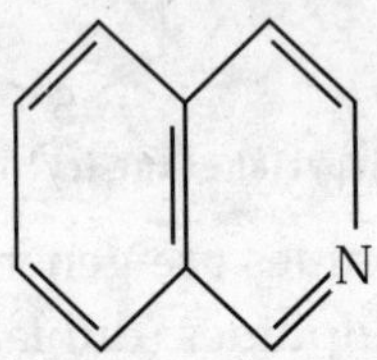

Isoquinoline

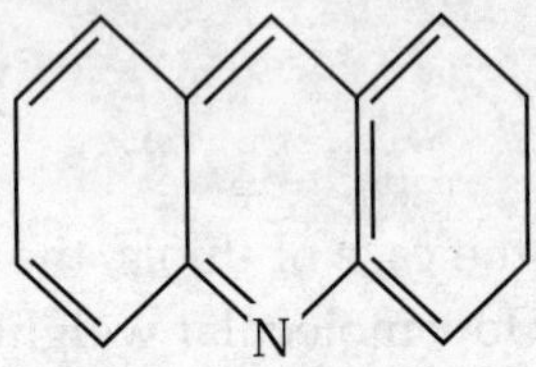

Acridine

The most abundant nitrogen compounds are non-basic and are found in the higher boiling fractions. Examples of non-basic nitrogen compounds are given below :

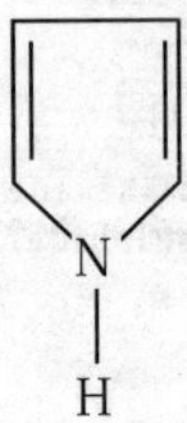

Pyrrole

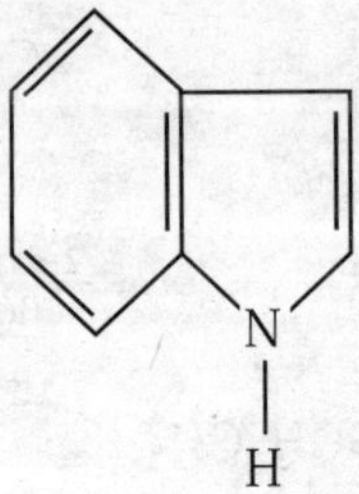

Indole

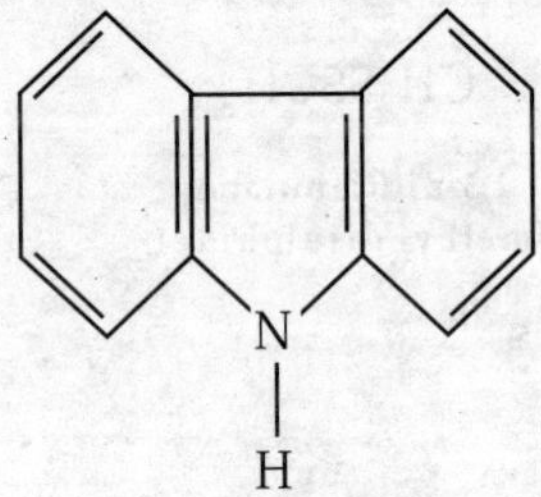

Carbazole

Additional nitrogen compounds such as anilines, phenazines and nitriles have been found in petroleum fractions after catalytic cracking.

Among nitrogen compounds, porphyrins occupy a special place. They may be present in crude oil either in free state (four pyrrole rings) as complexes containing nitrogen compounds and organic derivatives of vanadium and nickel.

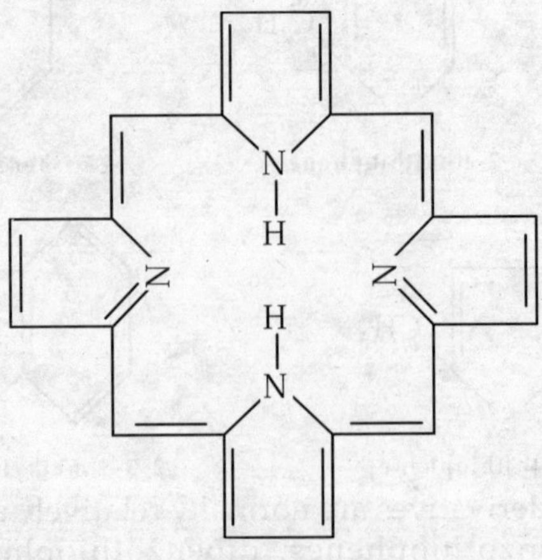

Under refining conditions, the nitrogen compounds are relatively stable, unlike some sulphur compounds. In the processing of crude oils, the concentration of nitrogen increases from lighter fraction to heavier one and the largest amount (65-75%) is concentrated in the residum. The presence of these compounds in trace amounts in distillates can however, lead to the loss in catalyst activity, colour instability of the products, and formation of gum.

Oxygen Compounds. The total amount of combined oxygen in crude oils is relatively low, varying from traces to a maximum of 2 per cent by weight. The oxygen concentration increases from lighter fractions to heavier ones; greater portion of oxygen containing constituents being concentrated in the residual oils, where values as high as 8 percent have been reported.

In the low- and medium-boiling range distillate fractions, oxygen is mainly present in the form of carboxylic acids and to a limited extent, phenols. A number of aliphatic monocarboxylic acids, dicarboxylic acids, and alicycli monocarboxylic acids have been reported in crude oils. The aliphatic acids range from formic acid to palmitic acid and include both the normal and isoalkyl chain structures. The dicarboxylic acids include dimethyl maleic acid and 1,2,2-trimethylcyclopropane-1, 3-dicarboxylic acid. The alicyclic acids, otherwise referred to as naphthenic acids, which is considered as derivatives of cyclopentane, and cyclohexane structures also occur. Phenols and substituted phenols are also found in crude oils, in spite of its solubility in water.

The determination of total oxygen content is customarily replaced by a determination of acidity of the fraction. In general, paraffinic crudes and their distillate fractions are low in acidity, whereas asphaltic crudes frequently contain substantial amounts of acids, of the order of 0.6 to 1.0 per cent by weight or higher.

Metallic Compounds. A large number of metallic elements such as iron, vanadium, nickel, aluminium, sodium, calcium, copper, magnesium, manganese, barium, silicon, chromium, strontium, lead, molybdenum, cobalt, gold, silver, titanium have been reported in crude oils.

The most abundant metal which are frequently found in crude oil is vanadium (upto 0.1 percent by weight), nickel and iron. Other trace metals are present in lower concentrations, i.e. less than one part per billion (1 ptb). They appear to be present in all crudes in the form of petroporphyrins and related complexes. They are undesirable in refining operations due to the fact that they produce undesirable reactions from refiner's point of view and their quantities and effects are cumulative.

Asphaltenes and Resins. These are present in the asphaltic residuum which remains after removal of distillable material from crude oil. Asphaltenes and resins contain in addition to carbon and hydrogen, some nitrogen, sulphur and oxygen. Some trace metals, such as nickel and vanadium are also present. About half of the total combined nitrogen and sulphur present in crude oils appears in the form of asphaltenes and resins. Most of these elements are present in the form of heterocyclic groups. The probability of all molecular species present containing one or more heteroatoms (N, S, or O) is high.

Asphaltenes are black amorphous solids. These are generally considered as highly aromatic materials of fairly high molecular weight ranging widely from 600 to 30000.

Resins are dark-coloured, solids or semi-solids. These possess lower molecular weights (800-2000) than asphaltenes and are generally less aromic in nature and comprise fewer condensed rings. Resins always seem to be more abundant than asphalenes and their ratio varies considerably depending on the crude oil.

3.3. TYPES OF CRUDE OIL

Although all crude oils consist mainly of hydrocarbons of the various series discussed earlier, oils from various sources differ widely in the proportion of the different hydrocarbon series which they contain. Thus, one type of crude oil may contain more of the paraffins, including the solid paraffin waxes, while another may contain more of the naphthenes. The products that are worth producing will therefore vary from crude to crude, and economic factors, such as market demand and accessibility, often have a determining influence on the products made. Thus, from some crude oils useful yields of lubricating oils and waxes may be obtained, while others may yield insignificant quantities of wax but may contain asphaltic bitumen together with lubricating oil of a different character due to different chemical nature. Other again may contain little lubricating oil but provide excellent stock for cracking. In a very general way, crude oils may be divided into three classes :

Paraffin Base Crudes

The crude oil which contain paraffin wax but little or no asphaltic matter. Such crude oils consist mainly of paraffin hydrocarbons and usually yield gasoline of low octane value and gas oil of high cetane value. They usually give good yields of paraffin wax and high grade lubricating oils.

Asphaltic Base Crudes

Crude oil containing little or no paraffin wax but usually asphaltic matter, often in quite large proportions. The hydrocarbons consist mainly of the naphthene series, and these crude oils yield lubricating oils.

Mixed Base Crudes

Mixed base are those crude oils which contain both paraffin wax and asphaltic matter. Both paraffin and naph-thenic hydrocarbons are present together with a certain proportion of aromatic hydrocarbons. The above classification of crude oils is a rough and ready division into types and should not be used too strictly. There is considerable overlapping between the types in the case of most crude oils.

Naphthenic base crudes are those crudes whose residue on atmospheric distillation contain asphaltic material with small amounts of paraffin wax.

3.4. BULK PETROLEUM PRODUCTS

Crude oil consists of a very complex range of different hydrocarbons. These form precursors for a variety of special products and the entire host of petrochemical products, apart from the bulk petroleum products. The bulk petroleum products, however, are mainly the following :

- Liquified Petroleum Gas (L.P.G.)
- Naphtha
- Motor Spirit (M.S.)
- Kerosene
- Aviation Turbine Fuel (A.T.F.)
- High Speed Diesel Oil (H.S.D)

- Light Diesel Oil (L.D.O.)
- Furnace Oil (F.O.)
- Bitumen

Some of the end-uses of the above bulk petroleum products are described below.

LPG (Liquified Petroleum Gas)

This is generally a mixture of propane and butane. At atmospheric temperature and pressure it is gas. It has therefore to be stored under pressure to keep it liquified, hence the name LPG. LPG is used largely as a domestic fuel. It is also invaluable for use as a fuel in certain industries such as glass manufacture due to its being a clean and a convenient fuel. LPG is also a good feedstock for the manufacture of various petrochemicals

Naphtha

The most important usage of Naphtha is in the production of Motor Spirit. It is partly blended direct into Motor Spirit. and partly processed in a catalytic reforming unit to produce high octane number reformate. Reformate is then blended into the motor spirit pool, Naphtha is also an important feedstock for fertilizer manufacture. Steam cracking of Naphtha produces a large variety of olefins including ethylene and propylene which are the precursors for various petrochemicals.

Motor Spirit

This is a blend of a variety of light hydrocarbons such as straight run naphtha, reformate, cracked gasoline from crackers, alkylates etc. Octane rating of motor spirit is an important anti-knock property which makes the modern high compression ratio internal combustion engines possible for automotive purpose.

Kerosene

While in India, kerosene is largely used for illumination elsewhere it is used for heating purposes. A special cut of Kerosene can be processed for the manufacture of Linear Alkyl Benzene used as a detergent.

A.T.F. (Aviation Turbine Fuel)

Aviation turbine fuel also called the jet fuel is becoming an increasingly important fuel with the development of aviation industry and with the phasing out of the earlier piston type of aero- engines. This is a kind of a kerosene cut with certain important specifications such as the freezing point of ATF should be below 47°C.

HSD (High Speed Diesel)

This is an important petroleum product used for automotive purposes in compression ignition engines. It is an important fuel used for road, rail and marine transport systems.

LDO (Light Diesel Oil)

This is an inferior kind of a gas oil used as fuel in low speed engines largely used in lift irrigation sector in India.

FO (Fuel Oil)

This is used in a variety of industrial applications and as fuel in furnaces.

Bitumen

The bulk grades of bitumen are used for road making.

The Composition of crude oil can be summarized below in **Fig. 3.1.**

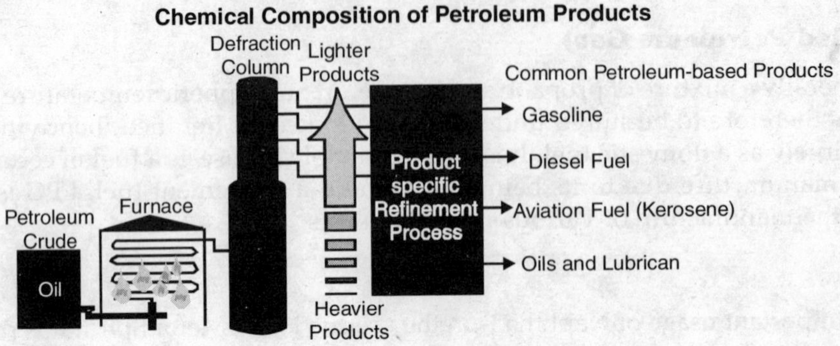

- When petroleum products. Or the waste associated with use of these products are released into the environment, they chemically and biologically interact with the soil, groundwater, and microorganisms.
- The ratios of these chemicals serve as a fingerprint for the type of product (diesel, gasoline etc.) Originally released into the environment.

Fig. 3.1. Composition of crude oil.

Some of the Chemical Found in Petroleum Products, Significant because of their Toxicity or Their Taste and Smell are :

Chemical	Symbol	Human Toxicity
Benzene	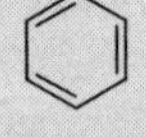	Probable carcinogen (cancer-causing agent). Central Nervous System (CNS) depression, Irregular heart rhythm.
Xylenes	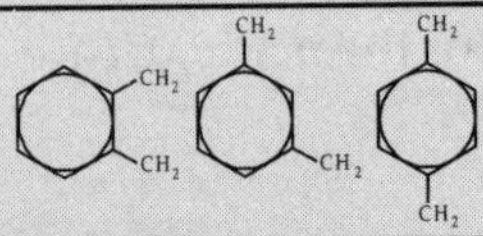	Abdominal pain, vomiting; CNS effects; Heart and liver damage; Bone marrow damage.
Toluene	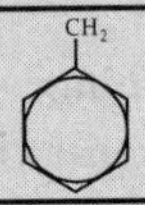	Mutagenic (causes mutations), CNS depression, Respiratory irritation, Liver, kidney damage
Ethylbenzene	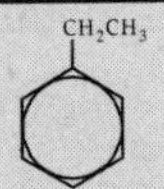	Narcosis, CNS effects, Skin and eye irritation
Naphthalene	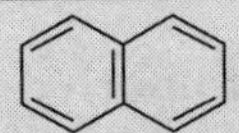	Abdominal Pain, Jaundice, Renal failure, Optical neuritis and corneal damage.
PAHs		General Class with varying degrees of toxicity, several of which are potential carcinogens. 17 specific PAHs are regulated under Florida's Petroleum Cleanup Program.
Minerals	Cadmium, Chromium, Lead, Arsenic	Varying degrees of toxicity. Arsenic is a carcinogen and has reproductive effects. Lead can cause kidney disease, malnutrition and weakness.

The petroleum industry generally classifies crude oil by the geographic location it is produced in (*e.g.* West Texas Intermediate, Brent, or Oman), its API gravity (an oil industry measure of density), and by its sulfur content. Crude oil may be considered *light* if it has low density or *heavy* if it has high density; and it may be referred to as sweet if it contains relatively little sulfur or *sour* if it contains substantial amounts of sulfur.

The geographic location is important because it affects transportation costs to the refinery. *Light* crude oil is more desirable than *heavy* oil since it produces a higher yield of gasoline, while *sweet* oil commands a higher price than *sour* oil because it has fewer environmental problems and requires less refining to meet sulfur standards imposed on fuels in consuming countries. Each crude oil has unique molecular characteristics which are understood by the use of crude oil assay analysis in petroleum laboratories.

Barrels from an area in which the crude oil's molecular characteristics have been determined and the oil has been classified are used as pricing references throughout the world. Some of the common reference crudes are:

- West Texas Intermediate (WTI), a very high-quality, sweet, light oil delivered at Cushing, Oklahoma for North American oil .
- Brent Blend, comprising 15 oils from fields in the Brent and Ninian systems in the East Shetland Basin of the North Sea. The oil is landed at Sullom Voe terminal in the Shetlands. Oil production from Europe, Africa and Middle Eastern oil flowing West tends to be priced off this oil, which forms a benchmark.
- Dubai-Oman, used as benchmark for Middle East sour crude oil flowing to the Asia-Pacific region.
- Tapis (from Malaysia, used as a reference for light Far East oil).
- Minas (from Indonesia, used as a reference for heavy Far East oil).
- The OPEC Reference Basket, a weighted average of oil blends from various OPEC (The Organization of the Petroleum Exporting Countries) countries.

There are declining amounts of these benchmark oils being produced each year, so other oils are more commonly what is actually delivered. While the reference price may be for West Texas Intermediate delivered at Cushing, the actual oil being traded may be a discounted Canadian heavy oil delivered at Hardisty, Alberta, and for a Brent Blend delivered at the Shetlands, it may be a Russian Export Blend delivered at the port of Primorsk.

REVIEW QUESTIONS

1. Discuss the composition of crude oil.
2. On the basis of hydrocarbons present, how will you classify the crude oil.
3. What is sweet and sour crude oil ?
4. What are ring hydrocarbons ? Discuss the ring hydrocarbons present in crude oil.
5. What are the other elements present in crude oil apart from hydrocarbon ? Give their composition by weight.
6. What are the bulk petroleum products obtained from crude oil ?
7. Give the composition of carbon and hydrogen in crude oil.
8. What are saturated and and unsaturated hydrocarbons ?
9. What are the average composition of hydrocarbons in crude oil ?
10. Illustrate some of chemicals found in crude oil.

❑❑❑

CHAPTER

4

CHARACTERISTICS OF PETROLEUM PRODUCTS

INSIDE THIS CHAPTER

INTRODUCTION

The petroleum products should meet certain specifications or characteristics for satisfactory performance, for the specific use of these products. Standard organisations have laid down specifications and methods of test. Bureau of Indian Standard (BIS), New Delhi is one such organisation in India which standardises procedures and issues specifications. Institute of Petroleum (IP), U.K. and American Society for Testing and Materials (ASTM), U.S.A. are the other two important organisations whose methods and specifications are widely followed. Apart from BIS specifications, Central Board of Revenue, New Delhi has certain other specifications for the purpose of excise levy. Some of the important limits set by them are for carbon residue, flame height, flash point, and viscosity.

4.1. LABORATORY TESTS

Most of the laboratory tests can be broadly classified based on the following characteristics :

- Volatility
- Combustion
- Viscosity and consistency
- Melting point

- Oxidation
- Corrosion and protection
- Miscellaneous tests.

Volatility is the major determinant of the tendency of a hydrocarbon to produce potentially explosive vapours. It is also critically important to an engine's start and warm-up. Volatility is assessed by the following tests :

(*a*) Distillation
(*b*) Vapour pressure
(*c*) Flash point and fire point

The combustion properties of petroleum products are evaluated by the following tests :

(*a*) Antiknock quality-Octane number, Performance number
(*b*) Ignition quality-Cetane number, Aniline point, Diesel index, Calculated cetane index
(*c*) Calorific value
(*d*) Burning quality-Smoke point, Char value

The determination of viscosity and consistency of petroleum products is done by the

(*a*) Kinematic viscosity (Redwood, Saybolt, Engler)
(*b*) Viscosit index
(*c*) Penetration tests

The tests designed to ascertain the tendency of certain petroleum products to melt or liquefy, to solidify or to precipitate wax-like materials are :

(*a*) Freezing point
(*b*) Cloud point and pour point
(*c*) Drop point of grease
(*d*) Melting and setting point of wax
(*e*) Softening point of bitumen

The storage stability and resistance to oxidation for gasoline and aviation turbine fuel is studied on the basis of :

(*a*) Induction period of gasoline
(*b*) Thermal stability of Jet fuels
(*c*) Gum content

Most crude oils are corrosive to greater or lesser extent, due to the presence of sulphur compounds, organic acids (mainly naphthenic acids) and traces of brine. Therefore tests methods have been designed to evaluate the corrosive potentialities of the petroleum products which are obtained by processing of crude oils. The following methods are available :

(*a*) Total sulphur
(*b*) Acidity and alkalinity
(*c*) Copper-strip corrosion test
(*d*) Silver-strip corrosion test for Aviation Turbine Fuels

The other characteristic test include

(*a*) Ash
(*b*) Carbon residue
(*c*) Colour
(*d*) Density and specific gravity
(*e*) Gas chromatography of petroleum gases and liquids
(*f*) Refractive index of hydrocarbon liquids
(*g*) Lead in gasoline
(*h*) Water separometer index (modified) (WSIM)
(*i*) Ductility

The various test processes are describe below :

4.2. DISTILLATION

The distillation apparatus consists of a round bottom flask fitted with a thermometer and metal steel condenser and a graduated collecting flask. 100 ml of sample is taken in the flask and vaporised which is condensed in ice-cooled tube and collected. Whereas an individual hydrocarbon would exhibit a single boiling point, commercial fuel blends boil over a range of temperatures. Corresponding readings of vapour temperature and condensate recovered are made at prescribed intervals and the results are plotted in the form of distillation curve. The initial boiling point (IBP) is taken as the temperature observed at the fall of the first drop of condensate, and the final boiling point (FBP) as the maximum temperature reached during the test. Due to small losses of vapour at the connections and openings in the apparatus and the residue remaining in the flask on completion of the test, the total recovery does not generally exceed about 97 percent. A maximum distillation temperature limit of 370°C has to be set, otherwise the heavier hydrocarbon molecules are liable to suffer from cracking into lighter molecules causing the distillation characteristics to change during their measurement. Thus fuels heavier than gas oils cannot be tested completely for distillation behaviour.

The safety of fuels and their performance on engine are governed by their distillation characteristics. Being a measure of volatility, they determine the system of fuel metering required (wick feed, carburation or atomization), and are indicative of the vaporisation behaviour of fuels in storage (vapour loss and vapour lock), and in piston-engine manifolds (uniformity of distribution to cylinders). The extent of the distillation range for any given fuel is representative of the availability of that fuel from the parent crude oil. For a fuel with a high demand, the aim of the supplier will be to extend the distillation range as far as practicable. However, due to the interrelationship between properties and the associated problems, minimum and maximum limits, respectively, may need to be set for the initial and final boiling points.

The significance of this test varies from product to product. In case of crude oil, the ASTM distillation data give some idea of the fractions that could be collected below 300°C. If it is a true boiling point (T.B.P.) distillation, the TBP curve reveals a lot of characteristics that are useful for the design of the refinery. The 10 vol.% of distillation for motor spirit is an indication of the ease with which the engine can be started. Too high a FBP will cause crankcase oil dilution.

4.3. VAPOUR PRESSURE

Vapour pressure of a liquid fuel may be defined as the pressure exerted by the vapour above the free surface of the liquid at the given temperature. For volatile, nonviscous petroleum products, it is determined by Reid method. Raid vapour pressure is a common measure of volatility of gasoline. This is the pressure exerted by vapour when it is in quilibriuum with the liquid under the conditions of test. For liquefied petroleum gas, the procedure is different and the determination should be done at 65°C. The conditions under which vapour pressures are determined give results which are not true because of the air which is invariably present in apparatus. The true vapour pressure is higher than the Reid vapour pressure by about 5 to 9 percent but this relationship varies widely.

The standard Reid apparatus consists of a fuel chamber connected to an air chamber of four times volume, and fitted with a pressure gauge.

This test is important with respect to safety in transport, vapour lock in the gasoline engines, types of storage tanks employed and the starting characteristics of motor fuels. High vapours pressure entails loss of the product in storage and transportation. In case of motor spirit, it may cause vapour lock in the gasoline engines.

4.4. FLASH POINT AND FIRE POINT

Flash point and fire point can be taken as indirect measure of volatility of the product. The flash point is the lowest temperature at which application of test flame causes the vapour above the oil to ignite. The fire point is the lowest temperature at which the oil ignites and continues to burn for 5 second.

The determination of flash point of petroleum products consists of heating a given volume of liquid at a standard rate of temperature rise until vapour is produced to such a degree as to give a flammable mixture with air in an enclosed space (i.e. closed flash point temperature) or with air in an open cup (i.e. the higher open flash point temperature), ignition resulting from the application of a small flame. At fire point, not only will the vapour air mixture flash but the liquid will continue to burn.

Abel apparatus is used for determining the closed cup flash point of petroleum products having flash points between 19°C and 49°C. Pensky-Martens apparatus is used for determing the flash point of fuel oils and lubricating oils, bitumen other than cutback bitumen having a flash point above 49°C. Cleveland apparatus is used for determining the flash and fire points of petroleum products except fuel oils and those products having an open cup flash point above 79°C.

Flash point measures the tendency of the fuel to form a flammable mixture with air under controlled laboratory conditions. This is the only property that must be considered in assessing the overall flammability hazard of a material. It is used in shipping and safety regulations that define flammable and combustible materials. Petroleum products having low flash points are potential to fire hazards. Flash point can indicate the possible presence of highly volatile and flammable materials in relatively nonvolatile or nonflammable material.

4.5. OCTANE NUMBER

This is an important test for measuring the antiknock quality of the gasoline (petrol or motor spirit). The knocking of the motor fuels is compared using blends of reference fuels. The standard reference fuels used for octane numbers below 100 are iso-octane and normal heptane which are assigned values of 100 and 0, respectively, on the octane number scale. The octane number of the fuel is defined as the volume percentage of iso-octane (2, 2, 4-trimethyl pentane) in a blend with n-heptane which is equal to the test fuel in knock intensity under standardised and closely controlled conditions of test in a single-cylinder, variable compression ratio engines, known as CFR engines. Thus, a fuel of 87 octane number has a CFR engine performance matching that obtainable with a blend of 87 volume percent iso-octane and 13 volume percent n-heptane. The rating can be done by either Research method or Motor method. The differences in the two methods are as follows : (Table 4.1).

TABLE 4.1.

	Research Method	Motor Method
rpm	600	900
Spark advance	13° btdc	Variable
Mixture heating	No	Yes

Octane number requirements of gasoline engines depend on their comparison ratio. If the fuel meets the minimum requirements in respect of octane number it ensures trouble free operation. Apart from being a nuisance, the knocking in an engine may result in loss of energy and at times may cause severe damage to the engine.

4.6. PERFORMANCE NUMBER

This is used to estimate knocking characteristics of aviation gasolines of octane number higher than 100. The standard reference fuels for knock ratings above 100 octane numbre are iso-octane and its blends with tetraethyl lead (TEL). The ratings of aviation gasoline above 100 octane number are normally expressed as performance number.

The performance number scale is based on engine power output. The performance number of an aviation fuel represents approximately the maximum knock-free power output. The performance number shows the percentage increase in aircraft engine power for addition or TEL to iso-octane and is given by

$$\text{Octane number} = 100 + \frac{(\text{Performance number} - 100)}{3}$$

On the performance number scale, 100 octane number equals 100 performance number. The ratings of aviation gasoline above 100 octane number can be done by

(*a*) Aviation method (lean mixture rating);

(*b*) Supercharge method (rich mixture rating); and

(*c*) Extended motor method.

In aviation method, the rating is done at 1200 rpm by comparing the combustion chamber temperature for the fuel with those of the blends of known performance number. This lean mixture rating gives us an idea of the availability of knock limited power in spark ignition type aircraft engines when the aircraft is under cruising conditions.

In supercharge method, the rating is done at 1800 rpm by comparing the knock limited power of the fuel with those for blends of iso-octane and isooctane plus TEL. This is done at constant compression ratio by measuring indicated mean effective pressure at enough points to define the mixture response curves for the sample and the reference fuels. When the knock limited power for the sample is bracketed between those for two adjacent reference fuels, the rating is calculated by interpolation. The rich mixture rating indicates the availability of knock limited power when the plane is under take-off conditions.

In extended motor method, the rating is done in a CFR engine normally used for rating of motor gasolines by motor method (rpm = 900). The knocking intensity of the fuel is bracketed between reference fuel prepared from iso-octane and TEL and the performance number is calculated by interpolation.

4.7. CETANE NUMBER

Cetane number is related to the ignition delay of a fuel in a diesel engine, *i.e.* how rapidly combustion begins after injection of the fuel into the combustion chamber. Hexadecane (C_{16} H_{34}, also called cetane) has a low ignition delay time and is the arbitrarily chosen high-quality reference compound for the petane scale. It has been assigned a cetane number of 100. The low quality reference compound is 2, 2, 4, 4, 6, 8, 8-heptamethyl nonane (C_{16} H_{34}), which has a long ignition delay time and has been assigned a cetane number of 15. The standards for the cetane scale show that long-chain, saturated, unbranched hydrocarbons have, high cetane numbers, while structural features such as branching or unsaturation lead to lower cetane numbers.

Cetane number, of diesel fuels is determined in a single cylinder CFR engine by comparing the ignition delay chltfacteristics of the diesel fuels with that of reference blends of known cetane number. Cetane number of a diesel fuel is denned as the whole number nearest to the value determined by calculation from the percentage by volume of normal cetane in a blend with heptamethyl nonane which matches tfteignition quality of the test fuel when compared by this method. The matching blend percentages to the first decimal are inserted in the following equation to obtain the cetane number :

Cetane number = % n-cetane + 0.15 (% heptamethyl nonane)

The shorter the ignition delay period, higher is the cetane number of the fuel.

Cetane number is the index of ignition quality of a fuel. High cetane number fuels will facilitate easy starting of compression ignition engines, particularly in cold weathers, and faster warm up. These also result in increased engine efficiency and power output, reduced exhaust smoke and odour and combustion noise. In the absence of test engine, the diesel index or the calculated cetane index will give an approximate idea of the ignition quality of the fuel. Cetaue number can also be roughly assessed by the formula :

Cetane Number = 0.72 × Diesel Index + 10

4.8. ANILINE POINT

Aniline is a poor solvent for aliphatic hydrocarbons and excellent one for aromatics. This property is used in the aniline point test, Aniline point of an oil is the lowest temperature at which the oil is completely miscible with an equal volume of aniline.

Equal volumes of the sample and aniline (5 ml each) are heated or cooled with stirring in a jacketed test tube and temperature at which complete miscibility occurs is noted.

High aniline point indicates that the fuel is highly paraffinic and nence has a high diesel index and very good ignition quality. In case of aromatics the aniline point is low and the ignition quality is poor.

4.9. DIESEL INDEX

Diesel index is an indication of the ignition quality of a diesel fuel. This is determined by calculation from the specific gravity and the aniline point of the sample. Although it is of the same order as the cetane number, it may differ widely from the cetane number. Higher the diesel index, better is the ignition quality of the diesel fuel. It is normally used as a guide to ignition quality of the diesel fuel in the absence of test engine for the direct measurement of cetane number.

The diesel index is calculated as follows :

(*a*) $\text{Diesel index} = \dfrac{\text{Aniline point, °F} \times \text{°API}}{100}$

(b) Diesel index = $\frac{\text{Aniline gravity constant}}{100}$

(c) Diesel index = $\frac{\text{Cetane number} - 10}{0.72}$

4.10. CALCULATED CETANE INDEX

Calculated Cetane Index (CCI) is based on specific gravity and the 10 percent, 50 percent and 90 percent distillation temperatures of the fuels and it gives numbers that correlate with the engine-testing method. The relationship is given by the following four-variable equation.

$$CCI = 45.2 + 0.0892T_{10N} + (0.131 + 0.901B)T_{50N} + (0.0523 - 0.42B)T_{90N} + 0.00049[(T_{10N})^2 - (T_{90N})^2] + 107B + 60B^2$$

where
$T_{10N} = T_{10} - 215\ °C$
T_{10} = 10 percent distillation temperature, °C
$T_{50N} = T_{50} - 260\ °C$
T_{50} = 50 percent distillation temperature, °C
$T_{90N} = T_{90} - 310\ °C$
T_{90} = 90 percent distillation temperature, °C
$B = e^{-3.5}(G - 0.85) - 1$
G = Specific gravity at 15°C

This CCI is useful for estimating cetane numbers when a test engine is not available for direct measurement, and it may be conveniently employed for estimating cetane number when the quantity of sample available is too small for an engine rating.

4.11. CALORIFIC VALUE

This is the quantity of heat released per unit quantity of fuel, when it is burned completely with oxygen and the products of combustion returned to ambient temperature. This quantity of heat will include the heat of condensation (latent heat) of the water vapour formed by the combustion of the hydrogen in the fuel, as it cools to ambient conditions. It is called the "gross calorific value" or "higher calorific value".

Most heating applications cannot recover the heat of the water vapour; it simply escapes with the vapour out of the chimney. The potential heat content is therefore more near indicated by subtracting this latent heat from the gross calorific value, and the resultant value is called the net calorific value, or lower calorific value.

A weighed quantity of the sample is burned in a bomb calorimeter under controlled conditions. The calorific value is calculated from the weight of the sample and the rise in temperature. It can also be calculated from the formulae

$$\text{Calorific value} = 12400 - 2100\,\rho^2$$

in which calorific value is in cal/gm and ρ is density at 15°C in gm/cm^3.

Calorific value is a measure of the energy available in a fuel. Thus a knowledge of the calorific value of the fuel, and the efficiency of the heating equipment, is essential to compare the merits and running costs of different fuels and energy costs. It is a critical property of fuel intended for use in weight-limited vehicles.

4.12. SMOKE POINT

Smoke point is the maximum flame height in mm at which the fuel will burn without smoking when determined in a smoke point apparatus under specified conditions.

Smoke point apparatus comprises four main parts—lamp body, candle socket, candle and stand. The lamp body with chimney is fitted on the inside with a polished black engraved SCALE which is marked in white. A gallery is secured in the lower part of the body. The candle socket assembly is designed to give a smooth rise and fall over the total distance of travel. To ensure interchangeability the candle is finished to close tolerances. The assembly is mounted on a stand. The sample is burned in a standard lamp with a specified wick for five minutes. The height of the flame is read when it leaves no smoky tail.

This is an important test for evaluation of illuminating oils (kerosines) for their ability to burn without producing smoke and the assessment of the burning quality of aviation fuels. Higher the smoke point better is its domestic use. It also serves as a guide to assess the aromatic content of kerosines.

4.13. CHAR VALUE

The amount and nature of the deposits (char) produced during combustion depend on the hydrocarbon composition of the fuel and also on the design of the appliances in which it is used. Char occurs as a result of the breakdown and decomposition of the kerosine under the local conditions existing at the wick surface, and these conditions will also determine what proportion of the decomposed products remains on the wick.

4.14. VISCOSITY

Kinematic viscosity is defined as the measure of the resistance to gravity flow of a fluid, the pressure head being proportional to the density. The time of flow of a fixed volume of fluid is directly proportional to its kinematic viscosity. The unit of kinematic viscosity is cm^2/s or Stoke. The unit most usually used in measurement of the kinematic viscosity of petroleum fuels is the centistoke (cSt) which is 10^{-2} Stoke.

Dynamic viscosity, also known as absolute viscosity, is the ratio of applied shear stress to rate of shear and thus a measure of the resistance of a fluid to flow. The unit of dynamic viscosity is gm/cm.s or Poise. Dynamic viscosity may be obtained from kinematic viscosity by multiplying it by the density of the fluid at the temperature at which measurement was made.

Kinematic viscosity may be measured as an absolute property of the fuel, or alternately as a conventional property that is dependent on the instrument and the method used. Both approaches depend on the efflux time of a given volume of sample flowing under its own head through a restriction. This follows because the force acting the laminar (low speed) flow of a fluid through a restriction is approximately proportional to the dynamic viscosity, whereas the force promoting the flow is that due to gravity, and is proportional to the density of the fluid. Hence the time taken for the gravity flow of a given volume of sample through a restriction is approximately proportional to the kinematic viscosity.

The conventional methods, which are generally simpler but less accurate, are represented by the Redwood instrument in the UK, Saybolt in the USA and Engler in continental Europe. They each comprise a sample cup fitted with a standard-sized orifice in the base and surrounded by a water jacket containing a heating device. When the temperature reaches the test level, the orifice is unsealed and the time of flow is determined for the given volume of sample. The result is reported as Redwood or Saybolt universal second or as Engler degree, given by the efflux time ratio for the sample and for water. When the efflux time exceeds a specified maximum-for example 2000 s-due to high viscosity, use is made of a Redwood No. 2, or a Saybolt Furol (fuel and road oils) instrument, incorporating a larger diameter orifice.

The absolute determination of kinematic viscosity generally employs a glass U-tube viscometer with a capillary tube built into one leg. The length-diameter ratio is such that end effects are negligible and the precision is therefore higher. The instrument is suspended vertically in a thermostatically controlled water bath, and the time is measured for a given volume of sample to flow through the capillary. This measured time period is inserted into an equation to give a direct measure of the kinematic viscosity in centistokes.

$$\nu = At - B/t$$

where A = instrument calibration constant;
B = instrument type constant, depending on the capillary diameter; and
t = efflux time, s (Engler degree for Engler viscometer)

Table 4.2 gives the values of A and B for Redwood, Saybolt and Engler viscometers.

Viscosity is an important characteristic of a fuel and it is used for the pump design. Pump clearance are adjusted according to the viscosity and if it is out of the range, it will result in pump seizer. Pump operation of an engine depends on the proper viscosity of the liquid fuel. The viscosity of liquid fuel is important to its flow through pipelines, injector nozzles, and orifices, and for atomization of fuel in the cylinder.

TABLE 4.2. Instrument Constant Values.

Instrument	Range of time in second/Engler degree,	A	B
Redwood No. 1	$34 \le t \le 100$ $t > 100$	0.00260 0.00247	1.79 0.5
Redwood No. 2	$32.5 \le t \le 90$ $t > 90$	0.02458 0.0244	1.0 0.0
Saybolt Universal	$t \le 100$ $t > 100$	0.00226 0.00220	1.95 1.35
Saybolt Furol	$25 \le t \le 40$ $t > 40$	0.0224 0.0216	1.84 0.6
Engler	$1.35 \le E° \le 3.2$ $E° > 3.2$	0.08 0.076	0.0864 0.04

4.15. VISCOSITY INDEX

Viscosity index (VI) is the most widely used way of characterizing the effect of change of temperature on the viscosity of any oil. Proposed by Dean and Davis, viscosity index is an empirical concept based on the behaviour of mineral oils. In this concept, an oil whose viscosity changes rapidly with change in temperature has a low VI. An oil with a minimum change in viscosity with change in temperature has a high VI. In this system, Pennsylvanian (paraffinic) oils of a selected type which show a desirable, relatively small change of viscosity with change in temperature, were assigned a VI of 100, while selected Texas Coastal oils showing less desirable viscosity-temperature characteristics were assigned a VI of 0.VI is governed by the type of hydrocarbons in the oil.

Dean and Davis prepared tables giving the kinematic viscosities at 40°C and 100°C of the Texas Coastal oils (L) and the Pennsylvanian oils (H). The values of kinematic viscosities of L and H are given in Table 4.2. The VI of an oil can be calculated from the equation

$$\mathrm{VI} = \frac{L-U}{L-H} \times 100$$

where U = kinematic viscosity at 40°C of the oil whose VI is to calculated
L = kinematic viscosity at 40°C of an oil of 0 VI
H = kinematic viscosity at 40°C of an oil of 100 VI

4.16. PENETRATION TESTS

Several standard grades of bitumen are commercially available, which are normally classified into different grades by penetration test. The sample of bitumen is placed in a suitable container and brought to a temperature of 25°C in a water bath. The weighted needle is brought to the surface and at the end of 5 seconds interval, the penetration of the needle into the bitumen, in units of 1/10 mm is termed the 'Penetration' of the bitumen.

The penetration at 25°C and the softening point, or penetrations at two different temperatures (for example, 25°C and 10°C) can be used to define the extent to which the consistency of a bitumen changes with temperature. This an important characteristics for bitumens, and determines the type of bitumen used for a particular application. Various factors have been used to define temperature dependence. The most commonly used factor, however, is the Penetration Index (PI), which is defined as follows :

TABLE 4.3. Values of H, L and D for Kinematic Viscosity at 100°C

(H = Kinematic viscosity at 40°C of an oil of 100 VI, cSt,
L = Kinematic viscosity at 40°C of an oil of 0 VI, cSt)

Kinematics viscosity at 100°C, cSt	H	L	D = L – H	Kinematic viscosity at 100°C, cSt	H	L	D = L – H
2.00	6.620	8.360	1.740	5.10	31.090	44.500	13.410
2.10	7.143	9.043	1.900	5.20	32.150	46.460	14.310
2.20	7.684	9.752	2.068	5.30	33.210	48.440	15.230
2.30	8.243	10.485	2.242	5.40	34.270	50.430	16.160
2.40	8.821	11.244	2.423	5.50	35.330	52.430	17.100
2.50	9.417	12.028	2.611	5.60	36.390	18.040	18.040
2.60	10.031	12.838	2.807	5.70	37.450	56.430	18.980
2.70	10.664	13.672	3.008	5.80	38.510	58.430	19.920
2.80	11.315	14.532	3.217	5.90	39.570	60.430	20.860
2.90	11.984	15.417	3.433	6.00	40.630	62.430	22.740
3.00	12.671	16.328	3.657	6.10	41.690	64.430	22.740
3.10	13.377	17.263	3.886	6.20	42.750	66.430	23.680
3.20	14.101	18.224	4,123	6.30	43.810	68.430	24.620
3.30	14.843	19.210	4,367	6.40	44.880	70.430	25.550
3.40	15.603	20.222	4.619	6.50	45.970	72.460	26.490
3.50	16.382	21.258	4.876	6.60	47.080	74.550	27.470
2.00	6.620	8.360	1.740	5.10	31.090	44.500	13.410
3.60	17.179	22.320	5.141	6.70	48.220	76.740	28.520
3.70	17.994	23.407	5.413	6.80	49.390	79.040	29.650
3.80	18.828	24.520	5.692	6.90	50.590	81.440	30.850
3.90	19.680	25.657	5.977	7.00	51.820	83.920	32.100
4.00	20.550	26.820	6.270	7.10	52.980	86.460	33.480
4.10	21.400	28.060	6.660	7.20	54.150	89.040	34.890
4.20	22.280	29.360	7.080	7.30	55.273	91.660	36.387
4.30	23.180	30.730	7.550	7.40	56.473	94.095	37.622
4.40	24.100	32.180	8.080	7.50	57.669	96.528	38.859
4.50	25.040	33.720	8.680	7.60	58.873	98.958	40.085
4.60	26.000	35.350	9.350	7.70	60.063	101.398	41.335
4.70	26.980	37.060	10.080	7.80	61.305	103.925	42.620
4.80	27.980	38.840	10.860	7.90	62.513	106.388	43.875
4.90	29.000	40.680	11.680	8.00	63.723	108.859	45.136
5.00	30.040	42.570	12.530	8.10	64.969	111.419	46.450

Kinematics viscosity at 100°C, cSt	H	L	D = L – H	Kinematic viscosity at 100°C, cSt	H	L	D = L – H
8.20	66.251	114.067	47.816	12.30	120.883	233.078	112.195
8.30	67.501	116.650	49.149	12.40	122.330	236.392	114.062
8.40	68.753	119.306	50.553	12.50	123.781	239.713	115.932
8.50	70.041	121.926	51.885	12.60	125.274	243.136	117.862
8.60	71.296	124.528	53.232	12.70	126.730	246.482	119.762
8.70	72.542	127.153	54.611	12.80	128.189	249.842	121.653
8.80	73.793	129.786	55.993	12.90	129.689	253.305	123.616
8.90	75.089	132.515	57.426	13.00	131.153	256.690	125.537
9.00	76.352	135.176	58.824	13.10	132.658	260.180	127.522
9.10	77.617	137.841	60.224	13.20	134.166	263.684	129.518
9.20	78.880	140.517	61.637	13.30	135.716	267.293	131.577
9.30	80.184	143.284	63.100	13.40	137.230	270.824	133.594
9.40	81.448	145.989	64.541	13.50	138.745	274.369	135.594
9.50	82.714	148.695	65.981	13.60	140.270	277.964	137.694
9.60	83.986	151.411	67.425	13.70	141.784	281.498	139.714
9.70	85.262	154.147	65.885	13.80	143.348	285.177	141.829
9.80	86.575	156.982	70.402	13.90	144.874	288.776	143.902
9.90	87.856	159.722	71.886	14.00	146.402	292.388	145.986
10.00	89.178	162.494	73.316	14.10	147.933	296.014	148.081
10.10	90.458	165.361	74.903	14.20	149.502	299.749	150.242
10.20	91.814	168.303	76.489	14.30	151.043	303.402	152.359
10.30	93.128	191.194	78.066	14.40	152.582	307.069	154.487
10.40	94.461	174.075	79.614	14.50	154.124	310.749	156.625
10.50	95.825	177.68	81.243	14.60	155.708	314.540	158.832
10.60	91.152	179.980	82.828	14.70	157.255	318.247	160.992
10.70	98.492	182.097	84.415	14.80	158.804	321.968	163.164
10.80	101.206	188.875	87.667	14.90	160.396	325.801	165.405
10.90	101.206	188.875	87.667	15.00	161.950	329.549	167.599
11.00	102.537	191.848	89.311	15.10	163.548	333.410	169.862
11.10	103.909	194.899	90.990	15.20	165.190	337.385	172.195
11.20	105.305	197.966	92.661	15.30	166.793	341.275	174.482
11.30	106.721	201.150	94.429	15.40	168.399	345.179	176.780
11.40	108.100	204.238	96.138	15.50	170.007	348.881	178.871
11.50	109.493	207.339	97.846	15.60	171.660	353.131	181.471
11.60	110.887	210.468	99.581	15.70	173.274	357.078	183.801
11.70	112.273	213.601	101.328	15.80	174.891	361.039	186.148
11.80	113.711	216.829	103.118	15.90	176.522	365.117	188.565
11.90	115.114	219.981	104.867	16.00	178.174	369.107	190.933
12.00	116.507	223.145	106.638	16.10	179.841	373.214	193.373
12.10	117.948	226.412	108.464	16.20	181.552	377.439	195.987
12.20	119.438	229.784	110.346	16.30	183.224	381.577	198.352

Kinematics viscosity at 100°C, cSt	H	L	D = L – H	Kinematic viscosity at 100°C, cSt	H	L	D = L – H
16.40	184.942	385.729	200.787	21.00	268.26	601.66	333.40
16.50	186.577	389.897	203 320	21.20	272.35	612.67	340.32
16.60	188.300	394.184	205.884	21.40	275.99	622.52	346.53
16.70	189.983	398.381	208.398	21.60	280.10	633.67	353.57
16.80	191.670	402.594	210.924	21.80	284.22	644.89	360.67
16.90	193.401	406.929	213.528	22.00	287.90	654.94	367.04
17.00	195.094	411.172	216.078	22.20	292.05	666.30	374.25
17.10	196.831	415.537	218.706	22.40	296.22	677.75	381.53
17.20	198.571	419.917	221.346	22.60	299.96	687.98	388.53
17.30	200.357	424.421	224.064	22.80	304.13	699.57	395.44
17.40	202.104	428.832	226.728	23.00	308.34	711.24	402.90
17.50	203.853	433.260	229.407	23.20	312.09	721.67	409.58
17.60	205.605	437.704	232.099	23.40	316.32	733.47	417.15
17.70	207.361	442.161	234.800	23.60	320.57	745.35	424.78
17.80	209.162	446.745	237.583	23.80	324.36	755.98	431.62
17.90	210.923	451.237	240.314	24.00	328.63	768.00	439.37
18.00	212.687	455.743	243.056	24.20	332.45	778.76	446.3
18.10	214.350	460.266	245.916	24.40	336.75	790.92	454.17
18.20	216.268	464.915	248.647	24.60	341.05	803.17	462.12
18.30	218.042	469.469	251.427	24.80	345.40	815.49	470.09
18.40	219.818	474.039	254.221	25.00	349.3	826.5	477.2
18.50	221.597	478.625	257.028	25.20	353.2	839.0	485.4
18.60	223.423	483.339	259.916	25.40	358.0	851.5	493.5
18.70	225.208	487.956	262.748	25.60	362.4	864.1	501.7
18.80	226.996	492.590	265.594	25.80	366.8	876.8	510.0
18.90	228.831	497.352	268.521	26.00	371.2	889.6	518.4
19.00	230.625	502.017	272.392	26.20	375.1	901.0	525.9
19.10	232.466	506.812	274.346	26.40	379.0	913.9	534.3
19.20	234.354	511.739	277.385	26.60	384.0	926.9	542.9
19.30	236.201	516.568	280.367	26.80	388.5	939.9	551.4
19.40	238.052	521.413	283.361	27.00	393.0	953.1	560.1
19.50	239.906	526.274	286.366	27.20	397.5	966.3	568.8
19.60	241.806	531.247	289.441	27.40	402.0	979.6	577.6
19.70	243.666	536.164	292.498	27.60	406.0	991.4	585.4
19.80	245.529	541.075	295.546	27.80	410.0	1004.0	594.3
19.90	247.440	546.120	298.680	28.00	415.1	1018.4	603.3
20.00	249.31	551.07	301.76	28.20	419.7	1032.0	612.3
20.20	253.10	561.12	308.02	28.40	424.3	1045.6	621.3
20.40	256.86	571.13	314.27	28.60	428.9	1059.4	630.5
20.60	260.59	581.08	320.49	28.80	433.5	1073.2	639.7
20.80	264.64	591.94	327.30	29.00	438.1	1087.0	648.9

Kinematics viscosity at 100°C, cSt	H	L	D = L – H	Kinematic viscosity at 100°C, cSt	H	L	D = L – H
29.20	442.8	1101.0	658.2	49.0	979.1	2879.6	1900.5
29.40	446.9	1113.5	666.6	49.5	994.8	2935.2	1940.4
29.60	451.6	1127.6	676.0	50.0	1010.4	2990.8	1980.4
29.80	456.2	1141.8	685.6	50.5	1026.2	3047.4	2021.2
30.0	460.9	1156.0	695.1	51.0	1042.0	3104.0	2062.0
30.5	472.8	1192.0	719.2	51.5	1058.4	3162.9	2104.5
31.0	484.1	1226.8	742.7	52.0	1074.7	3221.8	2147.1
31.5	496.1	1263.7	767.6	52.5	1090.9	3280.4	2189.5
32.0	508.2	1301.1	792.9	53.0	1107.1	3339.0	2231.9
32.5	520.4	1338.9	818.5	53.5	1123.9	3399.9	2276.0
33.0	532.6	1337.2	844.6	54.0	1140.6	3460.8	2320.2
33.5	544.9	1416.0	871.1	54.5	1157.2	3521.4	2364.2
34.0	557.3	1455.3	898.0	55.0	1173.7	3582.0	2408.3
34.5	569.9	1495.0	925.1	55.5	1190.5	3643.6	2453.6
35.0	582.4	1535.2	952.8	56.0	1207.2	3767.2	2498.0
35.5	595.8	1577.7	981.9	56.5	1224.2	3767.8	2543.6
36.0	608.5	1618.2	1010.4	57.0	1241.1	3830.4	2589.3
36.5	621.4	1660.6	1039.2	57.5	1258.6	3895.4	2636.8
37.0	634.3	1702.7	1068.4	58.0	1276.1	3960.4	2684.3
37.5	647.4	1745.3	1097.9	58.5	1293.5	4025.0	2731.5
38.0	660.5	1788.3	1127.8	59.0	1310.8	4089.6	2778.8
38.5	674.4	1833.9	1159.5	59.5	1328.3	4155.2	2826.9
39.0	687.7	1877.9	1190.0	60.0	1345.8	4220.7	2874.9
39.5	701.1	1922.4	1221.3	60.5	1363.9	4288.8	2924.9
40.0	714.6	1967.4	1252.8	61.0	1382.0	4356.8	2974.8
40.5	728.3	2013.1	1284.8	61.5	1399.9	4424.4	3024.5
41.0	741.9	2058.7	1316.8	62.0	1417.7	4492.0	3074.3
41.5	756.1	2106.4	1350.3	62.5	1435.8	4560.6	3124.8
42.0	770.2	2154.1	1383.9	63.0	1453.9	4629.1	3175.2
42.5	784.2	2201.7	1417.5	63.5	1472.6	4700.2	3227.6
43.0	798.2	2249.3	1451.1	64.0	1491.2	4771.3	3280.1
43.5	812.7	2299.0	1486.3	64.5	1509.7	4841.9	3332.2
44.0	827.2	2348.6	1521.4	65.0	1528.1	4912.4	3384.3
44.5	842.2	2400.4	1558.2	65.5	1546.8	4984.0	3437.2
45.0	857.2	2452.1	1594.9	66.6	1565.4	5055.5	3490.1
45.5	872.1	2503.8	1631.7	66.5	1574.7	5129.7	3545.0
46.0	886.9	2555.4	1668.5	67.0	1603.9	5203.8	3599.9
47.0	917.1	2660.7	1743.6	67.5	1622.9	5277.4	3654.5
47.5	932.7	2915.5	1782.8	68.0	1641.9	5350.9	3709.0
48.0	948.2	2770.3	1822.1	68.5	1661.1	5425.5	3764.4
48.5	963.7	2825.0	1861.3	69.0	1680.3	5500.0	3819.7

Kinematics viscosity at 100°C, cSt	H	L	D = L – H	Kinematic viscosity at 100°C, cSt	H	L	D = L – H
69.5	1700.1	5577.2	3877.1	73.0	1839.4	6123.1	4283.7
70.0	1719.9	5654.4	3934.5	73.5	1860.0	6205.4	4344.4
70.5	1739.5	5731.0	3991.5	74.0	1880.6	6285.6	4405.0
71.0	1759.1	5807.5	4048.4	74.5	1901.0	6366.2	4465.2
71.5	1779.3	5886.8	4107.5				
72.0	1799.5	5966.0	4166.5				
72.5	1819.5	6044.6	4225.1				

$$\frac{\log(800) - \log(PEN\ 25°C)}{T_{RB} - 25} = \left[\frac{20 - PI}{10 + PI}\right]\left[\frac{1}{50}\right]$$

where T_{RB} is the ring and ball softening point of the bitumen in °C. The penetration of *a* bitumenat the softening point temperature is about 800. Bitumens less affected by temperature change have positive values of PI and those more affected by temperature change negative values.

4.17. FREEZING POINT

Freezing point is the temperature at which crystals of hydrocarbons formed on cooling disappear when temperature of fuel is allowed to rise.

This method covers a procedure for the detection of separated solids in aviation reciprocating engine and turbine engine fuels at any temperature likely to be encountered during flight or on the ground.

4.18. CLOUD POINT AND POUR POINT

Cloud point of petroleum products is the temperature at which a cloud or haze of wax crystals appears at the bottom of the test jar when the oil is cooled under prescribed conditions It is generally determined for products that are transparent in a 40-mm thick layer and have cloud points below 49°C.

The cold filter plugging point test is used to determine the extent to which diesel fuel or gas oil will flow, even though the temperature is below that at which wax crystals normally appear, i.e. cloud point.

Cloud point gives a rough idea of the temperature above which the oil can be safely handled without any fear of congealing or filter clogging.

Pour point is the lowest temperature expressed in multiple of 3°C at which the oil is observed to flow when cooled and examined under prescribed conditions.

Pour point is a well-established test to estimate the temperature at which a sample of oil becomes sufficiently solid to prevent its movement by pumping. The pour point temperature depends to a large extent on the thermal history of the sample. Also, the pour point indicates the waxy nature of the oils.

4.19. DROP POINT OF GREASE

The standard drop point tests indicate that the temperature at which the thickener is so soluble in the base oil that the grease becomes substantially fluid. Clay and dye thickened greases have no measurable drop points.

The drop point can be used to assess whether a grease of known formulation has been properly made or to obtain an indication of the type of thickener which has been used in a grease of an unknown formulation. It cannot be used to measure- the upper operating temperature limit for a grease.

4.20. MELTING AND SETTING POINT OF WAX

The cooling curve method is used to determine the setting point of waxes. Molten wax is allowed to cool in a specified apparatus and the temperature is recorded at frequent intervals. The point at which the temperature remains within a range of 0.1°C for one minute is taken as the setting point. This method is not suitable for microcrystalline or intermediate waxes, or blends of paraffin waxes with these or any additives.

The congealing point of a petroleum wax or petrolatum is determined by applying a drop of molten wax to a thermometer bulb, and noting the temperature at which it congeals when the thermometer is rotated under standardised cooling conditions. This method is suitable for all waxes.

The drop melting point of wax or petrolatum is determined by recording the temperature at which a drop of the sample falls from the bulb of a thermometer when heated under standard conditions.

These are considered to be suitable for checking the constant quality of wax output in refineries,

4.21. SOFTENING POINT OF BITUMEN

Bituminous materials do not change from the solid state to the liquid state at any definite temperature, but gradually become softer and less viscous as the temperature rises. For this reason, the determination of the softening point must be made by a fixed arbitrary, and closely defined method.

The softening point is defined as the temperature at which a substance attains a particular degree of softness under specified conditions of test. A steel ball of specified weight is placed upon a disc of sample contained within a metal ring of specified dimensions. The assembly is heated at a constant rate and the temperature at which the sample becomes soft enough to allow the ball, enveloped in a bitumen, to fall to the specified distance is taken as the softening point.

The ring and ball test for softening point measures the temperature in °C at which a standard disc of bitumen confined in a metal ring softens to such an extent, when heated at a rate of 5°C/ minute, that it deforms under the loading imposed by a small steel ball which falls a distance of 2.54 cm.

The determination of the softening point of bitumen is regarded by some as an indication of viscosity, although from the point of view of the application of bitumen its use is limited to that of its title. The softening point is used in the designation of hard bitumens and oxidized bitumens.

4.22. INDUCTION PERIOD OF GASOLINE

Induction period of gasoline is the time elapsed between the placing of the bomb in the bath and break point at 100°C. Break point is the point in the pressure-time curve that is preceded by a pressure drop of exactly 2 psi within 15 minute and succeeded by a drop of not less than 2 psi in the next 15 minute.

Fifty millilitre of the sample is enclosed in a bomb with oxygen at 100 psi and heated in a water bath at 100°C. The pressure is then recorded either on a chart or read every 15 minutes. The test is continued until the break point is reached. The test result is reported as induction period in minute

This test is conducted to assess the stability of gasoline in storage. This test indicates the presence of unsaturated hydrocarbons in the fuel and hence its gum forming tendency. Higher the induction period, better is the storage stability of the fuel. An induction period of 360 minute under laboratory conditions ensures storage stability of at least six month However, this correlation may vary with different gasolines under different conditions.

4.23. THERMAL STABILITY OF JET FUELS

Jet fuel thermal oxidation tester (JFTOT) is used to measure the high temperature stability of gas turbine fuels. This subjects the test fuel to conditions which can be related to those occurring in gas turbine engine fuel systems. The fuel is pumped at a fixed volumetric flow rate through a heater after which it enters the stainless steel filter where fuel degradation products may become trapped. The apparatus requires 600 ml of test fuel for a 25 hour test The essential data derived are the amount of deposits on an aluminium heater tube, and the rate of plugging of filter located just downstream of the heater tube.

In the JFTOT a charge is placed in a reservoir and the whole system is pressurized to 3.45 MPa with nitrogen. This ensures a single-phase reaction in the heated section. The fuel passes from the reservoir through a 0.45 micron iilter, to remove trace particulate matter, and into the reactor section, where it passes upwards in an annular space over an aluminium tube and out via a 17 micron stainless steel iilter through a heat exchanger, to cool it, and back to the top of the reservoir. The used and unused fuel in the reservoir are separated by a floating piston. The fuel is rated by a visual tube rating or by placing the tube in a Tube Deposit Rater. In this the tube is rotated at a constant speed and its surface scanned by two light sources reflecting off the tube on to a photocell. The photocell gives a signal to a meter. Also, the differential pressure across the 17 micron filter is measured.

The test results are indicative of fuel performance during gas turbine operation and can be used to assess the level of deposits that can form when liquid fuel contacts a heated surface.

4.24. GUM CONTENT

The gum compounds which can be present or produced in the fuel are classified into two types for test evaluation. Existent gum may be already formed in the fuel and can be deposited from solution as the fuel evaporates. Potential gum may be formed under extended storage conditions during which unstable hydrocarbons are thereby polymerized and oxidized to form gums.

Existent gum is the amount of nonvolatile heptane insoluble residue left when the sample is evaporated in a jet of hot air at 160°C. For jet fuels, the evaporation is carried out in a jet of superheated steam at 232°C.

Potential gum is the amount of gum formed after the sample is aged in an oxidation stability bath and evaporated under specified conditions.

Gum is always troublesome in any fuel and it may cause piston ring sticking and deposits on engines. The amount of gum points to the presence of olefins which have very poor storage stability. The existent gum test is claimed to measure the amount of gum or gum-forming compounds existing in the fuel, while the potential gum test attempts to predict the tendency to form gum on storage and use. These gum tests are usually used as refinery control methods.

4.25. TOTAL SULPHUR

This is determined by lamp method or wickbold procedure for volatile petroleum products and by bomb method for heavier products. Sulphur in the sample is oxidized by combustion and is estimated volumetrically after absorption in H2O2 or by gravimetric methods after converting into barium sulphate.

Sulphur compounds pose a dual problem: they not only cause environmental pollution from their combustion products, but these products are also naturally corrosive and cause severe physical problems to engine parts. A knowledge of the sulphur content of petroleum products is therefore of importance to both refiner and user.

4.26. ACIDITY AND ALKALINITY

New and used petroleum products may contain acidic constituents present as additives or as degradation products, such as oxidation products, formed during service. Total acidity is a measure of the combined organic and inorganic acidity.

The acids in the sample are extracted in neutral alcohol and then titrated against standard alcoholic potassium hydroxide under hftt conditions. ,t :-

Total acidity is an indication of the corrosive properties of the product. Inorganic acidity is a measure of the mineral acid present. Organic acidity is obtained by deducting the inorganic acidity from the total acidity.

4.27. COPPER-STRIP CORROSION TEST

Petroleum products contain sulphur compounds, most of which are removed during refining. Of the sulphur compounds remaining in the petroleum product, however, some can have a corroding effect on various metals. This corrosivity is not necessarily directly related to the total sulphur content. The effect can vary according to other chemicals and types of sulphur compounds present.

A cleaned and smoothly polished copper strip is immersed in the sample, which is then maintained at the specified temperature for the specified length of time. This strip is removed from sample, washed with aromatic and sulphur free petroleum spirit and examined for evidence of etching, pitting or discolouration. It is then compared with ASTM copper-strip corrosion standard colour code. The classification cod«indicates that tire numbers 1, 2, H and 4 designate slight tarnish, moderate garnish, dark tarnish and corrosion, respectively. Subscripts a-e describe a standard colour reproduction in the standard chart. For example, the classification code la indicates slight tarnish with a light orange colour.

This test serves as a measure of possible difficulties with copper, brass, or bronze parts of the fuel systems.

4.28. SILVER-STRIP CORROSION TEST FOR AVIATION TURBINE FUELS

A polished silver strip is completely immersed in Aviation Turbine Fuel at 45 ± 1°C for a period of 16 hour. At the end of this period, the silver strip is removed from the sample, washed and evaluated for corrosion against the set of standard.

Since some parts of the fuel pumps in aircraft are made of silver, the corrosive tendency of the fuel for silver assumes special significance. The cumulative effect of corrosion on such a vital component in the aircraft is hazardous.

4.29. ASH

Ash can result from oil, water-soluble metallic compounds, or extraneous solids, such as dirt and rust.

A known amount of sample is ignited and the carbonaceous residue left is heated to 800^{U}C in a muffle furnace, cooled and weighed as ash.

A knowledge of the amount of a product's ash forming material can provide information on whether the product is suitable for use in a given application.

4.30. CARBON RESIDUE

Carbon residue can be defined as the amount of carbon residue left after evaporation and pyrolysis of an oil and is intended to provide some indication of relative coke forming tendency.

Either Conradson method or Ramsbottom method can be used to determine the carbon residue of petroleum products.

4.30.1. Conradson Method

A weighed quantity of sample is placed in a crucible and subjected to destructive distillation. The residue undergoes cracking and coking reactions during fixed period of severe heating. At the end of the specified heating period, the crucible containing the carbonaceous residue is cooled in a desiccator and weighed. The residue remaining is calculated as the percentage of the original sample and reported as conradson carbon residue.

4.30.2. Ramsbottom Method

The sample after being weighed into a special glass bulb having a capillary opening is placed in a metal furnace maintained at 550°C for 20 ± 1 minute. The sample is thus quickly heated to the point, at which all volatile matter is evaporated out of the bulb with or without decomposition while the heavier residue remaining in the bulb undergoes cracking and coking reactions. After a specified 20 minute heating period, the bulb is removed from the bath, cooled in a desiccator and again weighed. The residue remaining is calculated as the percentage of the original sample and imported ac ramsbnttom carbon residue.

Carbon residue gives a measure of the caibon depositing tendencies of a fuel oil when heated in a bulb under prescribed conditions. While not directly correlating with engine deposits, this property is considered as an approximation. For example, carbon residue value of diesel fuel correlates with the amount of carbonaceous deposits the fuel will form in the combustion chamber of the engine. The expected carbon deposits in the combustion chamber is greater for higher value of the carbon residue. Carbon residue is also used in design calculation of vessels.

4.31. COLOUR

Colour is an indication of the degree of refining of the products. Various test methods used for the measm-ement of colour of petroleum products along with their main items equipment and product application range are given in Table 4.4.

4.32. DENSITY AND SPECIFIC GRAVITY

Density of a fluid is its mass per unit volume. It is measured over a range of temperatures, usually for convenience at the temperature at which the fuel is to be stored.

TABLE 4.4. Colour Measurement Methods.

Name	Main equipment	Colour scale	Application
Saybolt chromometer	Chromometer	+30 to –16	White petroleum products
ASTM Colour	Colorimeter	0 to 8	Heavy petroleum products, lubricating oils
Colour of dyed aviation gasoline	Colour comparator	Permanent colour glass disc	Dyed aviation gasoline
Colour by the Lovibond tintometer	Livibond tintometer	Colour standards of different rating for red, yellow, blue and neutral tint	All petroleum products except black oils and bitumens

Specific gravity is the ratio of the density of a fluid to that of water at the same temperature. The temperature usually specified is 15.56°C. In the U&A, specific gravity of an oil is often expressed as degrees API. API gravity is an arbitrary figure related to the specific gravity of petroleum products in accordance with the formula :

$$\text{Degrees API} = \frac{141.5}{\text{Specific Gravity at } 15.56°\text{C} / 15.56°\text{C}} - 141.5$$

The specific gravity is an indication of the type of hydrocarbon present, being highest for aromatics and lowest for paraffins. The API gravity reverses this relationship.

The most accurate method of determining the specific gravity of an oil is to weigh a known volume in a specific-gravity bottle at 15.56°C. A correction may be applied by measuring the specific gravity at some conventional temperature near 15°C and adding or subtracting 0.00063 per °C above or below 15°C.

Another method for determining the specific gravity of the oil is by means of a set of hydrometers. A hydrometer is placed in the oil sample at 15.56°C and allowed to come to rest. The specific gravity is shown on the scale at the point coincident with the surface of the oil.

Accurate determinations of the density, specific gravity and API gravity of petroleum products are necessary for the conversion of measured volumes to volumes at standard temperature of 15.56°C. When the volume of oil is known, its mass can be calculated. These factors govern the quality of crude petroleum. These properties, however, are uncertain indications of fuel quality, unless correlated with other properties. A correlation of fuel density with particulate emissions indicates increasing particulate emissions with increasing density.

4.33. GAS CHROMATOGRAPHY OF PETROLEUM GASES AND LIQUIDS

A measured volume of the gas sample is introduced into a chromatographic column and transported through the column, the sample is split into various components, either by adsorption or partition, depending on the column packing. The composition of the sample is determined from the chromatograms by measuring the area under the peaks. An identification of the components is done by noting the elution time.

4.34. REFRACTIVE INDEX OF HYDROCARBON LIQUIDS

Refractive index is defined as the ratio of the velocity of light (of specified wavelength) in air to its velocity in the substance under examination. It may also be defined as the sine of the angle of incidence divided by sine of the angle of refraction, as light passes from air into the substance. The refractive index of liquids varies inversely with both wavelength and temperature. Refractive intercept is calculated by

$$\text{Refractive intercept} = n - \frac{\rho}{2}$$

where n is the refractive index at 20°C and p is the density at 20°C, gm/cm^3.

Naphthene content in naphthas can be easily calculated by knowing the refractive index and density of the saturates fractions as determined by the refractive intercept method. There is a relation between the molecular weight, aromatics and refractive index of hydrocarbons and hence the determination of refractive index gives an indication of the content of aromatics in the hydrocarbon fractions.

4.35. LEAD IN GASOLINE

The lead alkyl is converted to lead chloride and extracted from the gasoline by refluxing with concentrated hydrochloric acid. The acid extract is evaporated to dryness. Any organic material present is removed by oxidation with nitric acid and the lead is determined gravimetrically as lead chromate. The method covers the gravimetric determination of the total lead content of gasoline and other volatile distillates blended with lead alkyls (tetraethyl lead or tetramethyl lead, etc.).

Tetraethyl lead is added in gasoline to improve the octane number but it is highly poisonous. Hence, its concentration in gasoline is restricted and its handling is done with utmost precaution.

4.36. WATER SEPAROMETER INDEX (MODIFIED) (WSIM)

This is carried out with a water separometer. It measures the water separation characteristics of fuels expressed in terms of WSIM.

An emulsion of water and fuel is prepared and passed through a cell containing a standardized 'fibreglass' coalescer. The effluent from the ceil is examined for entrained water by light transmission. A numerical scale (0-100) rates the case with which the fuel releases emulsified water.

WSIM is a measure of fuel cleanliness relative to its freedom from surfactant materials. A higher WSIM rating indicates that the fuel is cleaner relative to surfactant materials.

4.37. DUCTILITY

Bituminous surfaces exposed to varying temperature conditions undergo a great deal of expansion and contraction. Therefore an important characteristic of the binder is its ductility and the degree of ductility has an effect on the cracking of bituminous surfaces caused by traffic stress. The ductility of bitumen is expressed as the distance in centimeter to which a standard briquette can be elongated before the thread formed breaks under the specified conditions.

A molten bitumen sample is poured into a standard mould, allowed to cool to room temperature and then placed in a water bath so that the briquette can be brought to test temperature before mounting in the testing machine. Ductility testing machine consists of a moving carriage moving over a lead screw. An electric motor driven reduction gear unit ensures smooth constant speed and continuous operation. The temperature is controlled thermostatically.

Bitumen having high ductility may be more temperature susceptible.

REVIEW QUESTIONS

1. What is flash point and fire point ? What are their significance ?
2. What is octane number and how it is determined.
3. Define cetane number ? Why is it necessary to determine the cetane number of a fuel ?
4. How do we calculate diesel index ? What is it significance ?
5. What is the advantage of determining the calculated cetane index of a fuel ? Give the relation to determine CCI.
6. Define viscosity ? Give the equation to determine it and the instrument used to determine conventional or kinematic viscosity.
7. What is the significance of viscosity index and how can it be determined ?
8. Why is it necessary to determine pour point or cloud point of a fuel ?
9. What is induction period of a gasoline ?
10. What are problems encountered when sulphur is present in petroleum ?
11. Explain the method to determine copper strip corrosion test. What is its relevance.
12. Explain the significance of aniline point.

❑❑❑

CHAPTER

5

REFINING OF CRUDE OIL

INSIDE THIS CHAPTER

5.1. REFINING OF CRUDE OIL

Cruide oil which reaches the surface, generally consists of a mixture of solid, liquid and gaseous hydrocarbons containing sand and water. Crude oil as such is of little importance. However, after the removal of dirt, water and much of the associated natural gas, the crude oil is separated into a number of useful fractions by fractional distillation. The resultant fractions are then subjected to purification or conversion processes, for the manufacture of various useful petroleum products. All these processes are referred to as **refining** of petroleum and the plant where these processes take place are called **refineries.**

Crude oil refining includes the following steps :

(*i*) **Demulsification :** The crude oil coming out from the well, is stable emulsion of oil and salt water, which is yellow to dark brown in colour. The emulsions are stabilized because of the presence of emulsifying agents in the crude oil (e.g. sodium salts of naphthenic acids, sulfoacids etc.) The emulsions are also stabilized due to the static electricity on the dispersed particles causing mutual repulsion.

The demulsification is achieved by **Cottrell's Process,** in which the water is removed from the oil by electrical process. The crude oil is subjected to an electrical field, when droplets of colloidal water coalesce to form large drops which separate out from the oil.

(*ii*) **Removal of Unwanted Salt :** Excessive salt content such as NaCl and $MgCl_2$ can corrode the refining equipment. These are removed by washing with water.

The objectionable sulphur compound are removed by treating the oil with copper oxide. The copper sulphide so formed is separated by filteration.

(*iii*) **Fractional Distillation :** The fractional distillation of petroleum is carried out in a specially designed tall fractioning tower or column made up of steel. Figure 5.1 shows the fractional distillation apparatus.

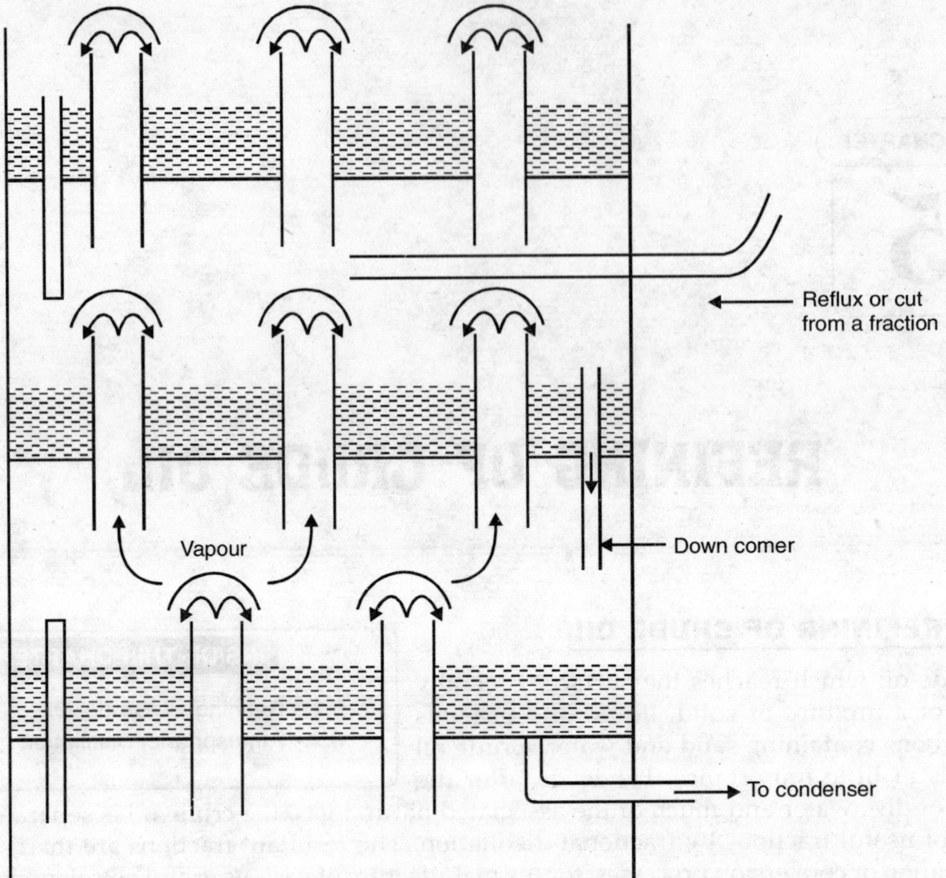

FIG. 5.1. SIMPLIFIED REPRESENTATION OF A BUBBLE-CAP FRACTIONATOR.

Early distillation processes were discontinuous or "batch proceses". In batch process, the crude petroleum was distilled from huge vessels till most of the oil separated into fractions. But this process is obsolete now a days, because due to increased operating temperatures, the cracking of various fractions take place. In modern refineries the batch process has been replaced by continuous process.

In continuous process, the crude oil is preheated to 350-380°C in specially designed tubular furnace known as pipe still. The temperature is kept within the above range so as to avoid thermal degradation of higher members into lower ones. The hot vapours from the crude are passed through a tall fractionating column, called bubble tower. Bubble tower consists of a number of horizontal trays, provided with a number of small chimneys, through which vapours rise. The chimneys are covered with loose caps, known as bubble caps. These bubble caps help to provide an intimate contact between the escaping vapours and down coming liquid.

The separation of various fractions is based on the fact that the hydrocarbons of petroleum boil at different temperatures. The temperature in the fractionating tower decreases gradually on moving upwards (About 300°C at the bottom and decreasing gradually to 125°C at the top). As the vapours of the crude oil go up, they become gradually cooler and fractional condensation takes place at different heights of column. Higher boiling fractions condense first at lower plates whereas lower boiling fractions condense turn by turn at upper plates. Thus, the crude oil gets separated into different fractions in the order of their boiling ranges and are collected at different heights in the columns with the help of condensers (Fig. 5.2).

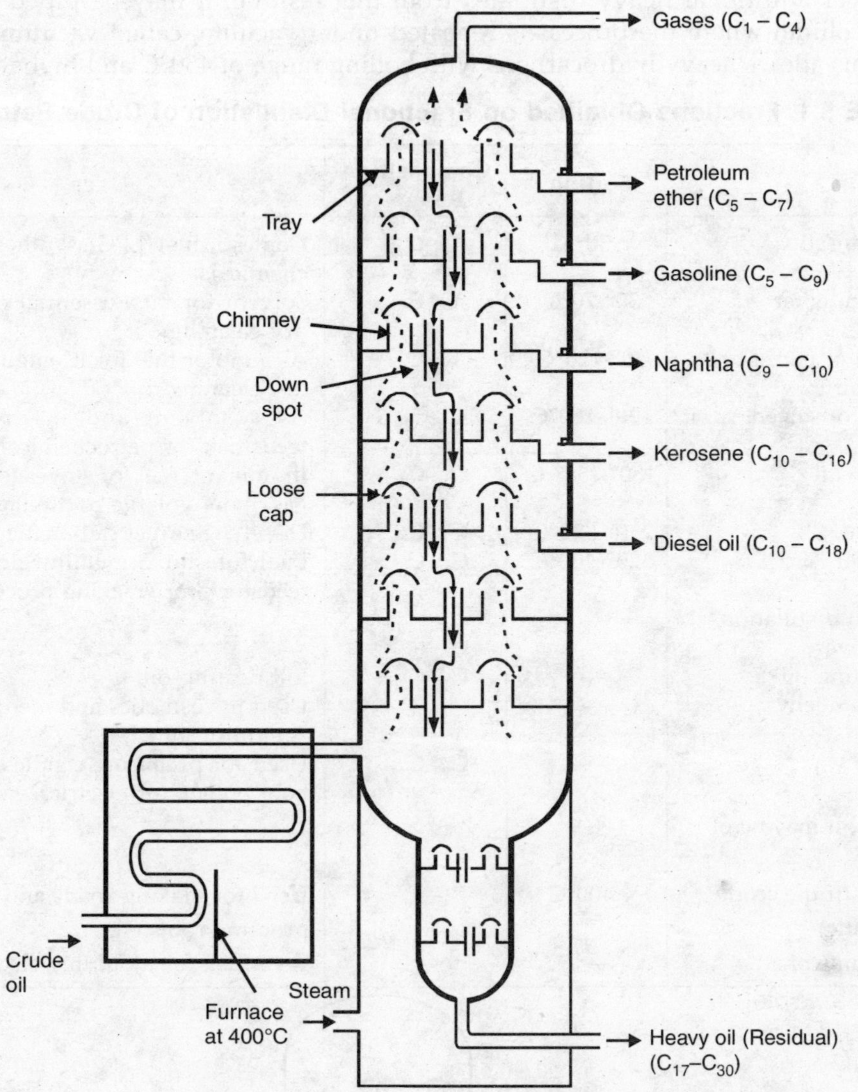

Fig. 5.2. Fractional Dislation of Cude Petroleum

The various primary fractions collected, their boiling ranges and uses are summarized in Table 5.1.

The atmospheric distillation takes place in the distilling column at or near atmospheric pressure. The crude oil is heated to 350-400°C and the vapour and liquid are piped into the distilling column. The liquid falls to the bottom and the vapour rises, passing through a series of perforated trays (sieve trays). Heavier hydrocarbons condense more quickly and settle on lower trays and higher hydrocarbons remains as a vapour longer and condense on higher trays. Liquid fractions are drawn from the trays and removed. In this way the light gases—methane, ethane, propane and butane pass out the top of the column, petrol is formed in the top trays, kerosene and gas oils in the middle and the fuel oils at the bottom. Residue drawn of the bottom may be burned as fuel, processed into lubricating oils, waxes and bilumens or used as feedstock for cracking units.

To recover additional heavy distillates from this residue, it may be piped to a second distillating column where the process is repeated under vacuum, called **vacuum distillation** (Fig. 5.3). This allows heavy hydrocarbons with boiling range of 450°C and higher to be.

TABLE 5.1. Fractions Obtained on Fractional Distillation of Crude Petroleum

Fraction	Boiling	Composition Range	Uses
1. Uncondensed	< 30° C	C_1-C_4	Domestic fuel (LPG), synthesis of organic chemicals.
2. Petroleum ether	30°-70°C	C_5-C_7	Solvent for fats, essential oils. Used in dry cleaning.
3. Gasoline or petrol or motor spirit	40°-120°C	C_5-C_9	As a motor fuel for IC engines, solvent, dry cleaning.
4. Naphtha or solvent spirit.	120°-180°C	C_9-C_{10}	As a solvent and in dry cleaning, feedstock for petrochemicals.
5. Kerosene oil	180°-250° C	C_{10}-C_{16}	Illuminant, fuel for stoves for making oil gas, more volatile portion as jet fuel.
6. Diesel Oil	250°-320°C	C_{10}-C_{18}	Diesel engine fuel, carburetting of water gas
7. Heavy Oil	320°-400°C	C_{17}-C_{30}	Fuel for ships, metallurgical furnaces, feedstock for cracking processes.
On Vacuum distillation, heavy oil gives :			
(a) Lubricating oil	—	C_{17}-C_{20}	Lubricating oil
(b) Petroleum jelly	—		Used in cosmetics and medicine
(c) Greases	—	—	As Lubricant
(d) Paraffin wax	—	C_{20}-C_{28}	Used for preparing candles, wax paper, boot polish, for electrical insulations.
8. Residue : It may be of two types			
(a) Asphalt (if the crude is asphaltic)	> 400°C	—	Used for making roads and water proofing roofs etc.
(b) Petroleum coke	—	—	As a fuel, for moulding electrode rods.

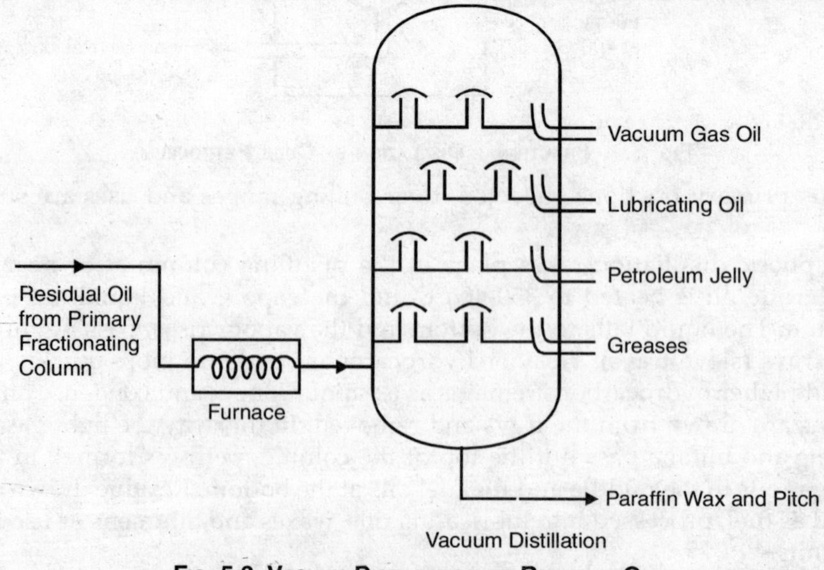

FIG. 5.3. VACUUM DISTILLATION OF RESIDUAL OIL

5.2. ATMOSPHERIC DISTILLATION

Atmospheric distillation takes place in a distilling column at or near atmospheric pressure. The crude oil is heated to 350-400°C and the vapour and liquid are piped into the distilling column. The liquid falls to the bottom and the vapour rises, passing through a series of perforated trays (sieve trays). Heavier hydrocarbons condense more quickly and settle on lower trays and lighter hydrocarbons remain as a vapour longer and condense on higher trays.

Liquid fractions are drawn from the trays and removed. In this way the light gases, methane, ethane, propane and butane pass out the top of the column, petrol is formed in the top trays, kerosene and gas oils in the middle, and fuel oils at the bottom. Residue drawn of the bottom may be burned as fuel, processed into lubricating oils, waxes and bitumen or used as feedstock for cracking units.

To recover additional heavy distillates from this residue, it may be piped to a second distillation column where the process is repeated under vacuum, called **vacuum distillation** as shown in Fig. 5.4. This allows heavy hydrocarbons with boiling points of 450°C and higher to be separated without them partly cracking into unwanted products such as coke and gas.

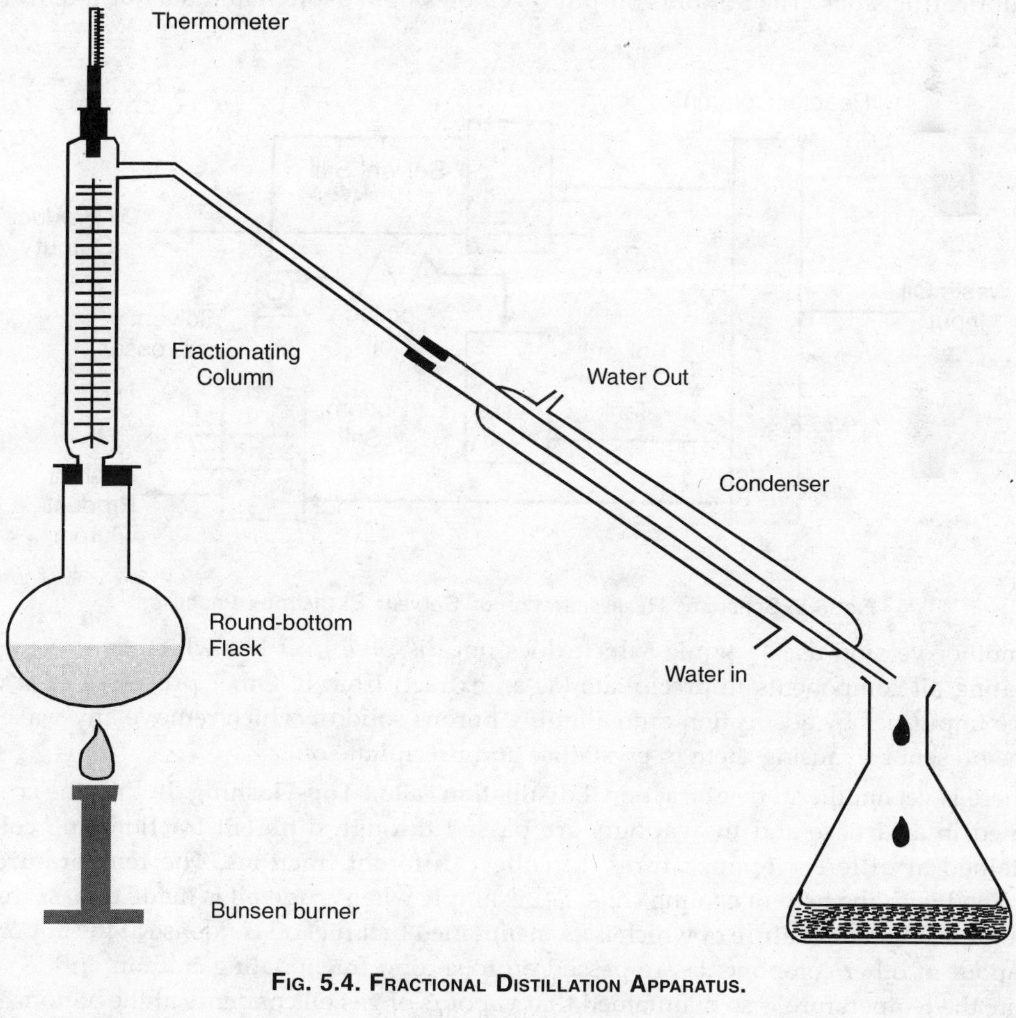

FIG. 5.4. FRACTIONAL DISTILLATION APPARATUS.

The heavy distillates recovered by vacuum distillation can be converted into lubricating oils by a variety of processes. The most common of these is called **solvent extraction**. In one version of this process the heavy distillate is washed with a liquid which does not dissolve in it but which dissolves (and so extracts) the non-lubricating oil components out of it. (Fig. 5.4)

The Fig. 5.5 presents a simplified, conceptual process flow diagram for the solvent extraction technology. Used oil is mixed in the Reactor Column with an aliphatic solvent such as liquefied propane. The solvent acts selectively, dissolving the oil fraction and leaving the impurities. The oil-laden solvent is transported from the top of the column to the Solvent Still, where the solvent fraction is separated from the oil for recycling within the process. The impurities (bottoms) slowly settle by gravity to the bottom of the reactor vessel where they are pumped to the Bottoms Still. Residual solvent is also separated there. The recovered solvent is liquefied through a compressor and cooling system and reinjected into the Reactor Column through the Solvent Recycle Tank to repeat the cycle. The oil product output, drawn from the Solvent Still, is considered suitable for blending with higher value fuel oils and with lubricating stock. The bottoms output resembles light asphalt and may be a marketable product.

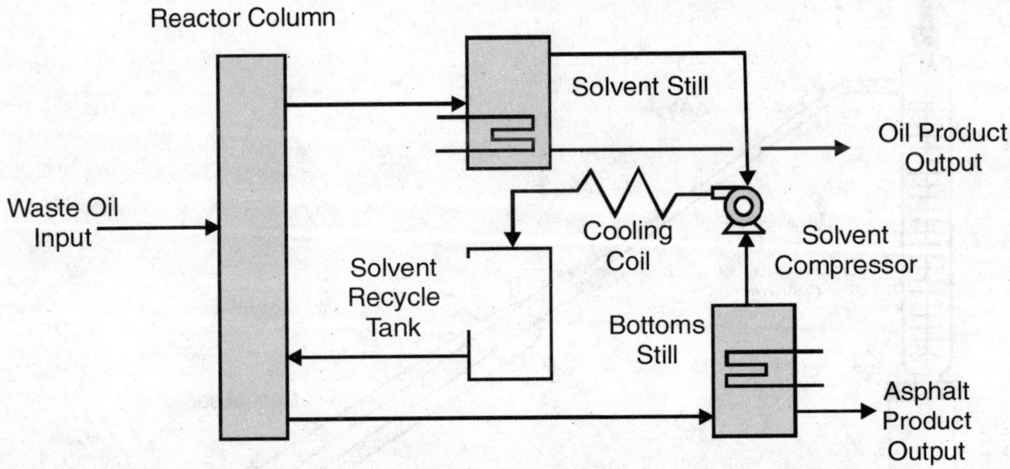

Fig. 5.5. Schematic Representation of Solvent Extraction Process.

Another version uses a liquid which does not dissolve in it but which causes the non-lubricating oil components to precipitate (as an extract) from it. Other processes exist which remove impurities by adsorption onto a highly porous solid or which remove any waxes that may be present by causing them to crystallise and precipitate out.

There is yet another type of fractional distillation called **Top-Flashing.** In this, the crude oil is heated in a furnace and the vapours are passed through different fractionating columns maintained at different temperatures, to collect different fractions. The temperatures are maintained with the help of cooling coils. For example when crude oil is made to pass through first column, the temperature of which is so maintained that fuel oil condense at the bottom and the vapour of other components are passed on to second fractionating column. In the second column, the temperature is so maintained that vapours of gas oil condense at the bottom and so on. Steam is injected at the bottom of each fractionating column for its partial pressure effect. Vapour of gasoline finally pass out from the top. (Fig. 5.6)

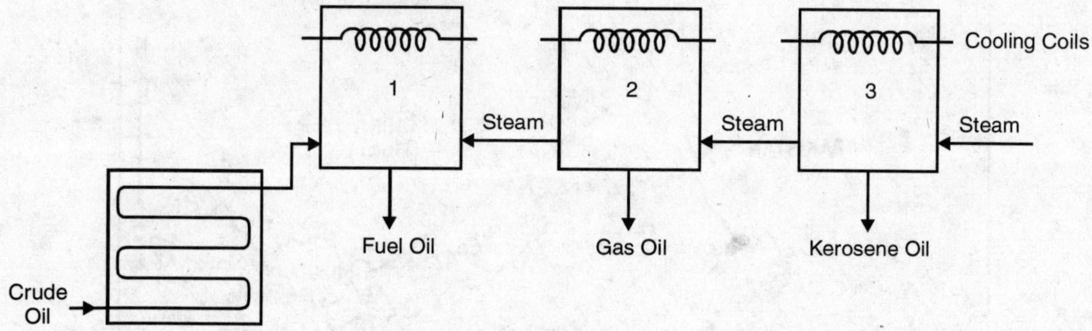

Fig. 5.6. Top Flashing

In top flashing, there is better control of product composition, but requires more pumps and instruments and hence is an expensive process.

The complete refining process and the product disposition is summarized below in Table 5.2.

TABLE 5.2. Refining Process of Crude Oil.

Crude Oil	Temperature	Product	Disposition
	<–90°	Butane and Lighter	Gas Processing
	90°–220°	Straight Run Gasoline	Motor Gasoline Blending
	220°–315°	Naphtha	Cat Reforming
	315°–450°	Kerosine	Hydrotreating
	450°–650°	Light Gas Oil	Distillate Fuel Blending
	650°–800°	Heavy Gas Oil	Cat Cracking
	800°+	Straight Run Residue	Flashing

Distilling crude and product disposition

The various refineries in India is depicted below in **Fig. 5.7.**

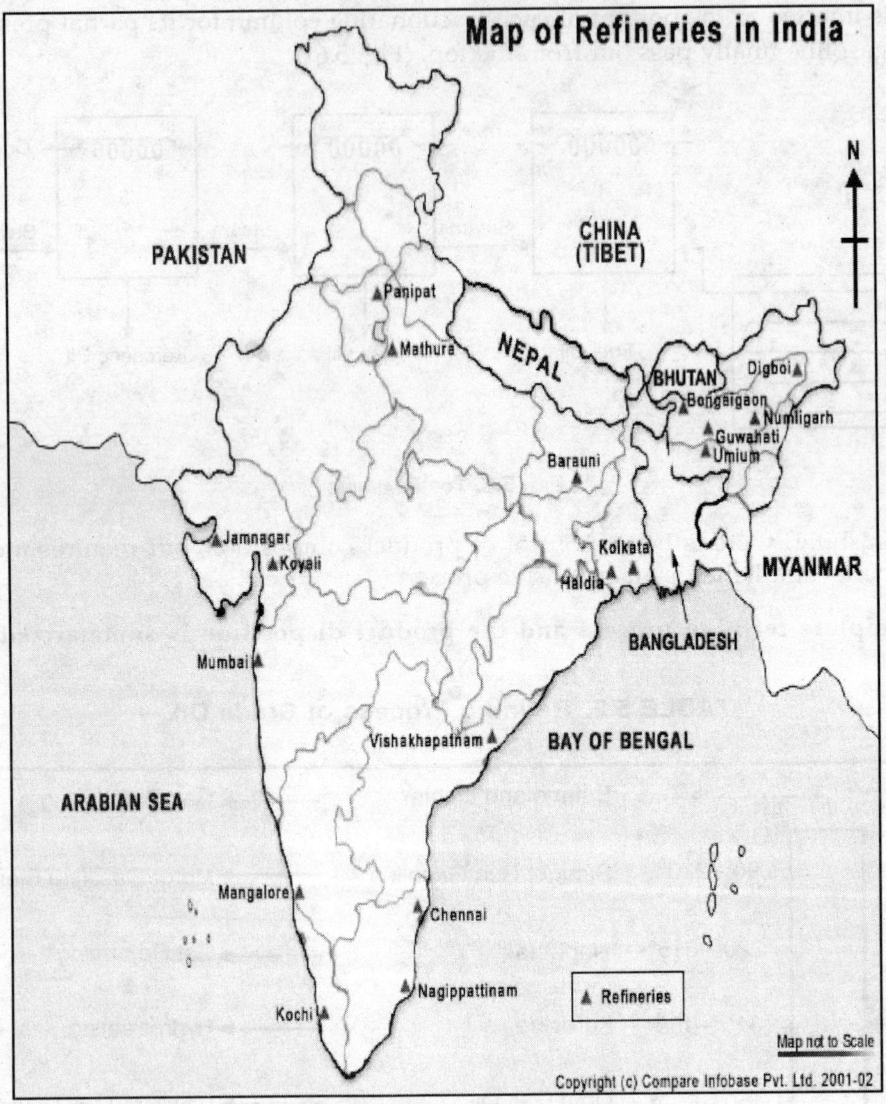

FIG. 5.7. VARIOUS REFINERIES IN INDIA.

REVIEW QUESTIONS

1. Explain the refining process of crude oil.
2. What is fractional distillation ?
3. Give a schematic diagram of various fraction obtained from crude oil.
4. What is top flashing of petroleum ?
5. Write short note on atmospheric distillation.

□□□

CHAPTER

6

CRACKING (CHEMISTRY)

6.1. INTRODUCTION

INSIDE THIS CHAPTER

In petroleum geology and chemistry, **cracking** is the process whereby complex organic molecules such as kerogens or heavy hydrocarbons are broken down into simpler molecules (*e.g.* light hydrocarbons) by the breaking of carbon-carbon bonds in the precursors. The rate of cracking and the end products are strongly dependent on the temperature and presence of any catalysts. Cracking, also referred to as pyrolysis, is the breakdown of a large alkane into smaller, more useful alkanes and an alkene. In other words, hydrocarbons cracking is the process of breaking long chain hydrocarbons into short ones. Cracking involves rupture of C-C and C-H bonds in the chains of higher molecular weight hydrocarbons

$$C_{10}H_{22} \text{ (Decane)} \longrightarrow C_5H_{12} \text{ (n pentane)} + C_5H_{10} \text{ (pentene)}$$

$$C_8H_{18} \longrightarrow C_5H_{12} + C_3H_6$$

The process of cracking involves following chemical changes :

- Higher hydrocarbons are converted to lower hydrocarbons by C-C cleavage. The products obtained on cracking have low boiling points than initial reactants.
- Formation of branched chain hydrocarbons takes place from straight chain alkanes
- Unsaturated hydrocarbons are formed from saturated hydrocarbons
- Cyclization may take place.

Cracking process takes place in following ways :

6.2. THERMAL CRACKING

Thermal Cracking uses heat to break down the residue from vacuum distillation. The lighter elements produced from this process can be made into distillate fuels and petrol. Cracked gases are converted to petrol blending components by alkylation or polymerisation. High temperatures and residence times were the variables used to determine yield of lighter products. Naphtha is upgraded to high quality petrol by reforming. Gas oil can be used as diesel fuel or can be converted to petrol by hydrocracking. The heavy residue is converted into residual oil or coke which is used in the manufacture of electrodes, graphite and carbides.

Modern high-pressure thermal cracking operates at absolute pressures of about 7,000 kPa. An overall process of disproportionation can be observed, where "*light*", hydrogen-rich products are formed at the expense of heavier molecules which condense and are depleted of hydrogen. The actual reaction is known as fission and produces alkenes, which are the basis for the economically important production of polymers.

Mechanism of Thermal Cracking

Thermal decomposition of the long chain. Oil molecules into shorter chains proceeds through a free radical mechanisms. It involves the following steps described below :

Initiation reactions, where a single molecule breaks apart into two free radicals. Only a small fraction of the feed molecules actually undergo initiation, but these reactions are necessary to produce the free radicals that drive the rest of the reactions. In steam cracking, initiation usually involves breaking a chemical bond between two carbon atoms, rather than the bond between a carbon and a hydrogen atom.

$$CH_3CH_3 \longrightarrow 2\,CH_3\bullet$$

Hydrogen abstraction, where a free radical removes a hydrogen atom from another molecule, turning the second molecule into a free radical.

$$CH_3\bullet + CH_3CH_3 \longrightarrow CH_4 + CH_3CH_2\bullet$$

Radical decomposition, where a free radical breaks apart into two molecules, one an alkene, the other a free radical. This is the process that results in the alkene.

$$CH_3CH_2\bullet \longrightarrow CH_2{=}CH_2 + H\bullet$$

Radical addition, the reverse of radical decomposition, in which a radical reacts with an alkene to form a single, larger free radical. These processes are involved in forming the aromatic products that result when heavier feedstocks are used.

$$CH_3CH_2\bullet + CH_2{=}CH_2 \longrightarrow CH_3CH_2CH_2CH_2\bullet$$

Termination reactions, which happen when two free radicals react with each other to produce products that are not free radicals. Two common forms of termination are **recombination,** where the two radicals combine to form one larger molecule, and **disproportionation,** where one radical transfers a hydrogen atom to the other, giving an alkene and an alkane.

$$CH_3\bullet + CH_3CH_2\bullet \longrightarrow CH_3CH_2CH_3$$

$$CH_3CH_2\bullet + CH_3CH_2\bullet \longrightarrow CH_2{=}CH_2 + CH_3CH_3$$

Thermal cracking is an example of a reaction whose energetics are dominated by entropy ($\Delta S°$) rather than by enthalpy ($\Delta H°$) in the Gibbs Free Energy equation $\Delta G° = \Delta H° - T\Delta S°$. Although the bond dissociation energy D for a carbon-carbon single bond is relatively high (about 375 kJ/mol) and cracking is highly endothermic, the large positive entropy change resulting from the fragmentation of one large molecule into several smaller pieces, together with the extremely high temperature, makes TΔS° term larger than the ΔH° term, thereby favouring the cracking reaction.

Here is an example of cracking with butane $CH_3—CH_2—CH_2—CH_3$

- **1st possibility (48%) :** Breaking is done on the CH_3-CH_2 bond.

$$CH_3{}^* \; / \; {}^*CH_2—CH_2—CH_3$$

Following product can be obtained

$$CH_4 + CH_2{=}CH—CH_3$$

- **2nd possibility (38%) :** Breaking is done on the $CH_2—CH_2$ bond.

$$CH_3—CH_2{}^* \; / \; {}^*CH_2—CH_3$$

An alkane and an alkene may be obtained

$$CH_3—CH_3 + CH_2{=}CH_2$$

- **3rd possibility (14%) :** Breaking of a C—H bond

An alkene and hydrogen gas can be obtained.

$CH_2{=}CH—CH_2—CH_3 + H_2$ this is very useful since the catalyst can be recycled.

Cracking reativity varies with the type of hydrocarbons and decrease in the following order :

n-paraffins > isoparaffins > cycloparaffins > aromatics > aromatics/naphthenes > poly nuclear aromatics.

Olefins crack to smaller olefins and di-olefins Naphthenes and aromatics with side chains undergo a shortening of side chains. Naphthene rings are stable below 490°C. Aromatics and polynuclear aromatics undergo condensation reaction with the formation of asphaltenic compounds. Some reactions are :

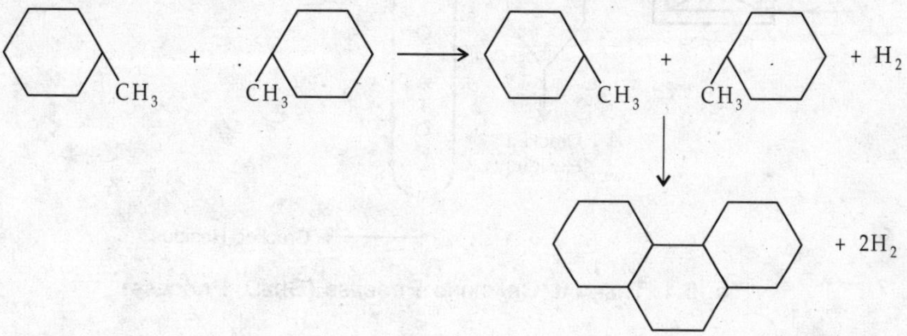

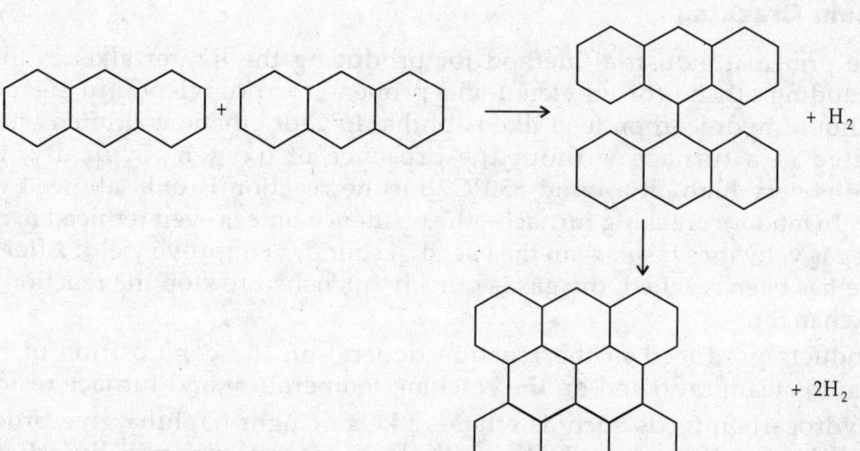

Types of Thermal Cracking includes

Fig. 6.1 shows the process of thermal cracking.

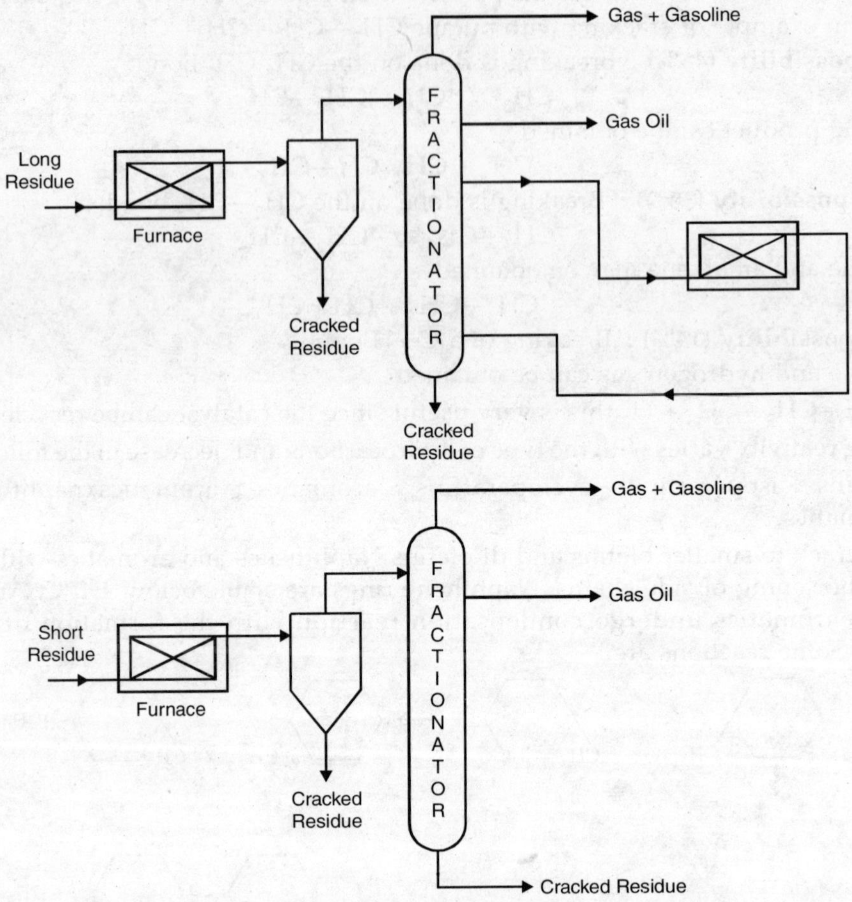

Fig. 6.1. Thermal Cracking Process (Shell Process).

6.2.1. Steam Cracking

It is the principal industrial method for producing the lighter alkenes (or commonly olefins), including ethene (or ethylene) and propene (or propylene). In steam cracking, a gaseous or liquid hydrocarbon feed like naphtha, LPG or ethane is diluted with steam and briefly heated in a furnace without the presence of oxygen. Typically, the reaction temperature is very high, at around 850°C, but the reaction is only allowed to take place very briefly. In modern cracking furnaces, the residence time is even reduced to milliseconds, resulting in gas velocities faster than the speed of sound, to improve yield. After the cracking temperature has been reached, the gas is quickly quenched to stop the reaction in a transfer line heat exchanger.

The products produced in the reaction depend on the composition of the feed, the hydrocarbon to steam ratio and on the cracking temperature and furnace residence time.

Light hydrocarbon feeds such as ethane, LPGs or light naphtha give product streams rich in the lighter alkenes, including ethylene, propylene, and butadiene. Heavier hydrocarbon (full range and heavy naphthas as well as other refinery products) feeds give some of these, but also give products rich in aromatic hydrocarbons and hydrocarbons suitable for inclusion in gasoline or fuel oil. The higher cracking temperature (also referred to as severity) favours the production of ethene and benzene, whereas lower severity produces relatively higher amounts of propene, C4-hydrocarbons and liquid products. The process also results in the slow deposition of coke, a form of carbon, on the reactor walls. This degrades the efficiency of the reactor, so reaction conditions are designed to minimize this. Nonetheless, a steam cracking furnace can usually only run for a few months at a time between de-cokings. Decokes require the furnace to be isolated from the process and then a flow of steam or a steam/air mixture is passed through the furnace coils. This converts the hard solid carbon layer to carbon monoxide and carbon dioxide. Once this reaction is complete, the furnace can be returned to service.

6.2.2. Visbreaking Process

Visbreaking, a mild form of thermal cracking, significantly lowers the viscosity of heavy crude-oil residue without affecting the boiling point range. Residual from the atmospheric distillation tower is heated (800°-950° F) at atmospheric pressure and mildly cracked in a heater (Fig. 6.2). It is then quenched with cool gas oil to control over cracking, and flashed in a distillation tower. Visbreaking is used to reduce the pour point of waxy residues and reduce the viscosity of residues used for blending with lighter fuel oils. Middle distillates may also be produced, depending on product demand. The thermally cracked residue tar, which accumulates in the bottom of the fractionation tower, is vacuum flashed in a stripper and the distillate recycled.

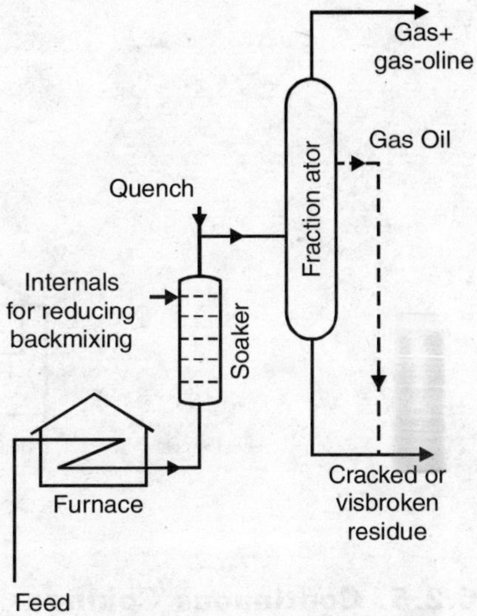

Fig. 6.2. Visbreaking.

6.2.3. Coking Processes

Coking is a severe method of thermal cracking used to upgrade heavy residuals into lighter products or distillates. Coking produces straight-run gasoline (coker naphtha) and various middle-distillate fractions used as catalytic cracking feedstock. The process so completely reduces hydrogen that the residue is a form of carbon called *"coke."* The two most common processes are delayed coking and continuous (contact or fluid) coking. Three typical types of coke are obtained (sponge coke, honeycomb coke, and needle coke) depending upon the reaction mechanism, time, temperature, and the crude feedstock.

6.2.4. Delayed Coking

In delayed coking the heated charge (typically residuum from atmospheric distillation towers) is transferred to large coke drums which provide the long residence time needed to allow the cracking reactions to proceed to completion. Initially the heavy feedstock is fed to a furnace which heats the residuum to high temperatures (900°-950° F) at low pressures (25-30 psi) and is designed and controlled to prevent premature coking in the heater tubes (**Figure 6.3**). The mixture is passed from the heater to one or more coker drums where the hot material is held approximately 24 hours (delayed) at pressures of 25-75 psi, until it cracks into lighter products. Vapors from the drums are returned to a fractionator where gas, naphtha, and gas oils are separated out. The heavier hydrocarbons produced in the fractionator are recycled through the furnace.

After the coke reaches a predetermined level in one drum, the flow is diverted to another drum to maintain continuous operation. The full drum is steamed to strip out uncracked hydrocarbons, cooled by water injection, and decoked by mechanical or hydraulic methods. The coke is mechanically removed by an auger rising from the bottom of the drum. Hydraulic decoking consists of fracturing the coke bed with high-pressure water ejected from a rotating cutter.

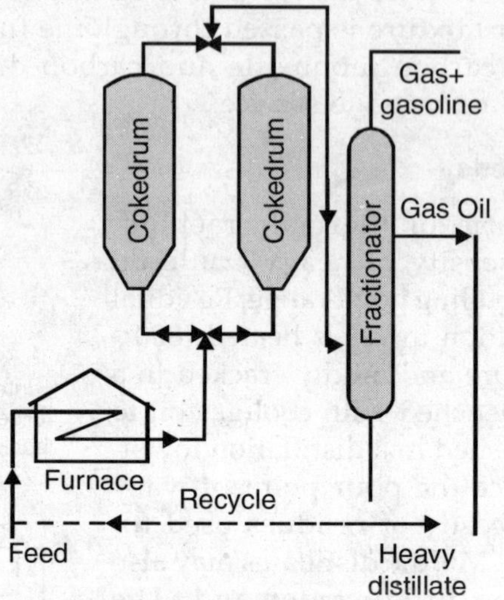

FIG. 6.3. DELAYED COKING.

6.2.5. Continuous Coking

Continuous (contact or fluid) coking is a moving-bed process that operates at temperatures higher than delayed coking. In continuous coking, thermal cracking occurs by using heat transferred from hot, recycled coke particles to feedstock in a radial mixer, called a reactor, at a pressure of 50 psi. Gases and vapours are taken from the reactor, quenched to stop any further reaction, and fractionated. The reacted coke enters a surge drum and is lifted to a feeder and classifier where the larger coke particles are removed as product. The remaining coke is dropped into the preheater for recycling with feedstock. Coking occurs both in the reactor and in the surge drum. The process is automatic in that there is a continuous flow of coke and feedstock.

6.3. CATALYTIC CRACKING

Catalytic cracking is used to convert heavy hydrocarbon fractions obtained by vacuum distillation into a mixture of more useful products such as petrol and light fuel oil. In this process, the feedstock undergoes a chemical breakdown, under controlled heat (450 - 500°C) and pressure, in the presence of a catalyst - a substance which promotes the reaction without itself being chemically changed. Small pellets of silica - alumina or silica - magnesia have proved to be the most effective catalysts. The cracking reaction yields petrol, LPG, unsaturated olefin compounds, cracked gas oils, a liquid residue called cycle oil, light gases and a solid coke residue. Cycle oil is recycled to cause further breakdown and the coke, which forms a layer on the catalyst, is removed by burning. The other products are passed through a fractionator to be separated and separately processed.

The catalytic cracking process involves the presence of acid catalysts (usually solid acids such as silica-alumina and zeolites) which promote a heterolytic (asymmetric) breakage of bonds yielding pairs of ions of opposite charges, usually a carbocation and the very unstable hydride anion. Carbon-localized free radicals and cations are both highly unstable and undergo processes of chain rearrangement, C—C scission in position beta (*i.e.*, cracking) and intra- and intermolecular hydrogen transfer or hydride transfer. In both types of processes, the corresponding reactive intermediates (radicals, ions) are permanently regenerated, and thus they proceed by a self-propagating chain mechanism. The chain of reactions is eventually terminated by radical or ion recombination.

Catalytic Cracking has an advantage over thermal cracking :

- High temperatures and pressures are not required in the presence of a catalyst and hence there is saving in the production cost.
- The use of catalyst not only accelerates the cracking reactions (decomposition, dehydrogenation, isomerization and polymerization), but also introduces new reactions which considerably modify the yield and the nature of the products.
- The yield of gasoline obtained by catalytic cracking contains low percentage of olefins and high percentage of isoparaffin and aromatic hydrocarbons. This increases the anti-knock properties of gasoline. Gasoline obtained by catalytic cracking has high octane number.
- No external fuel is required for cracking, since the heat required is obtained by burning off the coke deposited on the catalyst itself during regeneration process.
- The products contains very little amount of undesirable sulphur, because a major portion of it escapes out as H_2S gas, during cracking.
- Gasoline formed contains much less gum and gum forming compounds.
- Compared with thermal process, the catalytic process yield less coke, lesss gas and more liquid products.
- Catalysts are selective in action and hence cracking of only high boiling fractions takes place.

Depending upon the catalyst used catalytic cracking are as follows :

6.3.1. Fixed Bed Catalytic Reactor

In this type of cracking the catalyst is in the form of granules or pellets. These pellets or granules are used in the of fixed beds in catalyst towers or reactors. The heavy oil charge is passed through a heater and heated to a cracking temperature (420-450°C). The charge is then passed through a reactor containing catalyst. The catalyst used is either **silica alumina gel or bauxite mixed with clay and zirconium oxide.** The reactor is maintained at the temperature of 420-450°C and at the pressure of 1-5kg/m^2. Cracking of heavy oil results in low molecular weight hydrocarbons. The carbon formed during cracking process get adsorbed on the catalyst thereby deactivating it. The gasoline vapours are cooled and send to the separator were they are collected at the top and the heavy ends are collected at the bottom. The gas is send to the scrubber where the dissolved gases are separated and pure gasoline vapours are obtained and the dissolved gases make to makeup gases.

The process is shown below in Fig. 6.4

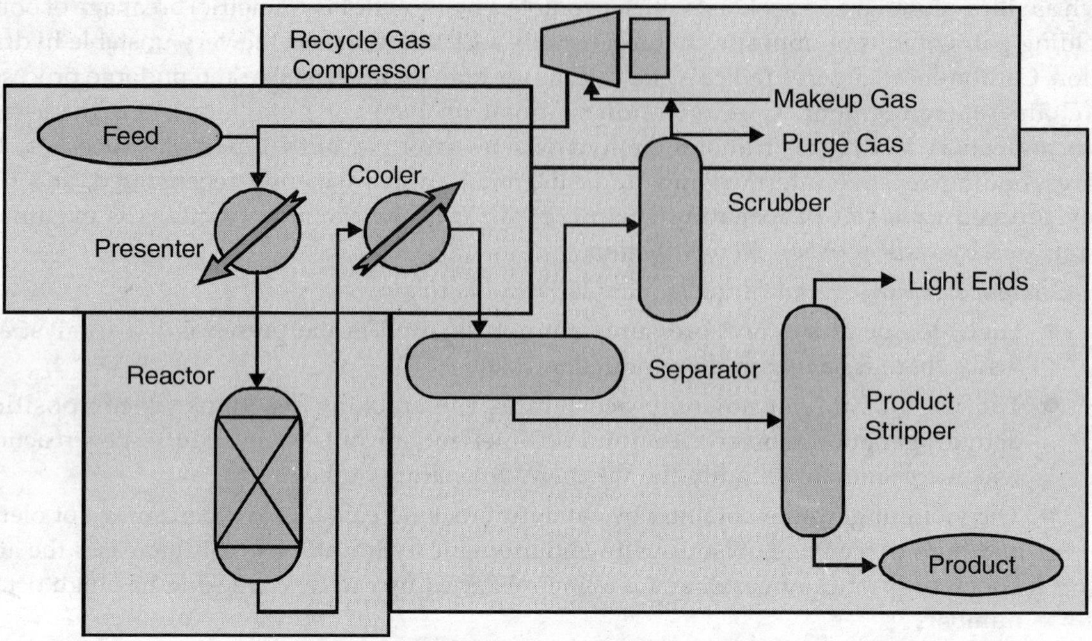

FIG. 6.4. FIXED BED CATALYTIC REACTOR.

6.3.2. Fluid Catalytic Cracking

Fluid Catalytic Cracking (FCC) is the most important conversion process used in petroleum refineries. It is widely used to convert the high-boiling, high-molecular weight hydrocarbon fractions of petroleum crude oils to more valuable gasoline, olefinic gases and other products. Cracking of petroleum hydrocarbons was originally done by thermal cracking which has been almost completely replaced by catalytic cracking because it produces more gasoline with a higher octane rating. It also produces byproduct gases that are more olefinic, and hence more valuable, than those produced by thermal cracking. The feedstock to an FCC is usually that portion of the crude oil that has an initial boiling point of 340°C or higher at atmospheric pressure and an average molecular weight ranging from about 200 to 600 or higher. This portion of crude oil is often referred to as heavy gas oil. The FCC process vaporizes and breaks the long-chain molecules of the high-boiling hydrocarbon liquids into much shorter molecules by contacting the feedstock, at high temperature and moderate pressure, with a fluidized powdered catalyst.

In effect, refineries use fluid catalytic cracking to correct the imbalance between the market demand for gasoline and the excess of heavy, high boiling range products resulting from the distillation of crude oil.

Basically, there are two different configurations for an FCC unit: **the "stacked" type** where the reactor and the catalyst regenerator are contained in a single vessel with the reactor above the catalyst regenerator and the **"side-by-side"** type where the reactor and catalyst regenerator are in two separate vessels. The fluid catalytic (Fig. 6.5) mechanism is described below.

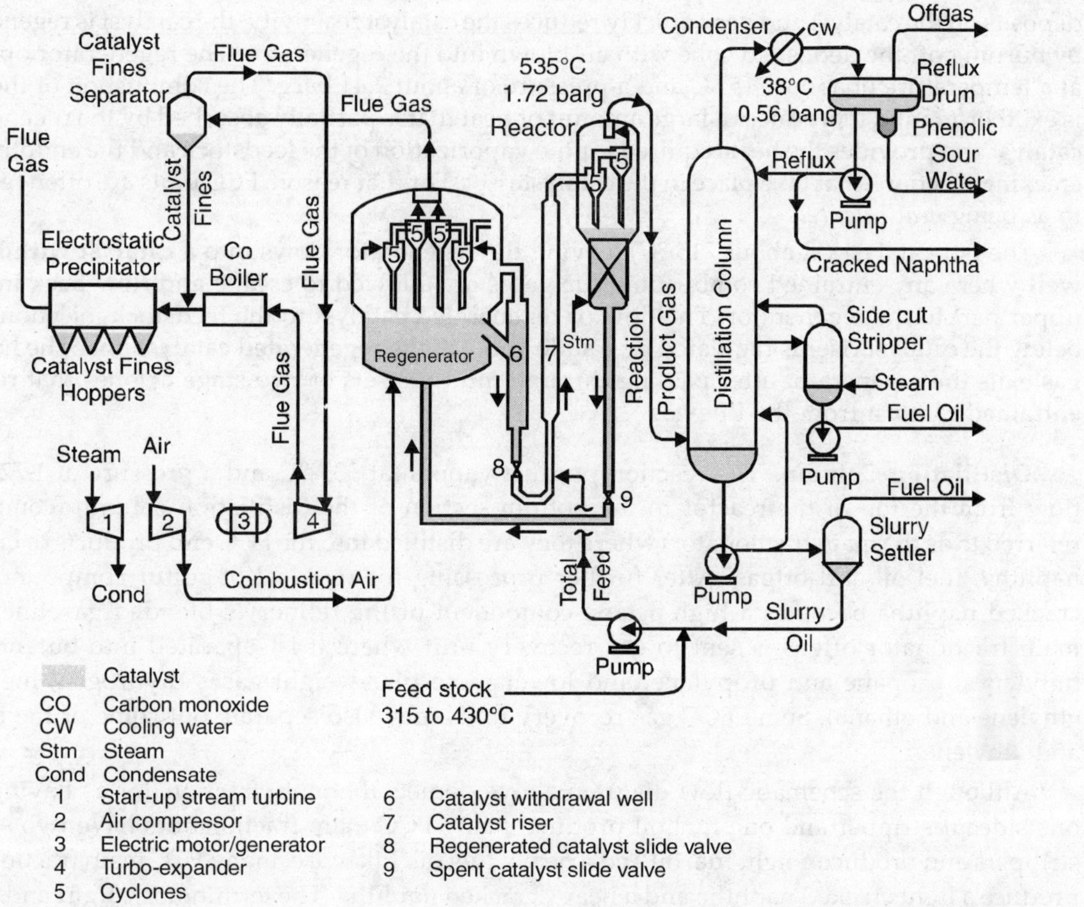

FIG. 6.5. A SCHEMATIC FLOW DIAGRAM OF A FLUID CATALYTIC CRACKING UNIT AS USED IN PETROLEUM REFINERIES.

Reactor and Regenerator : The schematic flow diagram of a typical modern FCC unit above is based upon the **"side-by-side"** configuration. The preheated high-boiling petroleum feedstock (at about 315 to 430°C) consisting of long-chain hydrocarbon molecules is combined with recycle slurry oil from the bottom of the distillation column and injected into the **catalyst riser** where it is vapourized and cracked into smaller molecules of vapor by contact and mixing with the very hot powdered catalyst from the regenerator. All of the cracking reactions take place in the catalyst riser. The hydrocarbon vapors *"fluidize"* the powdered catalyst and the mixture of hydrocarbon vapors and catalyst flows upward to enter the **reactor** at a temperature of about 535 °C and a pressure of about 1.72 barg. The reactor is in fact merely a vessel in which the cracked product vapors are :

- Separated from the so-called **spent catalyst** by flowing through a set of two-stage cyclones within the reactor and
- The **spent catalyst** flows downward through a steam stripping section to remove any hydrocarbon vapors before the spent catalyst returns to the **catalyst regenerator**. The flow of spent catalyst to the regenerator is regulated by a **slide valve** in the spent catalyst line.

Since the cracking reactions produce some carbonaceous material (referred to as coke) that deposits on the catalyst and very quickly reduces the catalyst reactivity, the catalyst is regenerated by burning off the deposited coke with air blown into the regenerator. The regenerator operates at a temperature of about 715 °C and a pressure of about 2.41 barg. The combustion of the coke is exothermic and it produces a large amount of heat that is partially absorbed by the regenerated catalyst and provides the heat required for the vaporization of the feedstock and the endothermic cracking reactions that take place in the catalyst riser. For that reason, FCC units are often referred to as being *heat balanced*.

The hot catalyst (at about 715 °C) leaving the regenerator flows into a **catalyst withdrawal well** where any entrained combustion flue gases are allowed to escape and flow back into the upper part to the regenerator. The flow of regenerated catalyst to the feedstock injection point below the catalyst riser is regulated by a slide valve in the regenerated catalyst line. The hot flue gas exits the regenerator after passing through multiple sets of two-stage cyclones that remove entrained catalyst from the flue gas.

Distillation Column. The reaction product vapors (at 535 °C and a pressure of 1.72 barg) flow from the top of the reactor to the bottom section of the distillation column (commonly referred to as the main fractionator) where they are distilled into the FCC end products of cracked naphtha, fuel oil and offgas. After further processing for removal of sulfur compounds, the cracked naphtha becomes a high-octane component of the refinery's blended gasolines. The main fractionator offgas is sent to **gas recovery unit** where it is separated into butanes and butylenes, propane and propylene, and lower molecular weight gases (hydrogen, methane, ethylene and ethane). Some FCC gas recovery units may also separate out some of the ethane and ethylene.

Although the schematic flow diagram above depicts the main fractionator as having only one sidecut stripper and one fuel oil product, many FCC main fractionators have two sidecut strippers and produce a light fuel oil and a heavy fuel oil. Likewise, many FCC main fractionators produce a light cracked naphtha and a heavy cracked naphtha. The terminology **light** and **heavy** in this context refers to the product boiling ranges, with light products having a lower boiling range than heavy products.

The bottom product oil from the main fractionator contains residual catalyst particles which were not completely removed by the cyclones in the top of the reactor. For that reason, the bottom product oil is referred to as a **slurry oil**. Part of that slurry oil is recycled back into the main fractionator above the entry point of the hot reaction product vapors so as to cool and partially condense the reaction product vapors as they enter the main fractionator. The remainder of the slurry oil is pumped through a slurry settler. The bottom oil from the slurry settler contains most of the slurry oil catalyst particles and is recycled back into the catalyst riser by combining it with the FCC feedstock oil. The so-called **clarified slurry oil** or **decant oil** is withdrawn from the top of slurry settler for use elsewhere in the refinery or as a heavy fuel oil blending component.

Regenerator Flue Gas. Depending on the choice of FCC design, the combustion in the regenerator of the coke on the spent catalyst may or may not be complete combustion to carbon dioxide (CO_2). The combustion air flow is controlled so as to provide the desired ratio of carbon monoxide (CO) to carbon dioxide for each specific FCC design. In the design shown in Figure the coke has only been partially combusted to CO_2. The combustion flue gas (containing CO and CO_2) at 715°C and at a pressure of 2.41 barg is routed through a secondary catalyst separator containing **swirl tubes** designed to remove 70 to 90 percent of the particulates in the flue gas leaving the regenerator. This is required to prevent erosion damage to the blades in the turbo-expander that the flue gas is next routed through. The expansion of flue gas through a turbo-expander provides sufficient power to drive the regenerator's combustion air compressor. The electrical motor-generator can consume or produce electrical power. If the expansion of the flue gas does not provide enough power to drive the air compressor, the electric motor/generator provides the needed additional power. If the flue gas expansion provides more power than needed to drive the air compressor, than the electric motor/generator converts the excess power into electric power and exports it to the refinery's electrical system. The expanded flue gas is then routed through a steam-generating boiler (referred to as a **CO boiler**) where the carbon monoxide in the flue gas is burned as fuel to provide steam for use in the refinery as well as to comply with any applicable environmental regulatory limits on carbon monoxide emissions.

The flue gas is finally processed through an electrostatic precipitator (ESP) to remove residual particulate matter to comply with any applicable environmental regulations regarding particulate emissions. The ESP removes particulates in the size range of 2 to 20 microns from the flue gas.The steam turbine in the flue gas processing system is used to drive the regenerator's combustion air compressor during start-ups of the FCC unit until there is sufficient conbustion flue gas to take over that task.

Chemistry Involved in Fluid Catalytic Cracking

Before delving into the chemistry involved in catalytic cracking, it will be helpful to briefly discuss the composition of petroleum crude oil. Petroleum crude oil consists primarily of a mixture of hydrocarbons with small amounts of other organic compounds containing sulfur, nitrogen and oxygen. The crude oil also contains small amounts of metals such as copper, iron, nickel and vanadium.

TABLE 6.1.

Composition	
Carbon	83-87%
Hydrogen	10-14%
Nitrogen	0.1-2%
Oxygen	0.1-1.5%
Sulfur	0.5-6%
Metals	< 0.1%

The elemental composition ranges of crude oil are summarized in Table 6.1 and the hydrocarbons in the crude oil can be classified into three types :

- **Paraffins or Alkanes :** Saturated straight-chain or branched hydrocarbons, without any ring structures.
- **Naphthenes or Cycloalkanes :** Saturated hydrocarbons having one or more ring structures with one or more side-chain paraffins.
- **Aromatics :** Hydrocarbons having one or more unsaturated ring structures such as benzene or unsaturated polycyclic ring structures such as naphthalene or phenanthrene, any of which may also have one or more side-chain paraffins.
- Olefins or alkenes, which are unsaturated straight-chain or branched hydrocarbons, do not occur naturally in crude oil.

The fluid catalytic cracking process breaks large hydrocarbon molecules into smaller molecules by contacting them with powdered catalyst at a high temperature and moderate pressure which first vapourizes the hydrocarbons and then breaks them. The cracking reactions occur in the vapor phase and start immediately when the feedstock is vapourized in the catalyst riser.

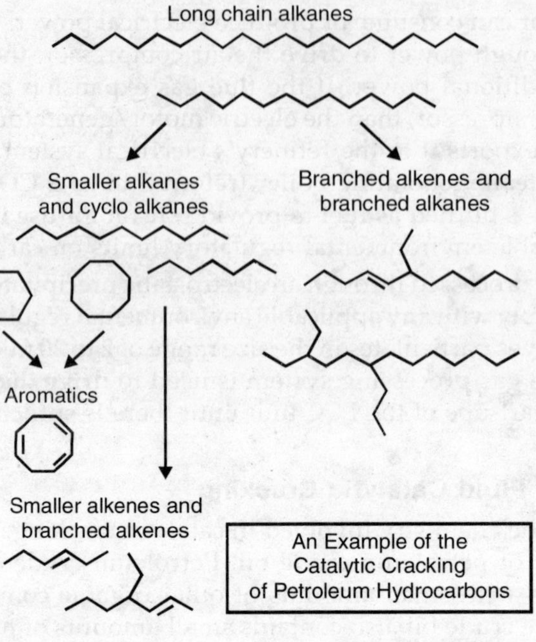

FIG. 6.6. DIAGRAMMATIC EXAMPLE OF THE CATALYTIC CRACKING OF PETROLEUM HYDROCARBONS.

Figure 6.6 is a very simplified schematic diagram that exemplifies how the process breaks high boiling, straight-chain alkane (paraffin) hydrocarbons into smaller straight-chain alkanes as well as branched-chain alkanes, branched alkenes (olefins) and cycloalkanes (naphthenes). The breaking of the large hydrocarbon molecules into smaller molecules is more technically referred to as **scission** of the carbon-to-carbon bonds. As depicted in Figure, some of the smaller alkanes are then broken and converted into even smaller alkenes and branched alkenes such as the gases ethylene, propylene, butylenes and isobutylenes. Those olefinic gases are valuable for use as petrochemical feedstocks. The propylene, butylene and isobutylene are also valuable feedstocks for certain petroleum refining processes that convert them into high-octane gasoline blending components. The cycloalkanes (naphthenes) formed by the initial breakup of the large molecules are further converted to aromatics such as benzene, toluene and xylenes which boil in the gasoline boiling range and have much higher octane ratings than alkanes. By no means does Figure include all the chemistry of the primary and secondary reactions taking place in the fluid catalytic process. There are a great many other reactions involved.

Catalysts : Modern FCC catalysts are fine powders with a bulk density of 0.80 to 0.96 g/cc and having a particle size distribution ranging from 10 to 150 μm and an average particle size of 60 to 100 μm. The design and operation of an FCC unit is largely dependent upon the chemical and physical properties of the catalyst. The desirable properties of an FCC catalyst are :

- Good stability to high temperature and to steam.
- High activity.
- Large pore sizes.
- Good resistance to attrition.
- Low coke production.

A modern FCC catalyst has four major components: crystalline **zeolite,** matrix, binder and filler. Zeolite is the primary active component and can range from about 15 to 50 weight percent of the catalyst. The zeolite used in FCC catalysts is referred to as **faujasite** and is comprised of silica and alumina tetrahedra with each tetrahedron having either an aluminum or a silicon atom at the center and four oxygen atoms at the corners. It is a molecular sieve with a distinctive lattice structure that allows only a certain size range of hydrocarbon molecules to enter the lattice. In general, the zeolite does not allow molecules larger than 8 to 10 nm (*i.e.*, 80 to 90 angstroms) to enter the lattice. The catalytic sites in the zeolite are strong acids (equivalent to 90% sulfuric acid) and provide most of the catalytic activity. The acidic sites are provided by the alumina tetrahedra. The aluminum atom at the center of each alumina tetrahedra is at a +3 oxidation state surrounded by four oxygen atoms at the corners which are shared by the neighboring tetrahedra. Thus, the net charge of the alumina tetrahedra is -1 which is balanced by a sodium ion during the production of the catalyst. The sodium ion is later replaced by an ammonium ion which is vapourized when the catalyst is subsequently dried, resulting in the formation of Lewis and Brønsted acidic sites. In some FCC catalysts, the Brønsted sites may be later replaced by rare earth metals such as cerium and lanthanum to provide alternative activity and stability levels. The matrix component of an FCC catalyst contains amorphous alumina which also provides catalytic activity sites and in larger pores that allows entry for larger molecules than does the zeolite. That enables the cracking of higher-boiling, larger feedstock molecules than are cracked by the zeolite.The binder and filler components provide the physical strength and integrity of the catalyst. The binder is usually silica sol and the filler is usually a clay (kaolin). Nickel, vanadium, iron, copper and other metal contaminants, present in FCC feedstocks in the parts per million range, all have detrimental effects on the catalyst activity and performance. Nickel and vanadium are particularly troublesome. There are a number of methods for mitigating the effects of the contaminant metals:

- **Avoid Feedstocks with High Metals Content :** This seriously hampers a refinery's flexibility to process various crude oils or purchased FCC feedstocks.
- **Feedstock Feed Pretreatment :** Hydrodesulfurization of the FCC feedstock removes some of the metals and also reduces the sulfur content of the FCC products. However, this is quite a costly option.
- **Increasing Fresh Catalyst Addition :** All FCC units withdraw some of the circulating **equilibrium catalyst** as spent catalyst and replaces it with fresh catalyst in order to maintain a desired level of activity. Increasing the rate of such exchange lowers the level of metals in the circulating **equilibrium catalyst**, but this is also quite a costly option.
- **Demetallization :** The **Demet Process** removes nickel and vanadium from the withdrawn spent catalyst. The nickel and vanadium are converted to chlorides which are then washed out of the catalyst. After drying, the demetallized catalyst is recycled into the circulating catalyst. Removals of about 95 percent nickel removal and 67 to 85 percent vanadium have been reported. Despite that, the use of the Demet process has not become widespread, perhaps because of the high capital expenditure required.
- **Metals passivation :** Certain materials can be used as additives which can be impregnated into the catalyst or added to the FCC feedstock in the form of metal-organic compounds. Such materials react with the metal contaminants and passivate the

contaminants by forming less harmful compounds that remain on the catalyst. *For example*, antimony and bismuth are effective in passivating nickel and tin is effective in passivating vanadium. A number of proprietary passivation processes are available and fairly widely used.

During the trip up the riser, the cracking catalyst is *"spent"* by reactions which deposit coke on the catalyst and greatly reduce activity and selectivity. The "spent" catalyst is disengaged from the cracked hydrocarbon vapours and sent to a stripper where it is contacted with steam to remove hydrocarbons remaining in the catalyst pores. The *"spent"* catalyst then flows into a fluidized-bed regenerator where air (or in some cases air plus oxygen) is used to burn off the coke to restore catalyst activity and also provide the necessary heat for the next reaction cycle, cracking being an endothermic reaction. The *"regenerated"* catalyst then flows to the base of the riser, repeating the cycle.

The gasoline produced in the FCC unit has an elevated octane rating but is less chemically stable compared to other gasoline components due to its olefinic profile. Olefins in gasoline are responsible for the formation of polymeric deposits in storage tanks, fuel ducts and injectors. The FCC LPG is an important source of C_3-C_4 olefins and isobutane that are essential feeds for the alkylation process and the production of polymers such as polypropylene

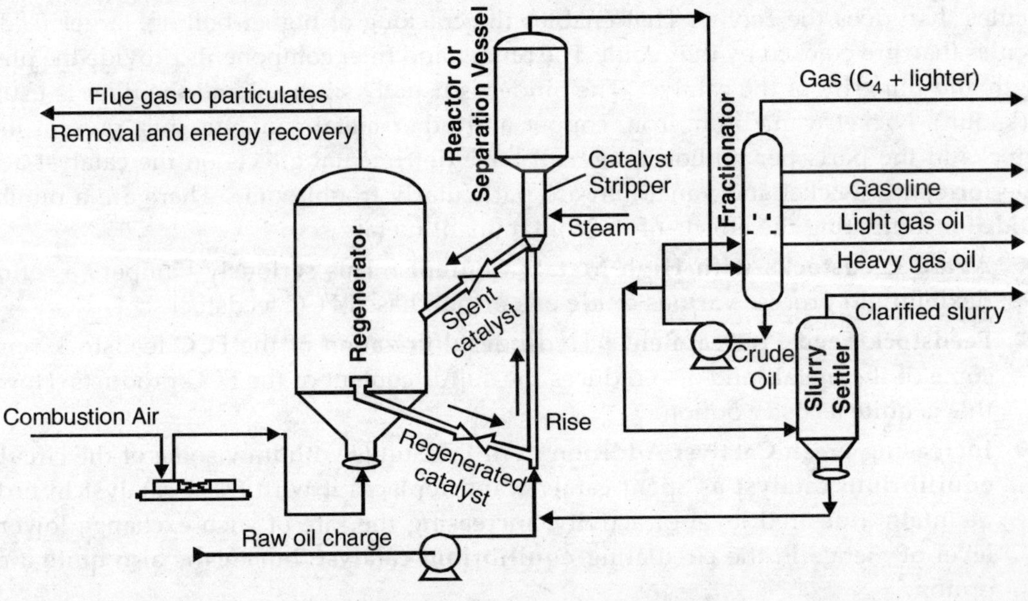

Fig. 6.7. Moving Bed Catalytic Reactor.

Moving bed catalytic cracking process is similar to the FCC process shown in **Fig. 6.7.** The catalyst is in the form of pellets that are moved continuously to the top of the unit by conveyor or pneumatic lift tubes to a storage hopper, then flow downward by gravity through the reactor, and finally to a regenerator. The regenerator and hopper are isolated from the reactor by steam seals. The cracked product is separated into recycle gas, oil, clarified oil, distillate, naphtha, and wet gas.

6.4. HYDROCRACKING

Hydrocracking can increase the yield of petrol components, as well as being used to produce light distillates. It produces no residues, only light oils. Hydrocracking is catalytic cracking in the presence of hydrogen. The extra hydrogen saturates, or hydrogenates, the chemical bonds of the cracked hydrocarbons and creates isomers with the desired characteristics. Hydrocracking is also a treating process, because the hydrogen combines with contaminants such as sulphur and nitrogen, allowing them to be removed. Gas oil feed is mixed with hydrogen, heated, and sent to a reactor vessel with a fixed bed catalyst, where cracking and hydrogenation take place.

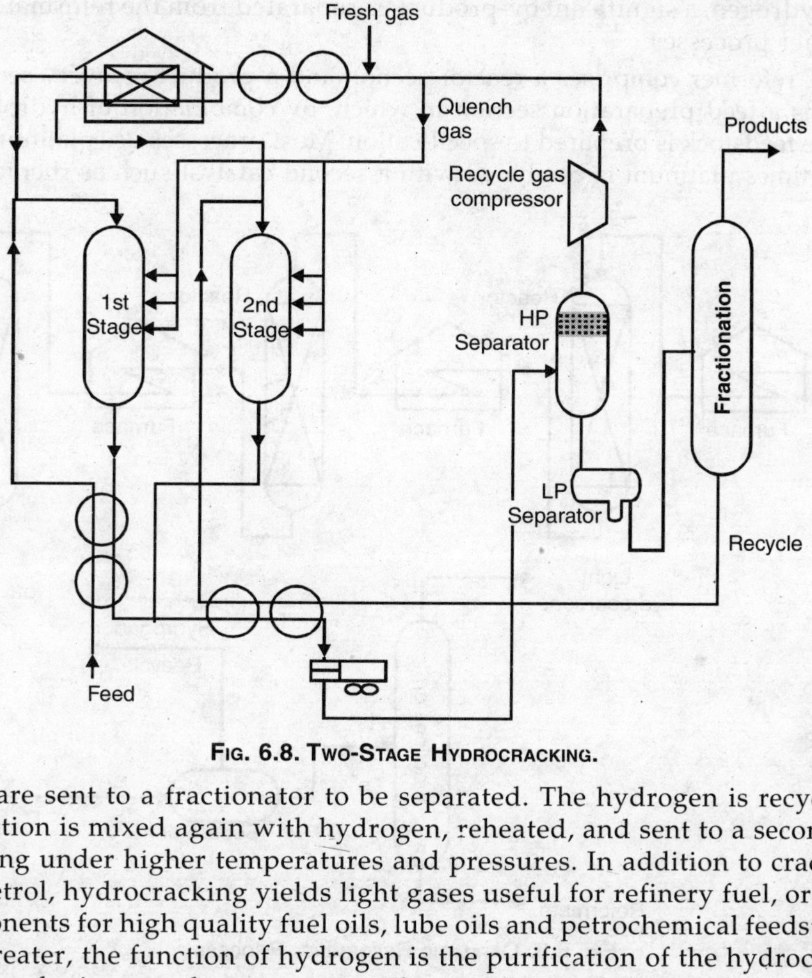

FIG. 6.8. TWO-STAGE HYDROCRACKING.

Products are sent to a fractionator to be separated. The hydrogen is recycled. Residue from this reaction is mixed again with hydrogen, reheated, and sent to a second reactor for further cracking under higher temperatures and pressures. In addition to cracked naphtha for making petrol, hydrocracking yields light gases useful for refinery fuel, or alkylation as well as components for high quality fuel oils, lube oils and petrochemical feedstocks. Similar to the hydrotreater, the function of hydrogen is the purification of the hydrocarbon stream from sulfur and nitrogen hetero-atoms.The products of this process are saturated hydrocarbons; depending on the reaction conditions (temperature, pressure, catalyst activity) these products range from ethane, LPG to heavier hydrocarbons comprising mostly of isoparaffins. Hydrocracking is normally facilitated by a bi functional catalyst that is capable of rearranging and breaking hydrocarbon chains as well as adding hydrogen to aromatics and olefins to produce naphthenes and alkanes. Major products from hydrocracking are jet fuel and diesel, while also relatively high octane rating gasoline fractions and LPG are produced. All these products have a very low content of sulfur and other contaminants. A two stage hydrocracker is shown below in **Figure 6.8.**

6.5. CATALYTIC REFORMING

Catalytic reforming is an important process used to convert low-octane naphthas into high-octane gasoline blending components called reformates. Reforming represents the total effect of numerous reactions such as cracking, polymerization, dehydrogenation, and isomerization taking place simultaneously. Depending on the properties of the naphtha feedstock (as measured by the paraffin, olefin, naphthenes, and aromatic content) and catalysts used, reformates can be produced with very high concentrations of toluene, benzene, xylene, and other aromatics useful in gasoline blending and petrochemical processing. Hydrogen, a significant by-product, is separated from the reformate for recycling and use in other processes.

A catalytic reformer comprises a reactor section and a product-recovery section More or less standard is a feed preparation section in which, by combination of hydrotreatment and distillation, the feedstock is prepared to specification. Most processes use platinum as the active catalyst. Sometimes platinum is combined with a second catalyst such as rhenium or another noble metal.

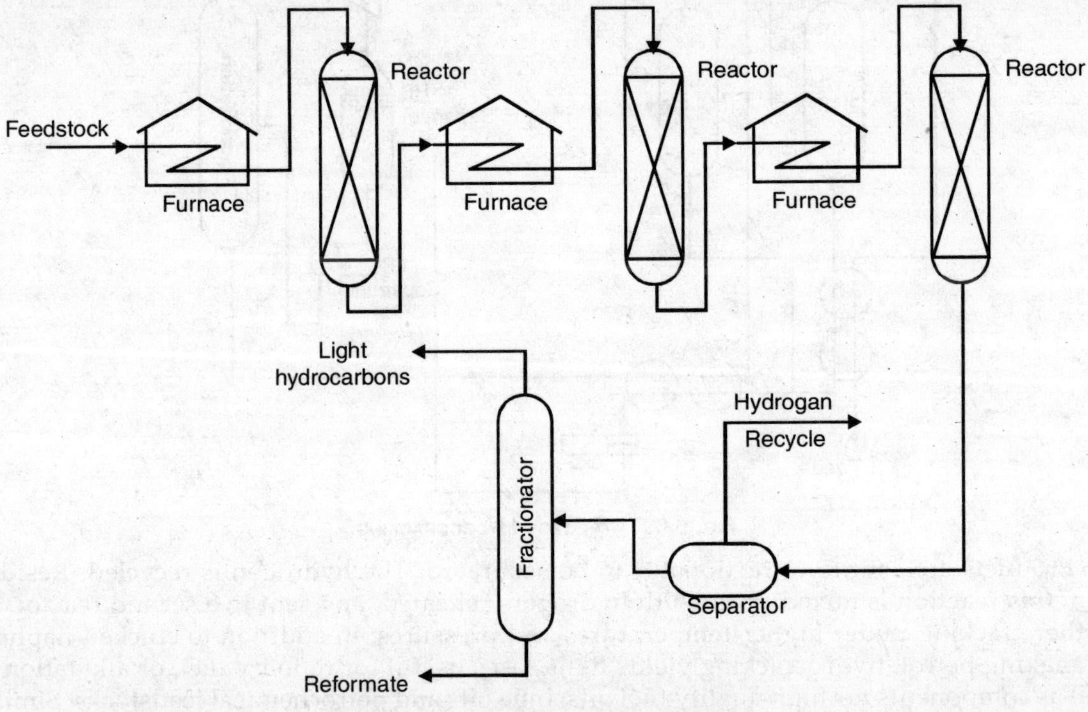

Fig. 6.9. Catalytic Reforming Process.

There are many different commercial catalytic reforming processes including platforming, powerforming, ultraforming and thermoform catalytic reforming. In the platforming process, the first step is preparation of the naphtha feed to remove impurities from the naphtha and reduce catalyst degradation. The naphtha feedstock is then mixed with hydrogen, vapourized, and passed through a series of alternating furnace and fixed-bed reactors containing a platinum catalyst. The effluent from the last reactor is cooled and sent to a separator to permit removal of the hydrogen-rich gas stream from the top of the separator for recycling. The liquid product from the bottom of the separator is sent to a fractionator called a stabilizer (butanizer). It makes a bottom product called reformate; butanes and lighter go overhead and are sent to the saturated gas plant.

Some catalytic reformers operate at low pressure (50-200 psi), and others operate at high pressures (up to 1,000 psi). Some catalytic reforming systems continuously regenerate the catalyst in other systems. One reactor at a time is taken off-stream for catalyst regeneration, and some facilities regenerate all of the reactors during turnarounds. **Figure 6.9** shows a functioning of catalytic reformer.

The Reaction Chemistry

The four major catalytic reforming reactions are :

- The dehydrogenation of naphthenes to convert them into aromatics as shown in the conversion methylcyclohexane (a naphthene) to toluene (an aromatic), as shown below:

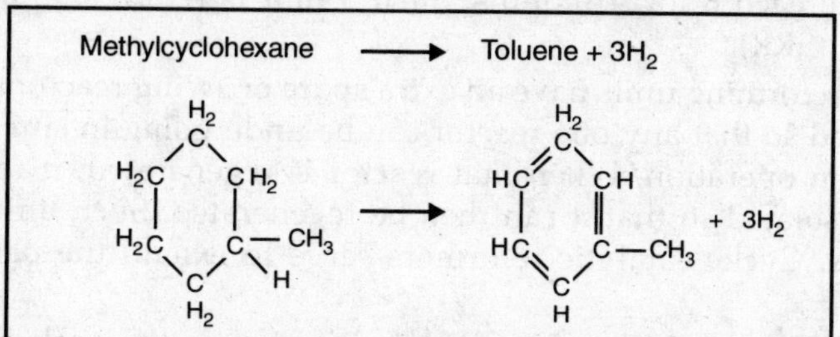

- The isomerization of normal paraffins to isoparaffins as exemplified in the conversion of normal octane to 2,5-Dimethylhexane (an isoparaffin), as shown below :

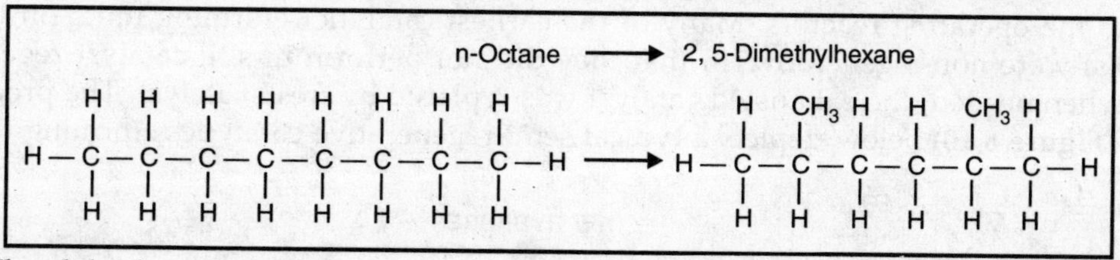

- The dehydrogenation and aromatization of paraffins to aromatics (commonly called dehydrocyclization) as depicted in the conversion of normal heptane to toluene, as shown below :

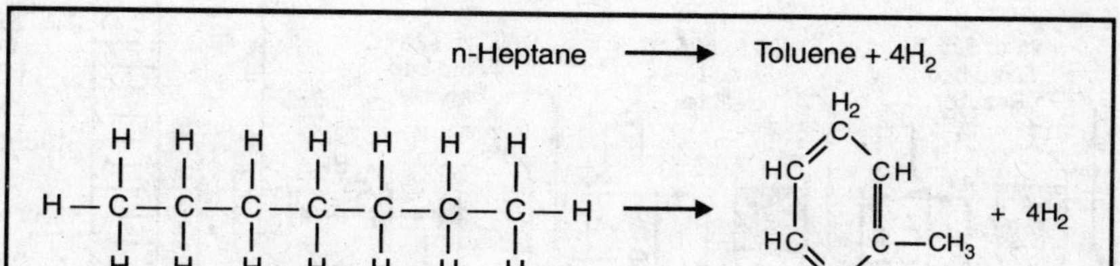

- The hydrocracking of paraffins into smaller molecules by the cracking of normal heptane into isopentane and ethane, as shown below :

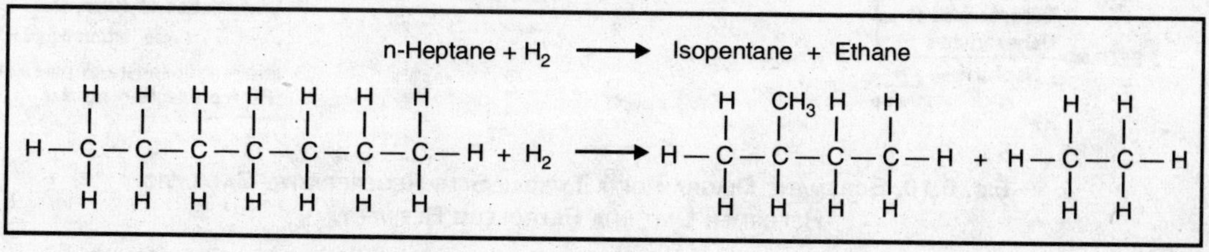

The hydrocracking of paraffins is the only one of the above four major reforming reactions that consumes hydrogen. The isomerization of normal paraffins does not consume or produce hydrogen. However, both the dehydrogenation of naphthenes and the dehydrocyclization of paraffins produce hydrogen. The overall net production of hydrogen in the catalytic reforming of petroleum naphthas ranges from about 50 to 200 cubic meters of hydrogen gas (at 0 °C and 1 atm) per cubic meter of liquid naphtha feedstock.

Semi-Regenerative Catalytic Reformer (SRR)

The catalytic reforming unit has three reactors, each with a fixed bed of catalyst, and all of the catalyst is regenerated *in situ* during routine catalyst regeneration shutdowns which occur approximately once each 6 to 24 months. Such a unit is referred to as a Semi-Regenerative Catalytic Reformer (SRR).

Some catalytic reforming units have an extra **spare or swing** reactor and each reactor can be individually isolated so that any one reactor can be undergoing in situ regeneration while the other reactors are in operation. When that reactor is regenerated, it replaces another reactor which, in turn, is isolated so that it can then be regenerated. Such units, referred to as **cyclic catalytic reformers.** Cyclic catalytic reformers serve to extend the period between required shutdowns.

The latest and most modern type of catalytic reformers are called **Continuous Catalyst Regeneration Reformers (CCR).** Such units are characterized by continuous in-situ regeneration of part of the catalyst in a special regenerator, and by continuous addition of the regenerated catalyst to the operating reactors. Many of the earliest catalytic reforming units (in the 1950s and 1960's) were non-regenerative in that they did not perform in situ catalyst regeneration. Instead, when needed, the exhausted catalyst was replaced by fresh catalyst. The process flow diagram (**Figure 6.10**) below depicts a typical semi-regenerative catalytic reforming unit.

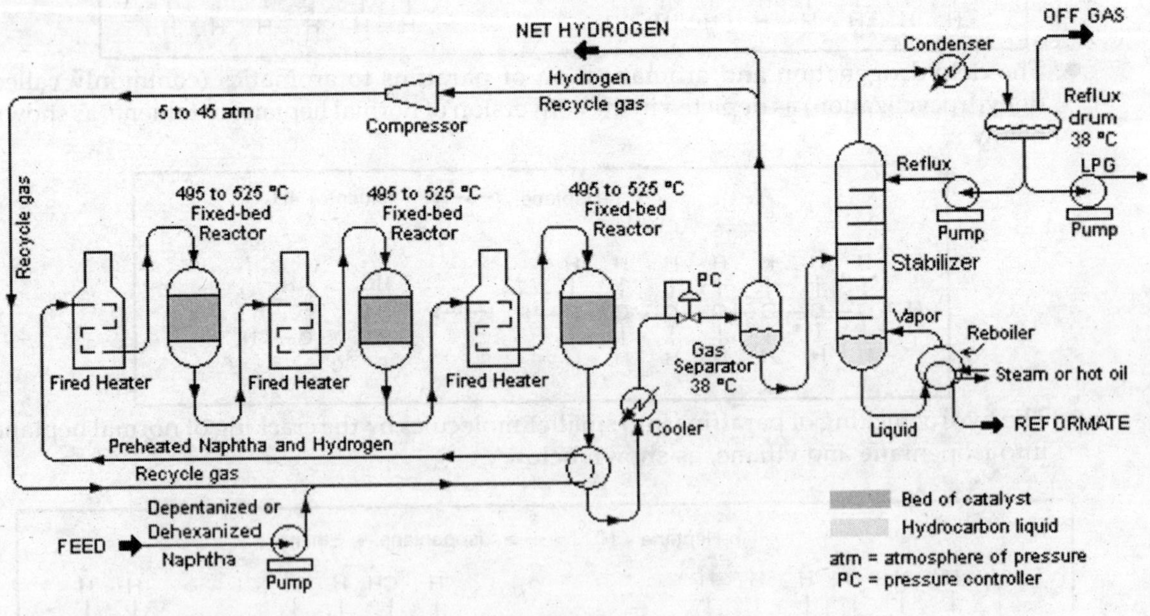

Fig. 6.10. Schematic Diagram of a Typical Semi-Regenerative Catalytic Reformer Unit in a Petroleum Refinery.

The liquid feed is pumped up to the reaction pressure (5 to 45 atm) and is joined by a stream of hydrogen-rich recycle gas. The resulting liquid-gas mixture is preheated by flowing through a heat exchanger. The preheated feed mixture is then totally vapourized and heated to the reaction temperature (495 to 520°C) before the vapourized reactants enter the first reactor. As the vapourized reactants flow through the fixed bed of catalyst in the reactor, the major reaction is the dehydrogenation of naphthenes to aromatics which is highly endothermic and results in a large temperature decrease between the inlet and outlet of the reactor. To maintain the required reaction temperature and the rate of reaction, the vapourized stream is reheated in the second fired heater before it flows through the second reactor. The temperature again decreases across the second reactor and the vapourized stream must again be reheated in the third fired heater before it flows through the third reactor. As the vapourized stream proceeds through the three reactors, the reaction rates decrease and the reactors therefore become larger. At the same time, the amount of reheat required between the reactors becomes smaller. Usually, three reactors are all that is required to provide the desired performance of the catalytic reforming unit. Some installations use three separate fired heaters as shown in the schematic diagram (**Figure 6.10**) and some installations use a single fired heater with three separate heating coils.

The hot reaction products from the third reactor are partially cooled by flowing through the heat exchanger where the feed to the first reactor is preheated and then flow through a water-cooled heat exchanger before flowing through the Pressure Controller (PC) into the gas separator.

Most of the hydrogen-rich gas from the gas separator vessel returns to the suction of the recycle hydrogen gas compressor and the net production of hydrogen-rich gas from the reforming reactions is exported for use in other the other refinery processes that consume hydrogen (such as hydrodesulfurization units and/or a hydrocracker unit).

The liquid from the gas separator vessel is routed into a fractionating column commonly called a **stabilizer**. The overhead offgas product from the stabilizer contains the byproduct methane, ethane, propane and butane gases produced by the hydrocracking reactions, and it may also contain some small amount of hydrogen. That offgas is routed to the refinery's central gas processing plant for removal and recovery of propane and butane. The residual gas after such processing becomes part of the refinery's fuel gas system. The bottoms product from the stabilizer is the high-octane liquid reformate that will become a component of the refinery's product gasoline.

Most catalytic reforming catalysts contain platinum or rhenium on a silica or silica-alumina support base, and some contain both platinum and rhenium. Fresh catalyst is chlorided (chlorinated) prior to use. The noble metals (platinum and rhenium) are considered to be catalytic sites for the dehydrogenation reactions and the chlorinated alumina provides the acid sites needed for isomerization, cyclization and hydrocracking reactions. The activity (*i.e.*, effectiveness) of the catalyst in a semi-regenerative catalytic reformer is reduced over time during operation by carbonaceous coke deposition and chloride loss. The activity of the catalyst can be periodically regenerated or restored by in situ high temperature oxidation of the coke followed by chlorination. As stated earlier herein, semi-regenerative catalytic reformers are regenerated about once per 6 to 24 months. Normally, the catalyst can be regenerated perhaps 3 or 4 times before it must be returned to the manufacturer for reclamation of the valuable platinum and/or rhenium content.

Following the cracking processes it is necessary to build or rearrange some of the lighter hydrocarbon molecules into high quality petrol or jet fuel blending components or into petrochemicals. The former can be achieved by several chemical process such as alkylation and isomerisation.

6.6. ALKYLATION

Olefins such as propylene and butylene are produced by catalytic and thermal cracking. Alkylation refers to the chemical bonding of these light molecules with isobutane to form larger branched-chain molecules (isoparaffins) that make high octane petrol. Olefins and isobutane are mixed with an acid catalyst and cooled. They react to form alkylate, plus some normal butane, isobutane and propane. The resulting liquid is neutralised and separated in a series of distillation columns. Isobutane is recycled as feed and butane and propane sold as Liquid Petroleum Gas (LPG).

For example, Isobutene (C_4H_{10}) + Butylene (C_4H_8) $\longrightarrow$ Isoactane (C_8H_{18})

6.7. ISOMERISATION

Isomerisation refers to chemical rearrangement of straight-chain hydrocarbons (paraffins), so that they contain branches attached to the main chain (isoparaffins). This is done for two reasons :

- They create extra isobutane feed for alkylation
- They improve the octane of straight run pentanes and hexanes and hence make them into better petrol blending components.

Isomerisation is achieved by mixing normal butane with a little hydrogen and chloride and allowed to react in the presence of a catalyst to form isobutane, plus a small amount of normal butane and some lighter gases. Products are separated in a fractionator. The lighter gases are used as refinery fuel and the butane recycled as feed. Pentanes and hexanes are the lighter components of petrol. Isomerisation can be used to improve petrol quality by converting these hydrocarbons to higher octane isomers. The process is the same as for butane isomerisation.

6.8. POLYMERISATION

Under pressure and temperature, over an acidic catalyst, light unsaturated hydrocarbon molecules react and combine with each other to form larger hydrocarbon molecules. Such process can be used to react butenes (olefin molecules with four carbon atoms) with iso-butane (branched paraffin molecules, or isoparaffins, with four carbon atoms) to obtain a high octane olefinic petrol blending component called polymer gasoline.

6.9. HYDROTREATING AND SULPHUR PLANTS

A number of contaminants are found in crude oil. As the fractions travel through the refinery processing units, these impurities can damage the equipment, the catalysts and the quality of the products. There are also legal limits on the contents of some impurities, like sulphur, in products. Hydrotreating is one way of removing many of the contaminants from many of the intermediate or final products. In the hydrotreating process, the entering feedstock is mixed with hydrogen and heated to 300-380°C. The oil combined with the hydrogen then enters a reactor loaded with a catalyst which promotes several reactions :

- Hydrogen combines with sulphur to form hydrogen sulphide (H_2S)
- Nitrogen compounds are converted to ammonia
- Any metals contained in the oil are deposited on the catalyst
- Some of the olefins, aromatics or naphthenes become saturated with hydrogen to become paraffins and some cracking takes place, causing the creation of some methane, ethane, propane and butanes.

Catalytic Hydrodesulfurization Process

Hydrotreating for sulfur removal is called hydrodesulfurization. In a typical catalytic hydrodesulfurization unit, the feedstock is deaerated and mixed with hydrogen, preheated in a fired heater (600°-800° F) and then charged under pressure (up to 1,000 psi) through a fixed-bed catalytic reactor. In the reactor, the sulfur and nitrogen compounds in the feedstock are converted into hydrogen sulfide (H_2S) and ammonia (NH_3). The reaction products leave the reactor and after cooling to a low temperature enter a liquid/gas separator. The hydrogen-rich gas from the high-pressure separation is recycled to combine with the feedstock, and the low-pressure gas stream rich in H_2S is sent to a gas treating unit where H_2S is removed. The clean gas is then suitable as fuel for the refinery furnaces. The liquid stream is the product from hydrotreating and is normally sent to a stripping column for removal of H_2S and other undesirable components. In cases where steam is used for stripping, the product is sent to a vacuum drier for removal of water. Hydrodesulfurized products are blended or used as catalytic reforming feedstock. **Figure 6.11** shows Catalytic Hydrodesulfurization Process.

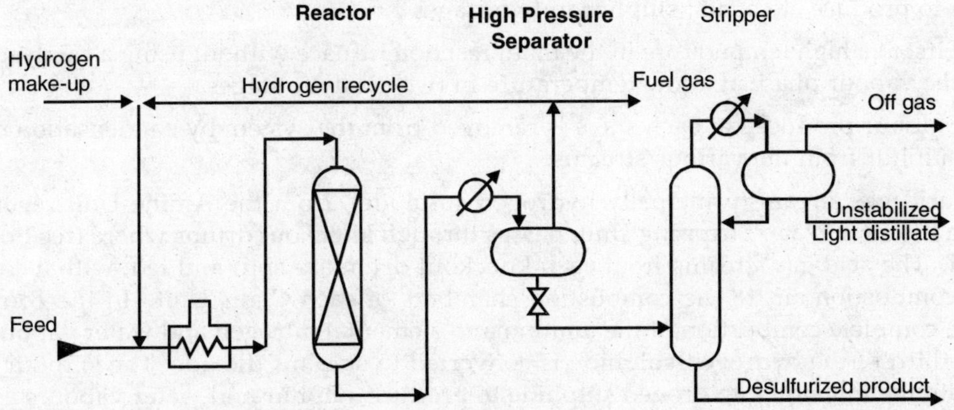

FIG. 6.11. CATALYTIC HYDRODESULFURIZATION PROCESS.

6.10. SULPHUR RECOVERY PLANTS

*The hydrogen sulphide created from hydrotreating is a **Toxic Gas** that needs further treatment. It involves two process :*

- Removal of hydrogen sulphide gas from the hydrocarbon stream.
- Conversion of hydrogen sulphide gas to elemental sulphur-a non-toxic and useful element .

Solvent extraction, using a solution of diethanolamine (DEA) dissolved in water, is applied to separate the hydrogen sulphide gas from the process stream. This stream is known as **Amine unit** .The hydrocarbon gas stream containing the hydrogen sulphide is bubbled through a solution of diethanolamine solution (DEA) under high pressure, such that the hydrogen sulphide gas dissolves in the DEA. The DEA and hydrogen mixture is the heated at a low pressure and the dissolved hydrogen sulphide is released as a concentrated gas stream which is sent to another plant for conversion into sulphur. Conversion of the concentrated hydrogen sulphide gas into sulphur occurs in two stages:

- Combustion of part of the H_2S stream in a furnace, producing sulphur dioxide (SO_2) water (H_2O) and sulphur (S).

$$2H_2S + 2O_2 \longrightarrow SO_2 + S + 2H_2O$$

- Reaction of the remainder of the H_2S with the combustion products in the presence of a catalyst. The H_2S reacts with the SO_2 to form sulphur.

$$2H_2S + 2O_2 \longrightarrow 3S + 2H_2O$$

As the reaction products are cooled the sulphur drops out of the reaction vessel in a molten state. Sulphur can be stored and shipped in either a molten or solid state upgrader sulphur recovery Claus unit is described below.

Upgrader Sulphur Recovery Claus Units

These units process sulphur containing gases from the Amine Unit(solution of diethanolamine (DEA)) and the Sour Water Stripping Unit. Two parallel Claus trains, each with a capacity to produce 70% of the normal sulphur production of the entire Upgrader are designed. Paralleling is employed to ensure the Upgrader can stay on stream at a reduced rate even if one sulphur recovery train fails to function. The Claus units partially oxidize hydrogen sulphide to produce elemental sulphur in two stages :

- First at a high temperature in the main reaction furnace without using a catalyst; then in the vapour phase at a low temperature in two catalytic stages.
- Sulphur produced at each stage is removed from the system by condensation of liquid sulphur from the various streams.

The acid gas stream (principally hydrogen sulphide), from the Amine Unit regenerators and from the Sour Water Stripping Unit, passes through knockout drums where free liquids are removed. The acid gas streams from each knockout drum are split and fed, with a controlled flow of combustion air, to the combustion chambers of each Claus Unit. In the combustion chamber, complete combustion of the ammonia to elemental nitrogen and water vapour occurs and one-third of the hydrogen sulphide is converted to sulphur dioxide. The sulphur dioxide reacts with the remaining hydrogen sulphide to produce sulphur and water vapours as shown above. The hot gases from the combustion chamber are cooled to condense the elemental liquid sulphur which is removed from the process by gravity. The partially cooled gases are reheated and fed to a converter where, in the presence of a catalyst, the reaction of sulphur dioxide and hydrogen sulphide is driven towards equilibrium. The sulphur produced is condensed and removed from the process in liquid form.

The gases are again reheated and fed to another converter where the reaction is again driven towards equilibrium, producing more sulphur, which is removed. The gases from the last sulphur condenser in each Claus Unit, containing less than 5% of the sulphur entering the units, is fed to a single train Tail Gas Clean-Up Unit for further processing. The condensed liquid sulphur is collected, degassed to remove dissolved hydrogen sulphide, and exported from the site on railcars or trucks as described in **Fig. 6.12.**

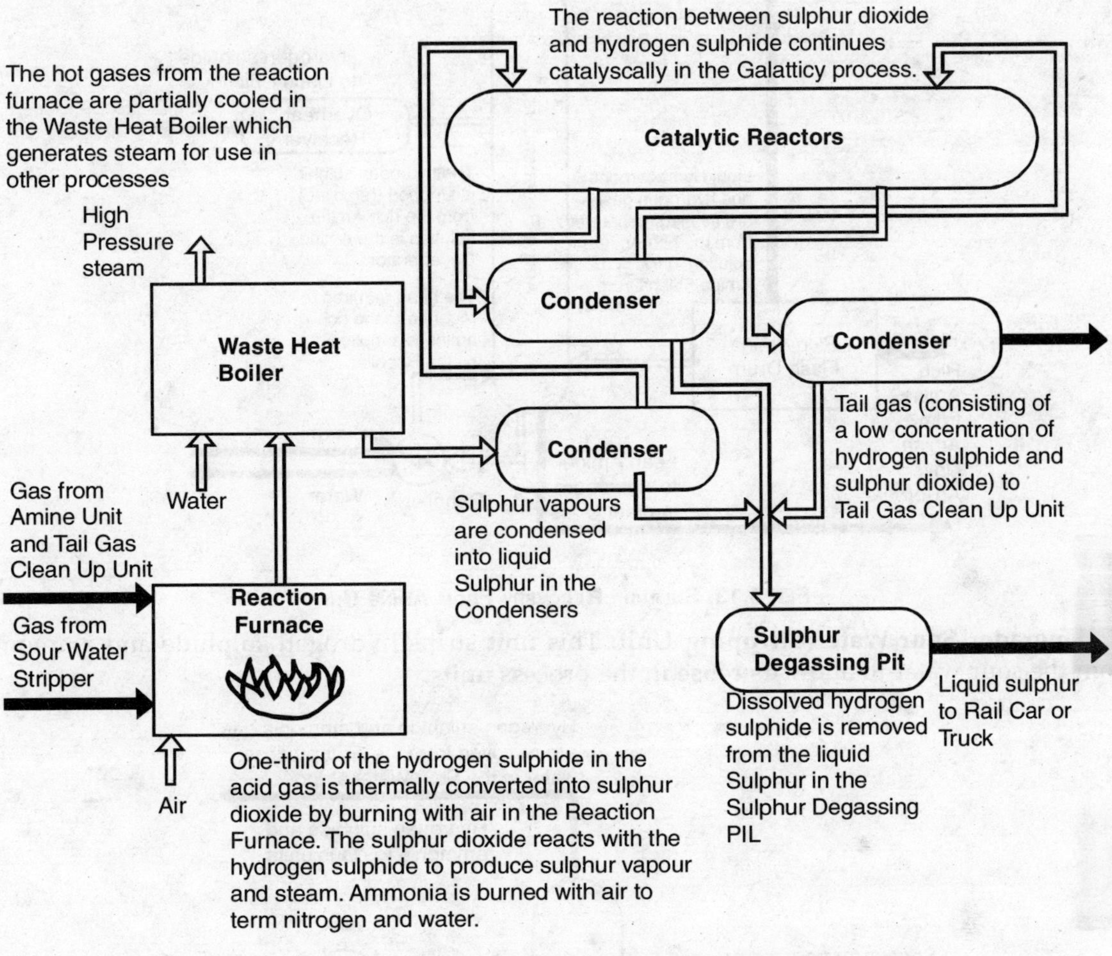

FIG. 6.12. CLAUS PROCESS FOR SULPHUR RECOVERY.

Upgrader Sulphur Recovery Amine Unit

Rich (hydrogen sulphide saturated) alkanolamine solution streams from the process units' amine absorbers are collected and routed into low pressure flash drums where dissolved hydrogen, hydrocarbon gases, and some hydrogen sulphide is flashed off. These flash gases are sweetened with a lean amine slipstream to remove the hydrogen sulphide present before routing them to fuel. The rich amine slipstream also enters the flash drums.

The resulting rich amine solution is preheated and introduced into two parallel regenerators where additional heat, in the form of striping steam, is applied to break the chemical bond between the amine molecules and hydrogen sulphide molecules. Subsequently, the hydrogen sulphide is removed from the solution and sent to the Claus Unit for conversion into elemental sulphur (**Figure 6.13**). The regenerated amine solution is cooled and returned to the absorbers for use.

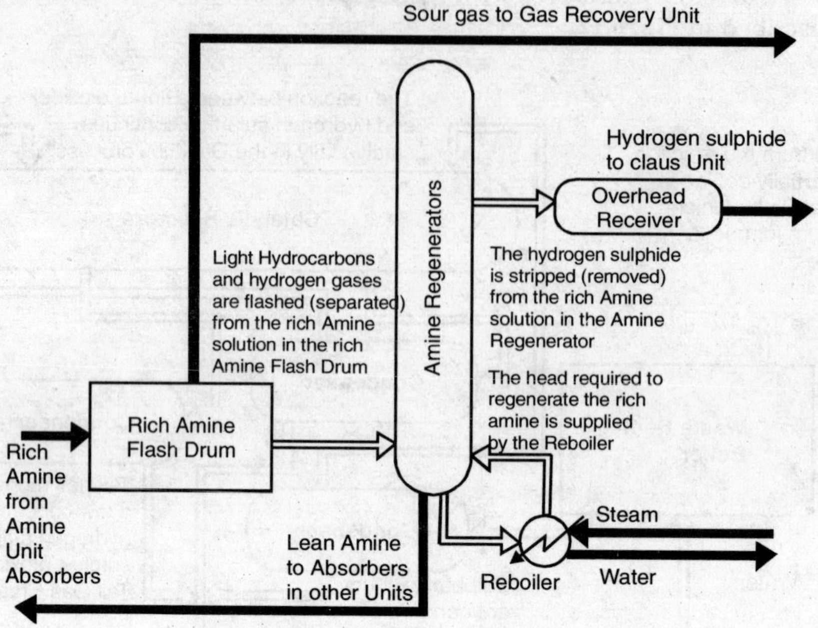

FIG. 6.13. SULPHUR RECOVERY FROM AMINE UNIT.

Upgrader Sour Water Stripping Unit: This unit strips hydrogen sulphide and ammonia from the sour water to allow its reuse in the process units.

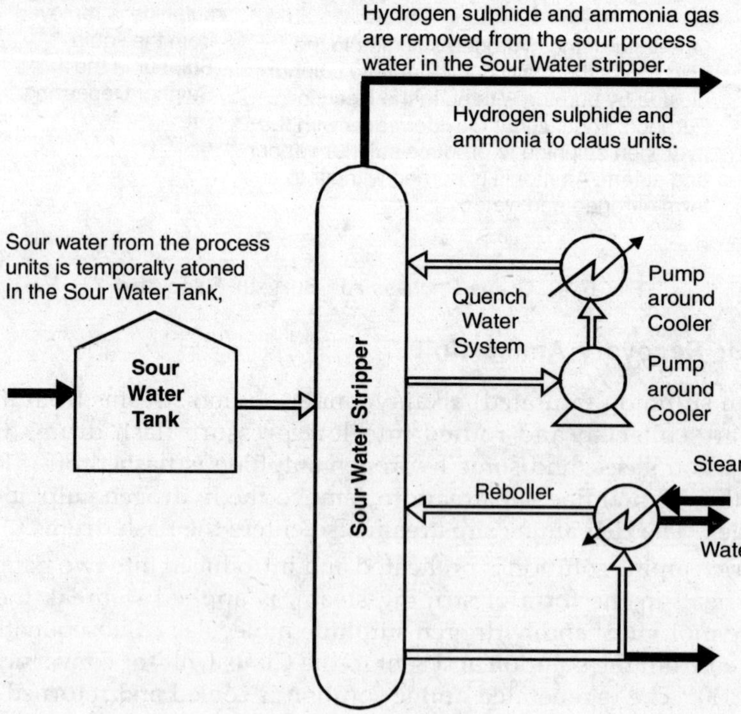

FIG. 6.14. WATER STRIPPING UNIT.

The sour water from the process units is fed to a stripping tower where heat, in the form of steam, is applied. The ammonia and hydrogen sulphide contained in the water is released by the heat and exits the top of the tower. Any excess steam is condensed by cooling the overhead stream and returning the condensate to the tower. The ammonia and hydrogen sulphide gases are directed to the Claus Units where they will be converted to sulphur and nitrogen. The stripped water from the tower bottom is returned to the process units for the reuse. The Waste Water Treatment Facility treats the stripped sour water biologically to remove phenols and other contaminants and enables the treated water to be recycled as described in Fig. 6.14.

6.11. APPLICATIONS

Oil refinery cracking processes allow the production of "light" products such as LPG and gasoline from heavier crude oil distillation fractions such as gas oils and residues. Fluid catalytic cracking produces a high yield of gasoline and LPG, while hydrocracking is a major source of jet fuel, diesel, naphtha and LPG. Thermal cracking is currently used to *"upgrade"* very heavy fractions (*"upgrading"*, *"visbreaking"*), or to produce light fractions or distillates, burner fuel and/or petroleum coke. Two extremes of the thermal cracking in terms of product range are represented by the high-temperature process called *"steam cracking"* or pyrolysis (750 to 900 °C or more) which produces valuable ethylene and other feedstocks for the petrochemical industry, and the milder-temperature delayed coking (500°C) which can produce, under the right conditions, valuable needle coke, a highly crystalline petroleum coke used in the production of electrodes for the steel and aluminium industries.

REVIEW QUESTIONS

1. Discuss the mechanism of thermal cracking.
2. What is cracking ? Explain the different type of cracking process.
3. Discuss catalytic cracking with respect to fixed bed catalytic cracking.
4. Whare are the advantage of catalytic cracking over thermal cracking ?
5. Give a diagrammatic example of the catalytic cracking of petroleum hydrocarbon.
6. Discuss hydrocracking.
7. Write short notes on hydro desulfurization process.
8. Give schematic representation of catalytic reforming and the reaction chemistry involved.
9. Discuss sulphur recovery claus units.
10. What are the various thermal cracking unit ?

❑❑❑

CHAPTER

7

PETROLEUM PRODUCTS

The petroleum products obtained from crude oil by various processes are discussed.

INSIDE THIS CHAPTER

7.1. NATURAL GAS

Natural gas is colorless and odorless in its pure form. Natural gas is combustible, and when burned it gives off a great deal of energy. Unlike other fosseil fuels, however, natural gas is clean burning and emits lower levels of potentially harmful air pollutants. In order to get energy constantly in the form of heat, electricity etc., natural gas has achieved a high level of importance in our society.

Composition of Natural Gas

Natural gas is a combustible mixture of hydrocarbon gases. While natural gas is formed primarily of methane, it can also include ethane, propane, butane and pentane. The composition of natural gas can vary widely. The Table 7.1 below shows the typical composition of natural gas.

TABLE 7.1. Composition of Natural Gas

Typical Composition of Natural Gas		
Methane	CH_4	70-90%
Ethane	C_2H_6	0-20%
Propane	C_3H_8	
Butane	C_4H_{10}	
Carbon Dioxide	CO_2	0-8%
Oxygen	O_2	0-0.2%
Nitrogen	N_2	0-5%
Hydrogen sulfide	H_2S	0-5%
Methane	CH_4	70-90%
Ethane	C_2H_6	0-20%

In its purest form, natural gas is almost pure methane. Methane is a molecule made up of one carbon atom and four hydrogen atoms, and is referred to as CH_4.

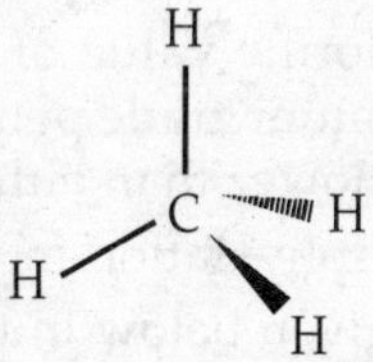

During most of the 19th century, natural gas was used almost exclusively as a source of light. Without a pipeline infrastructure, it was difficult to transport the gas very far, or into homes to be used for heating or cooking. Most of the natural gas produced in this era was manufactured from coal, as opposed to transported from a well. Near the end of the 19th century, with the rise of electricity, natural gas lights were converted to electric lights. This led producers of natural gas to look for new uses for their product.

In 1885, Robert Bunsen invented the Bunsen burner. He managed to create a device that mixed natural gas with air in the right proportions, creating a flame that could be safely used for cooking and heating. The invention of the Bunsen burner opened up new opportunities for the use of natural gas. The invention of temperature-regulating thermostatic devices allowed for better use of the heating potential of natural gas, allowing the temperature of the flame to be adjusted and monitored.

7.2. LIQUIFIED PETROLEUM GAS (LPG)

Liquefied petroleum gas (also called **LPG, GPL, LP Gas,** or autogas) is a mixture of hydrocarbon gases used as a fuel in cooking and vehicles and increasingly replacing chlorofluorocarbons as an aerosol propellant and a refrigerant to reduce damage to the ozone layer.

LPG consists of saturated Hydrocarbons - Propane (C_3H_8) and Butane (C_4H_{10}), which can be stored, transported separately or as a mixture. They exist as gases at normal room temperatures and atmospheric pressure. It is called liquefied petroleum gas because gases liquefy under moderate pressure. They readily vaporize upon release of pressure. It is this property that permits transportation of and storage of LP Gas in concentrated liquid form. They are derieved products and comes from two sources.

- It can be obtained from the refining of crude oil. When produced this way it is generally in pressurized form.
- LPG is also extracted from natural gas or crude oil streams coming from underground reservoirs.

60% of LPG in the world today is produced from natural gas whereas 40% of LPG is extracted from refining of crude oil. Commercial propane and butane are ideal products referred to as "propane" and "butane" are saturated hydrocarbons; but during the process of extraction/ production certain alowable unsaturated hydrocarbons like ethylene, propylene, butylenes etc. may be included in the mixture along with pure propane and butane. The presence of these in moderate amounts would not affect LPG in terms of combustion but may affect other properties slightly (such as corrosiveness or gum formation). The characteristics of LPG are :

- It is colorless and cannot be seen.
- It is odorless. Hence LPG is odorized by adding an odorant-ethanethiol prior to supply to the user, to aid the detection of any leaks.

- It is slightly heavier than air and hence if there is a leak it flows to lower lying areas.
- In liquid form, its density is half that of water and hence it floats initially before it is vaporized.
- LPG has a typical specific calorific value of 46.1 MJ/kg compared to 42.5 MJ/kg for diesel and 43.5 MJ/kg for premium grade petrol (gasoline). However, its energy density per volume unit of 26 MJ/l is lower than either that of petrol or diesel.
- It is non-toxic but can cause asphyxiation in very high concentrations in air.

Some of the other properties are given below in the Table 7.2.

TABLE 7.2. Properties of LPG.

Property	Units	Commercial Propane	Commercial Butane	Mixture 50% each
Specific gravity of liquid at 15 deg C (Water = 1)		0.504	0.582	0.543
Specific gravity of vapor at 15 deg C (Air = 1)		1.5	2.01	1.75
Vapor pressure at 38 deg C	Kg/sq.cm	13.8	2.6	8.0
Boiling point at atm pressure	Deg C	–42	9	+9 to –42
Ignition temperature in air	Dec C	495-605	480-535	480-605
Latent Heat of Vaporization	Btu/lb	184	167	175

LPG expands upon release and 1 liter of liquid will form approximately 250 liters of vapor. LPG is used as a fuel for domestic (cooking), industrial, horticultural, agricultural, heating and drying processes. LPG can be used as an automotive fuel or as a propellant for aerosols, in addition to other specialist applications. LPG can also be used to provide lighting through the use of pressure lanterns.

According to the 2001 Census of India, 17.5% of Indian households or 33.6 million Indian households used LPG as cooking fuel in 2001. 76.64% of such households were from urban India making up 48% of urban Indian households as compared to a usage of 5.7% only in rural Indian households. LPG is subsidised by the government.

The advantages of LPG are as follows :

- Because of its relatively fewer components, it is easy to achieve the correct fuel to air mix ratio that allows the complete combustion of the product. This gives LPG its clean burning characteristics.
- Both Propane and Butane are easily liquefied and stored in pressure containers. These properties make the fuel highly portable, and hence, can be easily transported in cylinders or tanks to end-users.
- LPG is a good substitute for petrol in spark ignition engines. Its clean burning properties, in a properly tuned engine, give reduced exhaust emissions, extended lubricant and spark plug life.
- As a replacement for aerosol propellants and refrigerants, LPG provides alternatives to fluorocarbons, which are known to cause deterioration of the earth's ozone layer.

The requirements for Liquified Petroleum Gases (IS : 4576–1978) is summarized below in the Table 7.3 and Fig. 7.1 shows the gas pipe line in India.

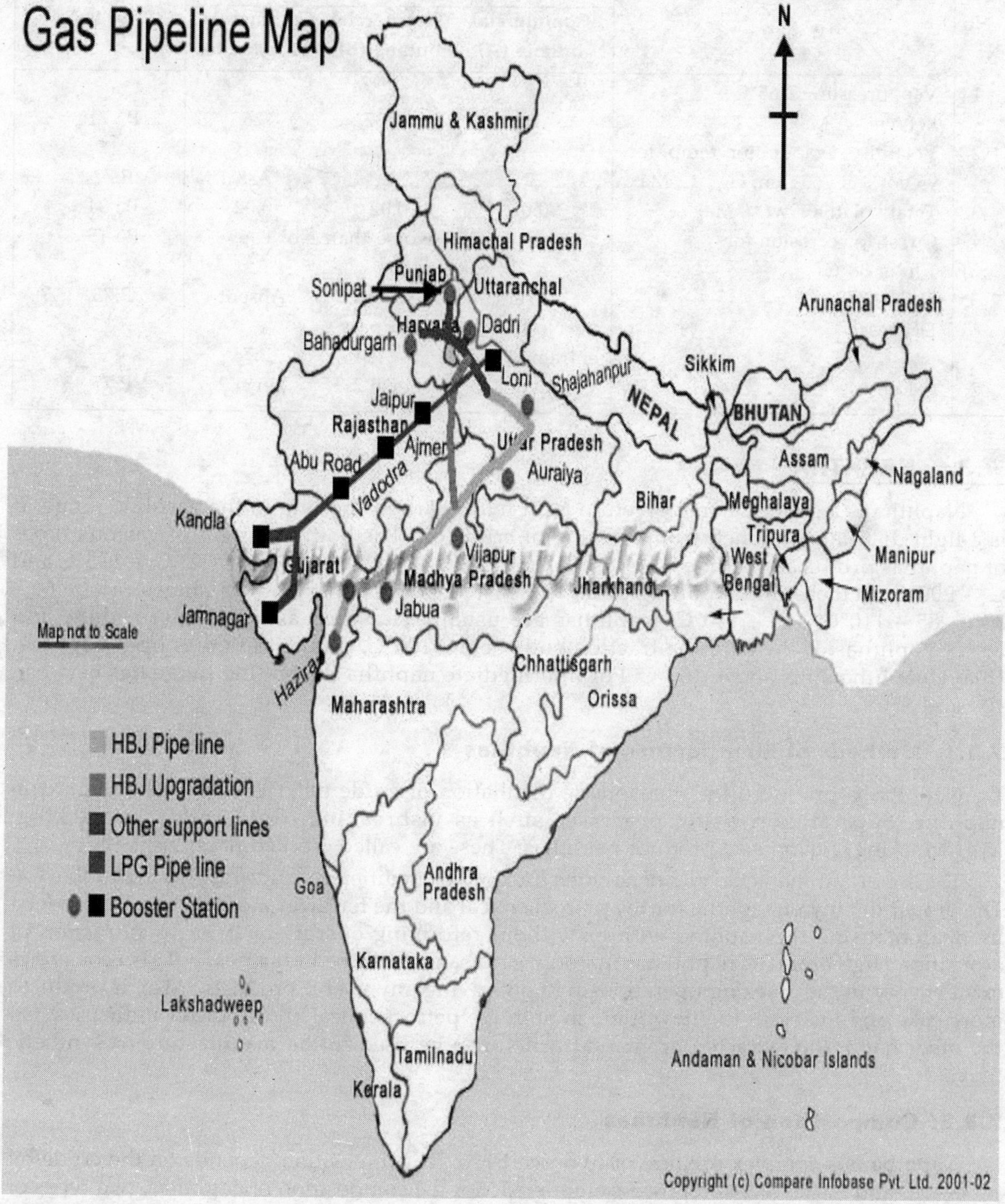

FIG. 7.1. GAS PIPELINE IN INDIA.

TABLE 7.3.

S. No.	Characteristics	Requirements for			Test method IS : 1448
		Commercial butane (A)	Commercial butane (B)	Commercial propane (c)	
1.	Vap. pressure @ 65°C kgf/cm^2 g. Max	10	16.87	26	P : 71
2.	Volatility, evaporation temp. for 95 vol% at 760 mm, Hg, °C, Max	2	2	–38	P:72
3.	Total volatile S wt % Max	0.02	0.02	0.02	P : 34
4.	Cu. strip corrosion for 1 hr at 38°C	Not worse than No. 1			P : 15
5.	H_2S	Absent	Absent	Absent	P:73
6.	Dryness	No free entramed H_2O	shall pass the list	P:74	
7.	Odour	Level 2	Level 2	Level 2	P:75

7.3. NAPHTHA

Naphtha is a generic name given to light hydrocarbons boiling in the gasoline range. It is a light distillate obtained from refining of crude oil. The boiling ranges of various types of naphthas produced include : C_5 – 85°C, C_5 – 110°C, C_5 – 140°C, C_5 – 160°C, C_5 – 175°C and C_5 – 200°C. In these initial boiling point (IBP) is constant. Other boiling ranges can be 60 – 85°C, 85 – 110°C, 110 – 140°C. Naphthas are usually classified as light, intermediate and heavy naphthas. If the naphtha fraction boils below 100°C, it is classified as light naphtha. Heavy naphtha boils above 150°C. For intermediate naphtha the boiling range lies between 100 and 150°C.

7.3.1. Methods of Manufacture of Naphthas

Naphtha is produced by atmospheric distillation of crude oil. This is called straight-run naphtha. Several conversion processes such as visbreaking, fluid catalytic cracking, hydrocracking, coking also produce naphthas. These are called cracked naphthas.

The important characteristics of naphtha fractions from different crudes are given in Table 7.4. The proper quality of naphtha for the petrochemical and the fertilizer industry can be achieved by dearomatising the naphtha with or without reforming operations prior to extraction of aromatics. High aromatic naphtha is not only a nuisance to these industries, but also consumes extra energy in the cracking operations without giving any useful products. Also, it produces more coke and increases the downtime in both the petrochemical and fertilizer industries. On the other hand, the extracted aromatic fraction can be used in the manufacture of synthetic fibres.

7.3.2. Composition of Naphthas

Naphtha is a complex mixture of hydrocarbons. Its composition depends on the crude oil processed and the conversion process employed. For the composition of naphthas, two types of analyses are usually carried out. These are :

TABLE 7.4. Important Characteristics of Naphthas from Different Crudes.

S. No.	Characteristics	Aghajari (34.3° API)		Light Arabian (34.5° API)		Kuwait (31.5° API)		Murban (39.3° API)		Darius (34° API)	
		C_5-175°C	C_5-200°C	C_5-175°C	C_5-200°C	C_5-175°C	C_5-200°C	C_5-175°C	C_5-200°C	C_5-175°C	C_5-200°C
1.	Gravity 15/15°C	0.73	0.74	0.72	0.735	0.72	0.730	0.73	0.74	0.718	0.727
2.	Yield on crude, wt. %	20.45	24.80	19.80	25.4	15.9	19.9	24.5	30.17	21	25
3.	Sulphur, wt %	0.096	0.101	0.03	0.04	0.044	0.075	0.018	0.022	0.08	0.103
4.	Total mercaptans, wt.%	0.0412	0.0356	0.00019	0.00010	0.015	0.014	0.013	0.013	—	—
5.	C/H ratio	5.64	5.68	5.84	5.97	5.72	5.75	6.1	6.2	5.56	5.62
6.	PONA*, vol. %										
	P	55.2	53.0	71.5	68.0	—	—	67.0	64.0	78.0	75.0
	N	29.0	30.6	20.0	20.0	—	—	18.0	19.0	14.0	15
	A	15.8	16.4	19.5	12.0	7.0	9.0	15.0	17.0	8.0	10.0

* Paraffin, Olefin, Naphthene, Aromatic

- Hydrocarbon type analysis
- Individual component wise analysis

The hydrocarbon type analysis determines the percentage of paraffins, olefins, naphthenes and aromatics. Different types of compounds found in naphtha fractions for various uses are given in Table 7.5. The approximate carbon number range of the products is also given in the same table. In Table 7.6, a summary of quality of the naphtha fractions from various indigenous and certain imported and worldwide available crudes with respect to the paraffin/naphthene/aromatic contents of their naphtha fractions are given. The value indicated is full range naphtha (C_5–140°C). The hydrocarbon type composition of naphthas from Indian crudes is given in Table 7.7 for boiling ranges of 60 – 85°C, 110 – 120°C, 120 – 130°C and 130 – 140°C.

7.3.3. Uses of Naphthas

The major end-uses of naphthas are listed in Table 7.8. The use of naphthas in fertilizer and petrochemical industries and as gasoline is also discussed.

TABLE 7.5. Types of Hydrocarbon in Naphthas

No. of atoms	5	6	7	8	9	10	11	12	13
Boiling point, °C	36	69	98	126	151	174	196	216	235
Types of compounds found									
Normal paraffins	P*	P	P	P	P	P	P	P	P
Branched paraffins	P	P	P	P	P	P	P	P	P
Saturated cyclic	P	P	P	P	P	P	P	P	P
Aromatics-Mono	—	P	P	P	P	P	P	P	P
Aromatics-Di	—	—	—	—	—	P	P	P	P
Aromatic naphthenes	—	—	—	—	P	P	P	P	P
Aromatic nonhydrocarbons	P	P	P	P	P	P	P	P	P
Aromatic olefins	—	—	—	PC	PC	PC	PC	PC	PC
Olefins cyclic	PC*	PC	PC	PC	PC	PC	PC	PC	PC
Approximate carbon number range of products	5	6	7	8	9	10	11	12	13
Ligroine									
Precipitation naphtha									
VM & P naphtha									
Mineral spirits									
Gasoline									
Kerosene									
Jet fuels									
Petrochemical feedstocks									

* P-Present, *PC-Present in cracked naphtha only

TABLE 7.6. Hydrocarbon Type Composition of Naphthas

Crude	Composition of naphtha (C_5 – 140°C)		
	Paraffin	Naphthene	Aromatic
Assam (Mix.), vol.%	32-40	52-43	16-17
North Gujarat, volt.%	52.5	42	5.5
Ankaleshwar, vol.%	70.8	25	4.2
Bombay High, vol.%	53.7	25	21.3
Iranian Light, wt.%	57.5	31.1	11.4
Iranian Heavy, wt.%	53.4	33.1	13.5
Basrah, vol.%	68.5	18.8	12.7
Kirkud blend, vol.%	67.5	22	10.5
Heavy Maya, vol.%	60.6	27	12.4
North Sea, wt.%	59	33	8
Russia export blend, wt.%	38.5	52	9.5

TABLE 7.7. Hydrocarbon type Composition of Naphthas from Some Indian Crudes

TBP Cut Range	Naharkatia vol. %			Lakwa-Rudrasagar vol. %			Ankleshwar vol. %			Kalol vol.%			Bombay High vol. %		
°C	A	N	P	A	N	P	A	N	P	A	N	P	A	N	P
60-85	13	50	37	8.6	50	41.4	3	42	55	6	38	56	22.3	29.1	48.6
110-120	18	36	36	9	63	28	15	41	44	—	—	—	—	—	—
120-130	4	51	45	12	66	22	4	38	58	9	50	41	30.9	20.7	48.4
130-140	50	28	22	22	52	15	33	27	40	—	—	—	—	—	—

A-Aromatics, N-Naphthenes, P-Paraffins

TABLE 7.8. Major End-Uses of Naphthas

Type of Naphtha	End-Uses
Light naphthas	(*a*) Gas making gasoline (*b*) Special gasoline
Intermediate naphthas	(*a*) Aviation gasoline (*b*) Motor gasoline (*c*) Marine gasoline (*d*) Commercial Solvents-Rubber, lacquer and pesticide diluents (*e*) Benzene - High-octane gasoline component, solvent, petrochemical manufacture (*f*) Toluene - Solvent, high-octane gasoline component, chemical intermediate, explosive (*g*) Xylenes - High-octane gasoline component, lacquer and enamels, chemical intermediate (*h*) Olefins and diolefins (*i*) Ammonia product
Heavy naphthas	(*a*) VM&P (Varnish manufacture and paints) naphtha (*b*) Thinner for paints, varnishes, lacquers (*c*) Stoddard Solvent - Special solvent for dry-cleaning trade (*d*) Mineral Spirits - thinner for paints and varnishes, turpentine substitute

Naphtha finds an extensive use as fertilizer feedstock where it is steam reformed to give synthesis gas, the starting material for fertilizers. Crudes which yield full range naphtha of high aromatic content of more than 10-12 percent aromatics are available and can be used by the fertilizer industry.

In petrochemical industry, naphtha (C_5-140°C) is used as feedstock for steam crackers, producing hydrogen, methane, ethylene, propylene, butenes, butadiene and to a lesser extent benzene and isoprene. These are the raw material for large number of petrochemicals. The petrochemical industry needs crude with as little aromatics percentage as possible with 4-6 percent as a 'benchmark' naphtha as per the supplies available in the past from the Ankaleshwar crude and from the imported Middle East crudes. The condensates from (LPG extraction in natural gas plants and the reformer raffinates can supply only a fraction (about 25 percent) of the total naphtha requirement of the petrochemical industry.

Catalytic reforming of straight-run naphtha yields a C_6-C_8 reformate rich in the aromatic hydrocarbons, benzene, toluene, ethylbenzene and the xylenes. The raffinate remaining after recovery of the aromatic hydrocarbons is sometimes used to augment straight-run naphtha as a steam-cracker feedstock. Naphtha-range products from hydrocracking of heavier oil fractions are also very occasionally used as steam-cracker feedstocks.

The steam-cracker feedstock is vaporized and steam is added in a ratio of about 0.3-0.5 tonne of steam per tonne of naphtha. The mixture is heated rapidly to the cracking temperature through tubular coils inside an oil- or gas-fired furnace. The temperature of operation varies from 750°C for medium/low severity to 900°C for high severity. To obtain the best yield of ethylene an important factor is a very short residence time, e.g. 0.25 s. This reduces the amount of coke produced. Longer residence time, e.g., 0.6 s, at low/medium severity increases the yield of aromatics.

The use of naphtha, as gasoline requires its high antiknock value or octane number. Depending on the crude source, the octane number of straight-run naphtha varies in te range of 50-70. Bombay High crude with its high aromatic content is more suitable for gasoline production because of distillation these crudes give high octane gasoline blending fractions. The octane number of cracked naphtha from fluid catalytic cracking is about 92-94. The present requirement of gasoline of 87 octane number for Indian vehicles is met by processing a large amount of naphtha in catalytic reformer to boost its octane number. Catalytic reforming of straight-run naphtha yields a reformate of high octane number in the range of 93-94.

7.4. GASOLINE

Gasoline is a most often produced by the fractional of distillation of crude oil within the boiling range of the crude oil is 30-180°C separated into fractions according to different boiling points of hydrocarbons of varying chain lengths. The fractional distillation process yields approximately 25% of straight run gasoline from each barrel of crude oil.

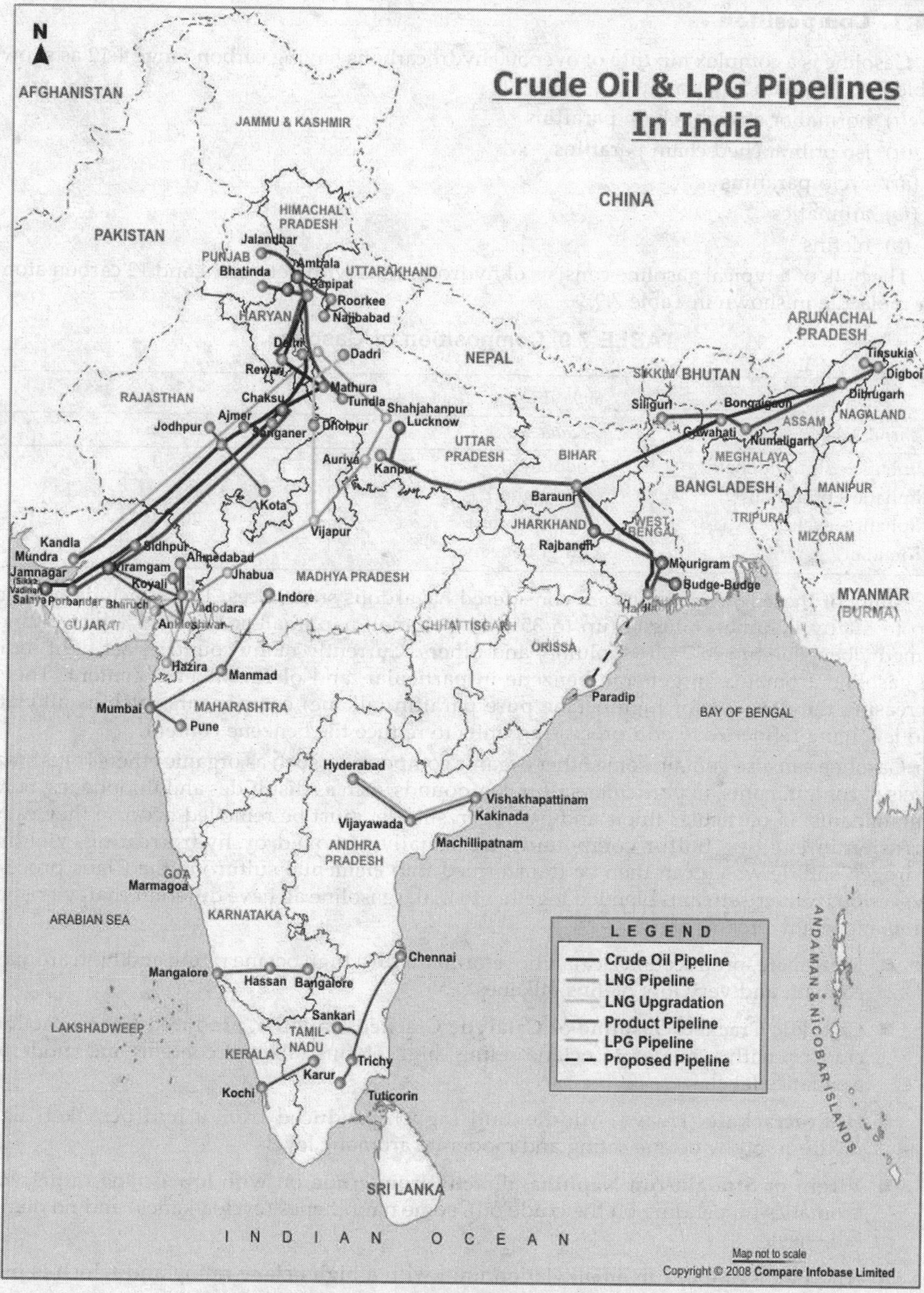

FIG. 7.2. CRUDE AND LPG PIPELINES IN INDIA.

7.4.1. Composition

Gasoline is a complex mixture of over 500 hydrocarbons having carbon range 4-12 as shown Table 7.9. The main types are :

(*i*) normal or straight chain paraffins
(*ii*) iso or branched chain paraffins
(*iii*) cyclo paraffins
(*iv*) aromatics
(*v*) olefins.

The bulk of a typical gasoline consists of hydrocarbons with between 4 and 12 carbon atoms per molecule in shown in Table 7.9.

TABLE 7.9. Composition of Gasoline.

Typical Composition of Gasoline		
General Name	*Examples*	*Percentage*
Aliphatic - straight chain	heptane	30-50
Aliphatic - branched	isooctane	
Aliphatic - cyclic	cyclopentane	20-30
Aromatic	ethyl benzene	20-30

Many of these hydrocarbons are considered hazardous substances. These include benzene (up to 5% by volume), toluene (up to 35% by volume), naphthalene (up to 1% by volume), trimethylbenzene (up to 7% by volume) and others. Currently many countries set tight limits on gasoline aromatics in general, benzene in particular, and olefin (alkene) content. This is increasing the demand for high octane pure paraffin (alkane) components, such as alkylate, and is forcing refineries to add processing units to reduce the benzene content.

Gasoline can also contain some other organic compounds: such as organic ethers (plus small levels of contaminants, in particular sulfur compounds such as disulfides and thiophenes. Some contaminants, in particular thiols and hydrogen sulfide, must be removed because they cause corrosion in engines. Sulfur compounds are usually removed by hydrotreating, yielding hydrogen sulfide which can then be transformed into elemental sulfur via the Claus process) The various refinery streams blended together to make gasoline all have different characteristics. Some important streams are:

- **Reformate**, produced in a catalytic reformer with a high octane rating and high aromatic content, and very low olefins (alkenes).
- **Catalytic Cracked Gasoline** or **Catalytic Cracked Naphtha**, produced from a catalytic cracker, with a moderate octane rating, high olefins (alkene) content, and moderate aromatics level.
- **Hydrocrackate** (Heavy, Middle, and Light), produced from a hydrocracker, with medium to low octane rating and moderate aromatic levels.
- **Virgin or Straight-run Naphtha**, directly from crude oil with low octane rating, low aromatics (depending on the crude oil), some naphthenes (cycloalkanes) and no olefins (alkenes).
- **Alkylate**, produced in an alkylation unit, with a high octane rating and which is pure paraffin (alkane), mainly branched chains.

Isomerate which is obtained by isomerising the pentane and overall a typical gasoline is predominantly a mixture of paraffins (alkanes), naphthenes (cycloalkanes), and olefins (alkenes). The exact ratios can depend on

- the oil refinery that makes the gasoline, as not all refineries have the same set of processing units.
- the crude oil feed used by the refinery.
- the grade of gasoline, in particular the octane rating.

7.4.2. Characteristics of Gasoline

Some of the characteristics of gosoline are discussed below and shown in Table 7.10.

TABLE 7.10.

Properties	Gasoline
Chemical formula	C4-C12
Mol. wt.	100-105
Composition wt.%	
Carbon	85-88
Hydrogen	12-15
Oxygen	0
Sp. gravity, 60°F/60°F	0.72-0.78
Density, 1b/gal @ 60°F	6-6.5
Boiling temp. °F	80-437
Reid vapor pressure (100°F) psi	8-15
Heating value	
Lower (Btu/gas)	116, 090
Lower (Btu/lb)	18, 676
Higher (Btu/gal)	124, 340
Higher (Btu/lb)	20,004
Octane Number	
Research Octane No. (RON)	88-98
Moles Octane No. (MON)	80-88
Freezing point, °F	–40
Viscosity	@ 140°F —
	@ 68°F 0.5-0.6
	@ –4°F 0.8-1
Flash point (closed cup, °F)	–45
Auto ignition temp, °F	495
Water solubility @ 70°F	
Fuel in water, vol%	Negligible
Water in fuel, vol%	Negligible
Flammability limits, vol%	
Lower	1.4
Higher	7.6
Specific heat Btu/lb °F	0.48
Latent heat of vaporization	
Btu/gal @ 60°F	~ 900
Btu/lb @ 60°F	~ 150
Stoichiometrican/fuel, weight	14.7
Vol. % fuel in vaporized stoichiometric mixture	2

Density

The specific density of gasoline ranges from 0.67–0.77, higher densities having a greater volume of aromatics. (0.026 lb/in^3; 719.7 kg/m^3; 6.073 lb/US gal; 7.29 lb/imp gal). Gasoline floats on water, so water cannot generally be used to extinguish a gasoline fire. Because of its specific density, and relative incompressibility even at extreme pressure, gasoline was chosen as the float fluid in the Bathyscaphe Trieste, the craft which reached a record-breaking depth of 10,900 metres (35,761 ft) in the deepest part of any ocean on Earth.

Volatility

Gasoline is more volatile than diesel oil, Jet-A or kerosene, not only because of the base constituents, but because of the additives that are put into it. The final control of volatility is often achieved by blending with butane. The Reid Vapor Pressure (RVP) test is used to measure the volatility of gasoline. The desired volatility depends on the ambient temperature: in hotter climates, gasoline components of higher molecular weight and lower volatility are used whereas in cold climates, too little volatility results in cars failing to start. In hot climates, excessive volatility results in **"vapor lock"** where combustion fails to occur, because the liquid fuel has changed to a gaseous fuel in the fuel lines, rendering the fuel pump ineffective and starving the engine of fuel. Volatility is regulated in order to reduce the emission of unburned hydrocarbons. In large cities, reformulated gasoline that is less prone to evaporation, among other properties, is required. In Australia summer petrol volatility limits are set by State Governments and vary between capital cities. The relaxed volatility standards may increase the atmospheric concentration of volatile organic compounds in warm weather, higher volatility gasoline effectively increases a nation's gasoline supply because the amount of butane in the gasoline pool is allowed to increase. Besides lowering the volatility of the fuel, other means of controlling the emission of unburned hydrocarbons, for environmental concerns, exist and are exercised. The vehicles sold should have a **fuel evaporative control system** (called an EVAP system) which collects expanding fuel vapor from the fuel tank in a charcoal-lined canister while the engine is stopped and then releases the collected vapors (through a "purge valve") into the engine intake for burning when the engine is running (usually only after it has reached normal operating temperature. The fuel evaporative control system is also required to include a gasketed filling cap which seals the fueling inlet to prevent vapors from escaping directly from the tank through it. Modern vehicles will turn on the MIL (Malfunction Indicator Light, *i.e.* "check engine" light) if it is detected that the gas cap is missing or loose and so not sealing.

Octane Rating

The **octane rating** is a measure of the resistance of gasoline and other fuels to detonation (engine knocking) in spark-ignition internal combustion engines. High-performance engines typically have higher compression ratios and are therefore more prone to detonation, so they require higher octane fuel. A lower-performance engine will not generally perform better with high-octane fuel, since the compression ratio is fixed by the engine design.

Isooctane or 2,2,4-trimethylpentane

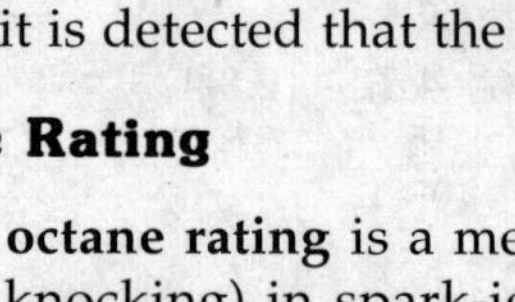

Octane number = 100

Propane
C_3H_8

Heptane

$CH_3CH_2CH_2CH_2CH_2CH_2CH_3$ Octane number = 0

Definition of Octane Rating

Octane rating of a spark ignition engine fuel is the detonation resistance (anti-knock rating) compared to a mixture of iso-octane (2,2,4-trimethylpentane, an isomer of octane) and n-heptane. By definition, iso-octane is assigned an octane rating of 100, and heptane is assigned an octane rating of zero. An 90-octane gasoline, for example, possesses the same anti-knock rating of a mixture of 90% (by volume) iso-octane, and 10% (by volume) n-heptane. This does not mean, however, that the gasoline actually contains these hydrocarbons in these actual proportions. It simply means that it has the same detonation resistance properties as the described 'standard' defined mixture. Octane rating does not relate to the energy content of the fuel. It is only a measure of the fuel's tendency to burn in a controlled manner, rather than exploding in an uncontrolled manner.

Research Octane Number (RON)

The most common type of octane rating worldwide is the Research Octane Number (RON). RON is determined by running the fuel in a test engine with a variable compression ratio under controlled conditions, and comparing the results with those for mixtures of iso-octane and n-heptane.

It is possible for a fuel to have a Research Octane Number (RON) greater than 100, because iso-octane is not the most knock-resistant substance available. Racing fuels, Aviation Gas, liquefied petroleum gas (LPG), and alcohol fuels such as methanol or ethanol may have octane ratings of 110 or significantly higher — ethanol's RON is 129 (102 MON, 116 AKI). Typical "octane booster" gasoline additives include tetra-ethyl lead, MTBE, and toluene. Tetra-ethyl lead (the additive used in leaded gasoline) is easily decomposed to its component radicals, which react with the radicals from the fuel and oxygen that start the combustion, thereby delaying ignition and leading to an increased octane number. However, tetra-ethyl lead and its byproducts are toxic; moreover, the use of tetra-ethyl lead creates an environmental hazard. Its use is now generally currently limited to being an additive to aviation gasoline.

Motor Octane Number (MON)

There is another type of octane rating, called Motor Octane Number (MON), or the aviation lean octane rating, which is a better measure of how the fuel behaves when under load as it is done at 900 rpm instead of the 600 rpm of the RON. MON testing uses a similar test engine to that used in RON testing, but with a preheated fuel mixture, a higher engine speed, and variable ignition timing to further stress the fuel's knock resistance. Depending on the composition of the fuel, the MON of a modern gasoline will be about 8 to 10 points lower than the RON. Normally, fuel specifications require both a minimum RON and a minimum MON

Anti-Knock Index (AKI)

It is also called the Road Octane Number (RdON), Pump Octane Number (PON), or (R+M)/2 (average of RON and MON).

Examples of Octane Ratings

The RON of n-heptane and iso-octane are exactly 0 and 100, by definition. Table 7.11 lists octane ratings for various other fuels.

TABLE 7.11. Octane Ratings of Various Fuels.

Fuel	RON
hexadecane	< -30
n-octane	-10
n-heptane (RON by definition)	0
diesel fuel	15–25
2-methylheptane	23
n-hexane	25
2-methylhexane	44
1-heptene	60
n-pentane	62
requirement for a typical two-stroke outboard engine	69
1-pentene	84
n-butanol	87
n-butane	91
t-butanol	97
cyclohexane	97
iso-octane (RON by definition)	100
benzene	101
methane	107
ethane	108
propane	110
methanol	113
toluene	114
ethanol	116
xylene	117
hydrogen*	> 130

*Hydrogen does not fit well into the normal definitions of octane number. It has a very high RON and a low MON so that it has low knock resistance in practice, due to its low ignition energy (primarily due to its low dissociation energy) and extremely high flame speed. These traits are highly desirable in rocket engines, but undesirable in Otto-cycle engines. However, as a minor blending component (e.g. in a bi-fuel vehicle), hydrogen raises overall knock resistance. Flame speed is limited by the rest of the component species; hydrogen may reduce knock by contributing its high thermal conductivity

Effects of Octane Rating

Higher octane ratings correlate to higher activation energies. Activation energy is the amount of energy necessary to start a chemical reaction. Since higher octane fuels have higher activation energies, it is less likely that a given compression will cause detonation. The fuels with higher octane ratings are used in more powerful engines, since such fuels explode less easily. However, an explosion is not desired in an internal combustion engine. An explosion will cause the pressure in the cylinder to rise far beyond the cylinder's design limits, before the force of the expanding gases can be absorbed by the piston traveling downward. This actually reduces power output, because much of the energy of combustion is absorbed as strain and heat in parts of the engine, rather than being converted to torque at the crankshaft. A fuel with a higher octane rating can be run at a higher compression ratio without detonating. Compression is directly related to power so engines that require higher octane usually deliver more motive power. Engine power is a function of the fuel, as well as the engine design, and is related to octane rating of the fuel. Power is limited by the maximum amount of fuel-air mixture that can be forced into the combustion chamber. When the throttle is partially open, only a small fraction of the total available power is produced because the manifold is operating at pressures far below atmospheric. In this case, the octane requirement is far lower than when the throttle is opened fully and the manifold pressure increases to atmospheric pressure, or higher in the case of supercharged or turbocharged engines. Many high-performance engines are designed to operate with a high maximum compression, and thus demand high-octane premium gasoline. A common misconception is that power output or fuel mileage can be improved by burning higher octane fuel than a particular engine was designed for. The power output of an engine depends in part on the energy density of its fuel, but similar fuels with different octane ratings have similar density. Since switching to a higher octane fuel does not add any more hydrocarbon content or oxygen, the engine cannot produce more power. However, burning fuel with a lower octane rating than required by the engine often reduces power output and efficiency one way or another. Power and efficiency for the reasons stated above. Many modern car engines feature a knock sensor – a small piezoelectric microphone which detects knock, and then sends a signal to the engine control unit to retard the ignition timing. Retarding the ignition timing reduces the tendency to detonate, but also reduces power output and fuel efficiency. Most fuel stations have two storage tanks (even those offering 3 or 4 octane levels), and give a mixture of the higher and lower octane fuel.

Octane ratings can vary greatly from region to region. In higher-altitude areas, a typical combustion engine draws in less air per cycle due to the reduced density of the atmosphere. This directly translates to reduced absolute compression in the cylinder, therefore deterring knock. It is safe to fill up a car with a carburetor that normally takes 87 AKI(Anti knock Index) fuel at sea level with 85 AKI fuel in the mountains, but at sea level the fuel may cause damage to the engine. A disadvantage to this strategy is that most turbocharged vehicles are unable to produce full power, even when using the "premium" 91 AKI fuel.

Energy Content

Gasoline contains about 32.0 MJ/l (9.67 kWh/l, 132 MJ/US gal or 36.6 kWh/US gal). This is an average; gasoline blends differ and therefore actual energy content varies from season to season and from batch to batch, by up to 4% more or less than the average, according to the US EPA. On average, about 19.5 US gallons (16.2 imp gal; 74 L) of gasoline are available from a 42-US-gallon (35 imp gal; 160 L) barrel of crude oil (about 46% by volume), varying due to quality of crude and grade of gasoline. The remaining residue comes off as products ranging from tar to naptha.

A high octane fuel such as Liquefied petroleum gas (LPG) has a lower energy content than lower octane gasoline, resulting in an overall lower power output at the regular compression ratio an engine ran at on gasoline. However, with an engine tuned to the use of LPG (i.e. via higher compression ratios such as 12:1 instead of 8:1), this lower power output can be overcome. This is because higher-octane fuels allow for a higher compression ratio - this means less space in a cylinder on its combustion stroke, hence a higher cylinder temperature which improves efficiency according to Carnot's theorem, along with fewer wasted hydrocarbons (therefore less pollution and wasted energy), bringing higher power levels coupled with less pollution overall because of the greater efficiency.

The main reason for the lower energy content (per litre) of LPG in comparison to gasoline is that it has a lower density. Energy content per kilogram is higher than for gasoline (higher hydrogen to carbon ratio). The weight-density of gasoline is about 740 kg/m^3 (6.175 lb/US gal; 7.416 lb/imp gal).

Different countries have some variation in RON (Research Octane Number) is standard for gasoline, or petrol. In the UK, ordinary regular unleaded petrol is 91 RON (not commonly available), premium unleaded petrol is always 95 RON, and super unleaded is usually 97-98 RON. However both Shell and BP produce fuel at 102 RON for cars with hi-performance engines, and the supermarket chain Tesco began in 2006 to sell super unleaded petrol rated at 99 RON. In the US, octane ratings in unleaded fuels can vary between 86-87 AKI (91-92 RON) for regular, through 89-90 AKI (94-95 RON) for mid-grade (European Premium), up to 90-94 AKI (95-99 RON) for premium (European Super).

7.4.3. Petrol Engine

Petrol Engine

A petrol engine (known as a gasoline engine, is an internal combustion engine with spark-ignition, designed to run on petrol (gasoline) and similar volatile fuels.

Petrol engine was introduced by German scientist - Gottlieb Daimler and Karl Benz in 1885. It is considered as one of biggest achievement in the automotive field. It uses petrol called as gasoline in USA as a fuel. Within the engine burning of fuel mixed with air causes hot gases to expand against parts of the engine and force them to move. So petrol engines are called internal-combustion engines.

Petrol engines are compact and light in weight for the power they produce. The rate at which it produces work is usually measured in horsepower or watts.

Kinds of Petrol Engines

Reciprocating petrol engines are distinguished in a number of ways. Some of them are as follows:

Type of Compression

As a piston moves from the bottom to the top of a cylinder, it compresses the air and petrol mixture. A number, called the compression ratio, tells how much the mixture is compressed. A high-compression engine may have a compression ratio of 10 to 1. Such an engine compresses the mixture to a tenth of its original volume. A low-compression engine may have a ratio of 8 to 1. High-compression engines burn petrol more efficiently than low-compression engines.

But high-compression engines require high-octane petrol. Earlier the octane level of petrol depended on the amount of lead additives—the more lead, the higher the octane. Later on the car manufactured were equipped with catalytic converters that reduce the pollutants in car exhausts. Lead was found to interfere with the effectiveness of catalytic converters. Cars with catalytic converters had to use low-octane petrol because high-octane lead-free petrol was costly to produce. As a result, compression ratios were reduced so that engines could burn lower octane lead-free fuels efficiently. The **compression ratio** of an internal-combustion engine or external combustion engine is a value that represents the ratio of the volume of its combustion chamber; from its largest capacity to its smallest capacity. It is a fundamental specification for many common combustion engines. In a piston engine it is the ratio between the volume of the cylinder and combustion chamber when the piston is at the bottom of its stroke, and the volume of the combustion chamber when the piston is at the top of its stroke. A high compression ratio is desirable because it allows an engine to extract more mechanical energy from a given mass of air-fuel mixture due to its higher thermal efficiency. High ratios place the available oxygen and fuel molecules into a reduced space along with the adiabatic heat of compression - causing better mixing and evaporation of the fuel droplets. Thus they allow increased power at the moment of ignition and the extraction of more useful work from that power by expanding the hot gas to a greater degree. Higher compression ratios will however make gasoline engines subject to engine knocking, also known as detonation and this can reduce an engine's efficiency or even physically damage it. The ratio is calculated by the following formula:

$$\mathrm{CR} = \frac{\frac{\pi}{4}b^2 s + V_c}{V_c}$$

where b = Cylinder bore (diameter)

s = Piston stroke length

V_c = Volume of the combustion chamber (including head gasket). This is the minimum volume of the space at the end of the compression stroke. Because of the complex shape of this space, it is usually measured directly rather than calculated.

Valve Arrangement

The two most common valve arrangements are :

(1) L-head

(2) I-head

An L-head, or underhead, valve engine has the intake and exhaust valves side by side in the cylinder block. The intake valve admits the air-fuel mixture into the cylinder and the exhaust valve lets out the exhaust gases. An I-head, or overhead, valve engine has the two valves side by side in the cylinder head, the cylinder block's top cover. In some cars, each cylinder has four valves—two intake valves and two exhaust valves.

The Way They are Cooled

The burning fuel-air mixture in a cylinder produces gas temperatures of about 2500 °C. Therefore, the metal parts of the engine must be cooled or they would melt. Most automotive petrol engines are liquid cooled. A liquid, usually water, is circulated around the cylinders to cool the metal. The heated liquid is then pumped through a radiator. A fan driven by the engine or by an electric motor draws air through the radiator to cool the liquid. Petrol engines may be air-cooled, with fins (to increase the surface area on the cylinders) and cylinder head; or liquid-cooled, by a water jacket and radiator. The coolant was formerly water, but is now usually a mixture of water and ethylene glycol. This mixture has a lower freezing-point and a higher boiling-point than pure water, and also prevents corrosion, with modern antifreezes also containing lubricants and other additives to protect water pump seals and bearings. The cooling system is usually slightly pressurised to raise the boiling point of the coolant. Most aircraft reciprocating engines are air cooled to reduce weight. Air is not as effective a coolant as liquids, so the outsides of the cylinders have many metal fins. These fins conduct heat out of the cylinder and offer a large surface area for the air to sweep over, thus ensuring effective cooling.

The Way They are Supplied With Air and Fuel

Fuel may be sent, to the cylinders by either a carburettor or an injection system. Therefore, reciprocating engines are also classified as carburetted or as fuel-injected engines. Because combustion depends upon both air and fuel, the power of an engine is limited by the amount of air reaching the cylinders. To increase power, an engine may be supercharged or turbocharged. A supercharger is an engine-driven pump, and a turbocharger is an exhaust-driven pump. Both pumps force extra air into the cylinders, increasing the engine power. The air needed to burn 1 unit of petrol weighs about 15 times as much as the petrol.

Number of Piston Strokes per Cycle

The engine is being classified based on number of stokes per cycle :

(*a*) Two Stroke

(*b*) Four Stroke

Four Stroke Petrol Engine

The four stroke engine is called so because the working of an internal combustion engine is divided into four stages called four strokes of the engine. The Figs. 7.3 and 7.4 below shows the four stroke of an engine and a typical model of engine.

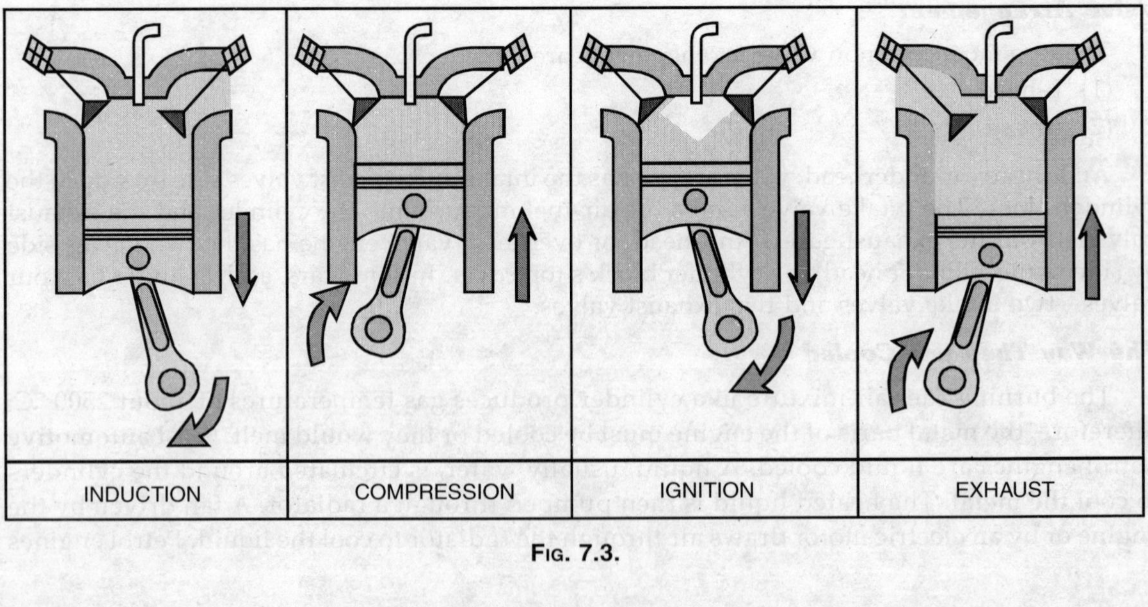

FIG. 7.3.

Intake stroke : When the engine starts, the piston moves downwards in the cylinder, because of which a region of low pressure is created in the cylinder, above the piston. At this moment, the intake valve opens and the fuel mixture(petrol vapour and air mixture) is sucked into the cylinder from the carburettor. (Fig. 7.5)

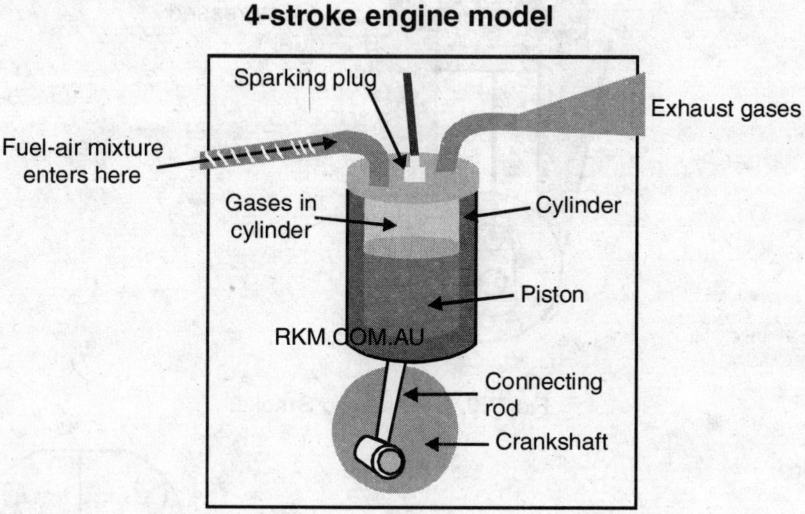

FIG. 7.4. A FOUR STROKE ENGINE MODEL.

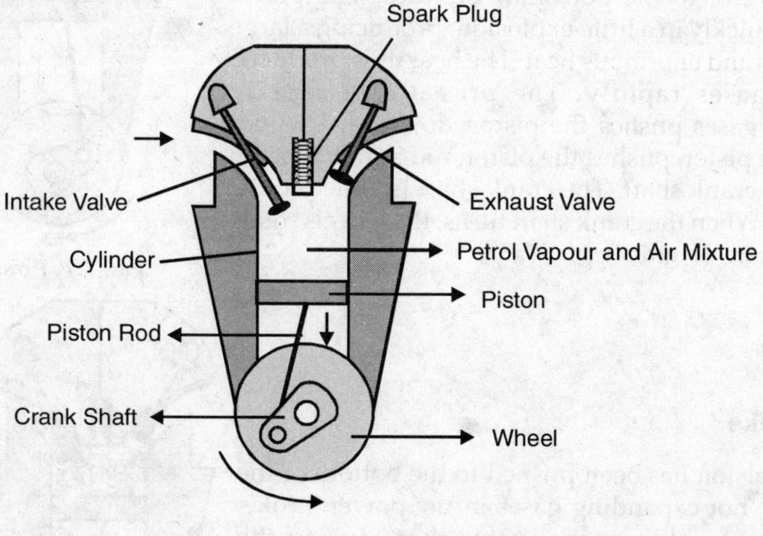

FIG. 7.5. INTAKE STROKE.

Compression Stroke

When the sufficient amount of the fuel mixture (petrol vapour and air mixture) has entered the cylinder, the intake valve gets closed. The piston is then forced to move upwards which compresses the fuel-mixture to about one-eighth of its original volume. Higher the compression ratio, more will be the efficiency of the engine. (Fig. 7.6)

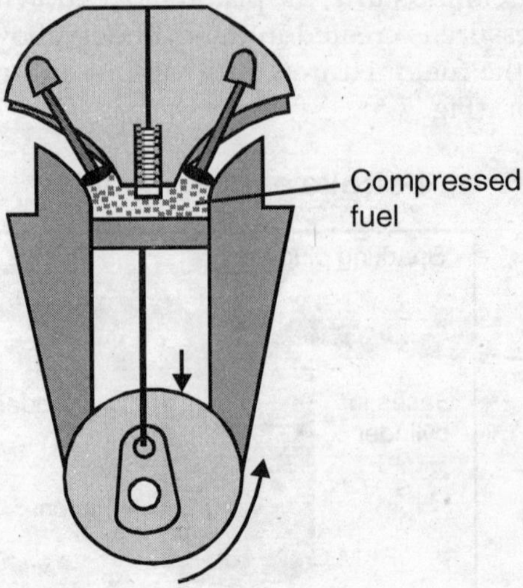

Fig. 7.6. Compressed Stroke.

Power Stroke

Before the piston completes its upward movement, compressing the petrol vapour and air mixture, the spark plug produces a little electric spark inside the cylinder and this spark sets fire to the petrol-air mixture. The petrol vapour burns quickly in a little explosion, producing a large volume of gases and enormous heat. The heat thus produced expands the gases rapidly. The pressure of rapidly expanding hot gases pushes the piston downward with a great force. The piston pushes the piston rod and the piston rod pushes the crank shaft. The crank shaft is joined to the wheels of a car. When the crank shaft turns, the wheels rotate and move the car. (Fig. 7.7)

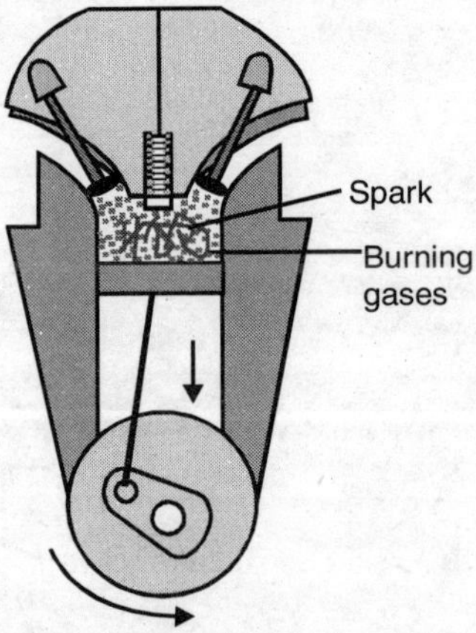

Fig. 7.7. Power Stroke.

Exhaust Stroke

When the piston has been pushed to the bottom of the cylinder by the hot expanding gases in the power stroke, then the exhaust valve opens. After that, due to the momentum gained by the wheels, the piston is pushed upwards. The upward movement of the piston, expels the spent gases through the exhaust valve into the atmosphere, carrying away the unused heat. The exhaust valve then closes, the intake valve opens up, and the above four strokes of the engine are repeated again and again. (Fig. 7.8)

Figure 7.9 shows the typical petrol engine.

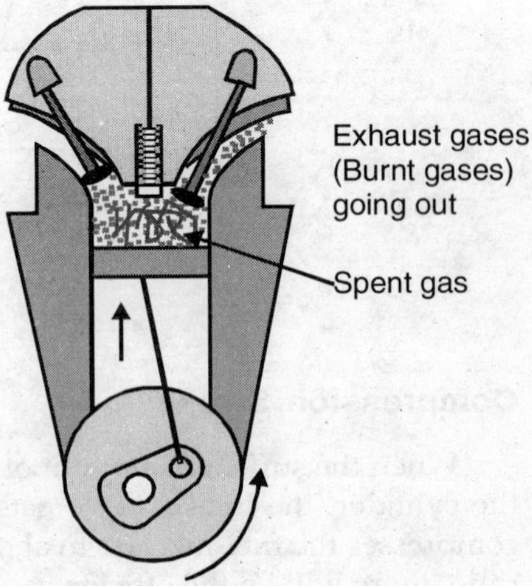

Fig. 7.8. Exhaust Stroke.

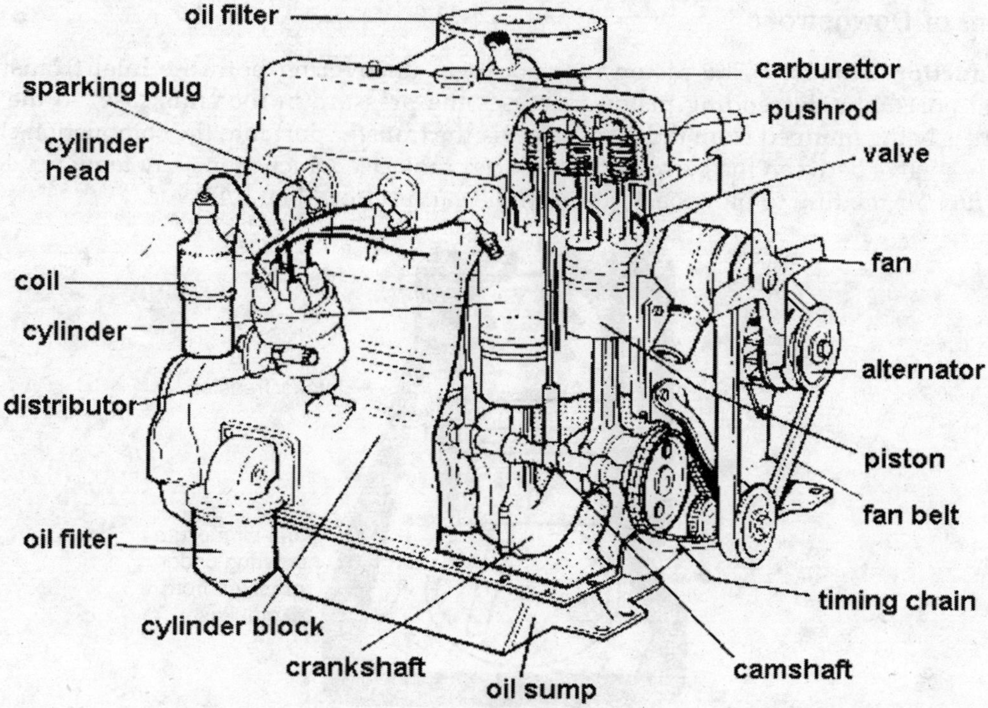

FIG. 7.9. A TYPICAL PETROL ENGINE.

Two Stroke Petrol Engine

In this type of engine ignites fuel at every upward stroke, so there are two strokes for every ignition of fuel. They are called upward and downward stroke. As the piston moves in upward direction from bottom to top in the first stroke the air and fuel mixture gets compressed and ignited by spark plug as upward stroke comes to end. This results in an explosion of mixture which forces the piston to move downwards thereby producing power. Since they have less moving parts they are light in weight. Big size bulky two-stroke cycle engines have lubrication systems like that of four-stroke cycle engines.

In two-stroke engine there are no valves, camshafts. The piston is connected by the con-rod to the crankshaft. Lubrication is achieved by mixing oil with the fuel. The complete cycle takes only one upstroke and one downstroke of the piston. The fuel and air are mixed in the carburettor and the mixture enters the sealed crankcase, i.e. the space underneath the piston. The crankcase is connected to the combustion chamber in the cylinder by an **inlet port**, sometimes known as a transfer port. Opposite the inlet port there is another port, the **exhaust port**. Both the inlet and exhaust port are uncovered by the piston in the bottom part of its stroke, thus replicating the function of the valves in the four-stroke engine and allowing gases to enter and leave the cylinder. In the top part of the stroke both ports are covered, sealing the cylinder. This allows compression to occur at the top of the upstroke and allows the power of the expanding gases to be harnessed at the top of the downstroke. Figure 7.10-7.12 explains the working of two stroke engine.

Bottom of Downstroke

Induction/Exhaust : The piston moves down, uncovering both the inlet (transfer) and exhaust ports. The descending piston increases the pressure in the crankcase, so the fuel-air mixture is being pumped from the crankcase via the transfer port into the combustion chamber. There is a valve between the carburettor and the crankcase which stops any tendency for some of the fuel-air mixture to blow back through the carburettor. (Fig. 7.10)

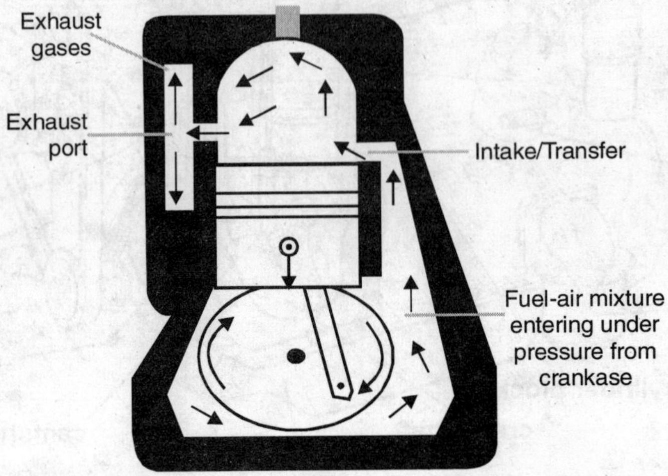

FIG. 7.10. INDUCTION PROCESS IN TWO-STROKE ENGINE.

Top of Upstroke–Compression

The piston has passed Bottom Dead Centre (BDC) and is now rising. The inlet and exhaust ports are both covered, so the fuel-air mixture in the combustion chamber is being pressurised and is heating up. At the same time, a vacuum is developing in the crankcase, so a fresh charge of fuel-air is being drawn into the crankcase from the carburettor. As the piston rounds BDC, a high voltage discharge from the spark plug ignites the mixture in the combustion chamber. (Fig. 7.11)

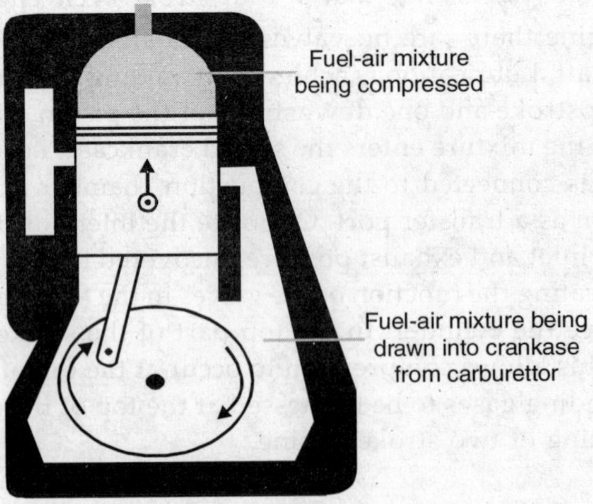

FIG. 7.11. COMPRESSION PROCESS.

Top of Downstroke-Ignition/Power

With both ports still closed, the pressure of the expanding gases forces the piston down again. The pressure in the crankcase is already rising. Later in the downstroke the exhaust port will be uncovered, allowing the spent gases to escape. Very shortly after that the inlet/transfer port will also be uncovered, which takes the engine back to where it was at the start working of two-stroke engine of the cycle. (Fig. 7.12)

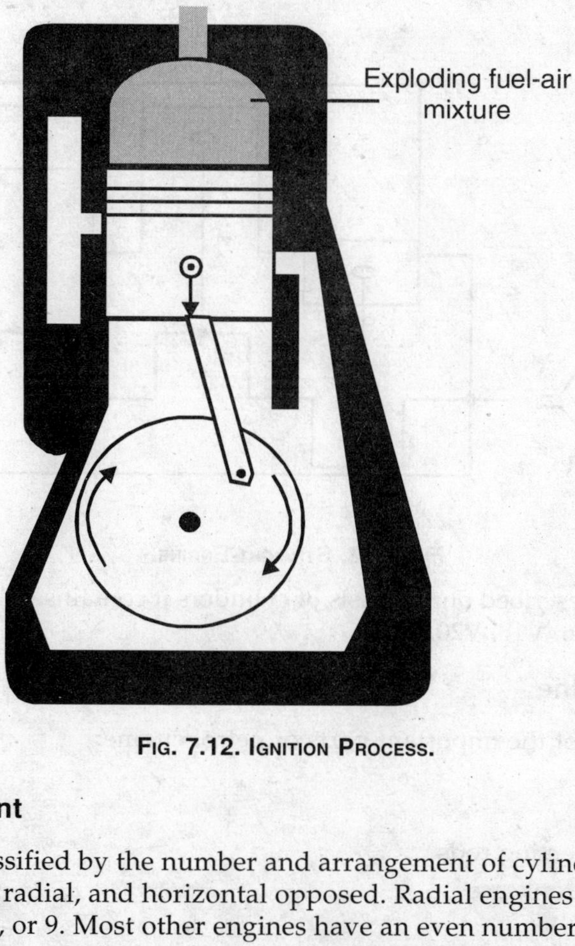

Fig. 7.12. Ignition Process.

Cylinder Arrangement

Engines are also classified by the number and arrangement of cylinders. The most common types include in-line, V, radial, and horizontal opposed. Radial engines have an odd number of cylinders, such as 3, 5, 7, or 9. Most other engines have an even number of cylinders—4, 6, 8, or 12.Common cylinder arrangements are from 1 to 6 cylinders in-line (Usually found in four- and six-cylinder configurations, the **straight engine**, or **inline engine** (Figure 7.13) is an internal-combustion engine with all cylinders aligned in one row or plane, with no offset. A straight engine is considerably easier to build than an otherwise equivalent horizontally opposed or V engine, because both the cylinder bank and crankshaft can be milled from a single metal casting, and it requires fewer cylinder heads and camshafts. In-line engines are also much smaller in overall physical dimensions than designs like the radial, and can be mounted in any direction. Straight configurations are simpler than their V-shaped counterparts. They have a support bearing between each piston as compared to "flat and V" engines which have support bearings between every two pistons. Although six-cylinder engines are inherently balanced, the four-cylinder models are inherently off balance and rough, unlike 90 degree V four sand horizontally opposed 'boxer' 4 cylinders or from 2 to 16 cylinders in V-formation. (A **V engine**, or **Vee engine** is a common configuration for an internal combustion engine. The cylinders and pistons are aligned, in two separate planes, so that they appear to be in a **"V"** when viewed along the axis of the crankshaft. The V configuration generally reduces the overall engine length, height and weight compared to an equivalent inline configuration. Various cylinder bank angles of V are used in different engines; depending on the number of cylinders, there may be angles that work better than others for stability. Very narrow angles of V combine some of the advantages of the V engine and the straight engine (primarily in the form of compactness) as well as disadvantages.

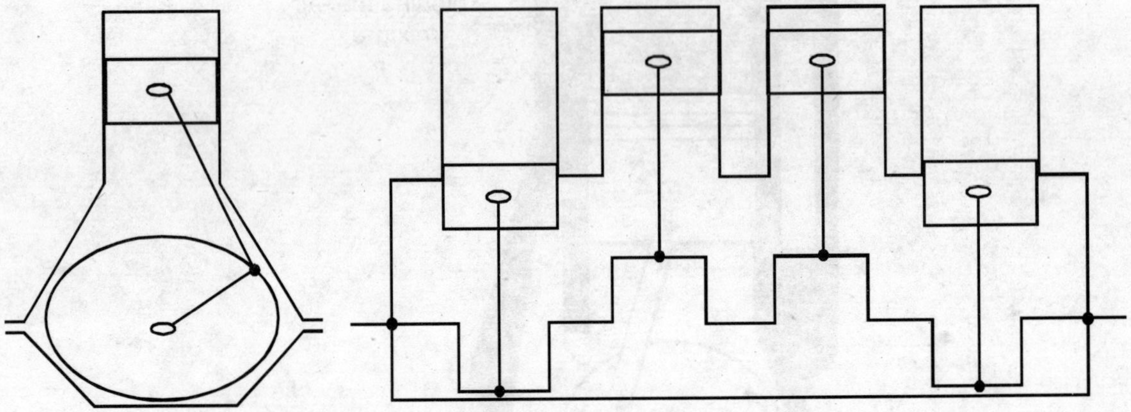

FIG. 7.13. STRAIGHT ENGINE.

The V engines are described on the basis of cylinders it contains. These may be V-twin, V4, V5, V6, V8, V10, V12, V16, V18, V20, V24

Parts of a Petrol Engine

Following are some of the important parts of petrol engine:

(*i*) Cylinders
(*ii*) Cylinder block
(*iii*) Piston and Connecting rods
(*iv*) Cylinder head Crankcase
(*v*) Valves
(*vi*) Crank shaft Flywheel
(*vii*) Exhaust system
(*viii*) Camshaft Fuel system
(*ix*) Lubrication system
(*x*) Ignition system

Ignition

Petrol engines use spark ignition and high voltage current for the spark may be provided by a magneto or an ignition coil. In modern car engines the ignition timing is managed by an electronic Engine Control Unit.

Carburetor

It is the heart of gasoline engine. They meter the fuel and mix it with the air in precise proportions. Old carburetors do spark advance by measuring the difference in pressure between the outside and inside of the carburetor. The amount of throttle advance is also measured. The engine's remains which may be carbon monoxide or unburned hydrocarbons shows how well the carburetor is working.

These engines are widely used in vehicles, portable power plants to supply the power to run pumps and other machinery on farms. Many small boats, aero planes, trucks and buses also use it.

7.4.4. Additives

Lead

Gasoline, when used in high compression internal combustion engines, has a tendency to autoignite (*detonation*) causing a damaging "engine knocking" (also called "pinging") noise. The discovery that lead additives modified this behavior led to the widespread adoption of their use in the 1920s and therefore more powerful higher compression engines. The most popular additive was tetra-ethyl lead. However, with the discovery of the environmental and health damage caused by the lead, and the incompatibility of lead with catalytic converters Most countries are phasing out leaded fuel; different additives have replaced the lead compounds. The most popular additives include aromatic hydrocarbons, ethers and alcohol (usually ethanol or methanol). The Clean Air Act banned the sale of leaded fuel for use in on-road vehicles. Possession and use of leaded gasoline in a regular on-road vehicle now carries a maximum $10,000 fine in the US. However, fuel containing lead may continue to be sold for off-road uses, including aircraft, racing cars, farm equipment, and marine engines. The ban on leaded gasoline led to thousands of tons of lead not being released in the air by automobiles. Similar bans in other countries have resulted in lowering levels of lead in people's bloodstreams. A side effect of the lead additives was protection of the valve seats from erosion. Many classic cars' engines have needed modification to use lead-free fuels since leaded fuels became unavailable. Gasoline, as delivered at the pump, also contains additives to reduce internal engine carbon buildups, improve combustion, and to allow easier starting in cold climates.

Tetra-Ethyl Lead

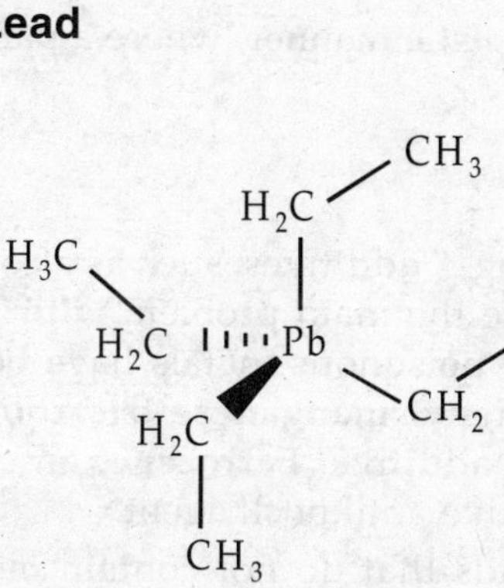

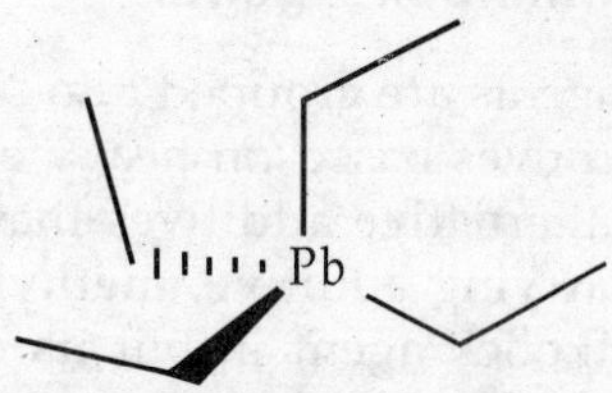

Tetra-ethyl lead, abbreviated **TEL**, is an organometallic compound with the formula $(CH_3CH_2)_4Pb$. Once a common antiknock additive in gasoline (petrol), TEL usage was largely discontinued because of the toxicity of lead and its deleterious effect on catalytic converters. It is still used as an additive in aviation fuel for piston engine-powered aircraft.

TEL is produced by reacting chloroethane with a sodium–lead alloy.

$$4\ NaPb + 4\ CH_3CH_2Cl \longrightarrow (CH_3CH_2)_4Pb + 4\ NaCl + 3\ Pb$$

TEL, is a viscous colorless liquid. Because TEL is charge neutral and contains an exterior of alkyl groups, it is highly lipophilic and soluble in petrol (gasoline).

TEL has four C–Pb bonds. At the temperatures found in internal combustion engines $(CH_3CH_2)_4Pb$ decomposes completely into lead and lead oxides and combustible, short-lived ethyl radicals. Lead and lead oxide scavenge radical intermediates in combustion reactions. This prevents ignition of unburnt fuel during the engine's exhaust stroke. Lead itself is the reactive antiknock agent, and TEL serves as a gasoline-soluble lead carrier. When $(CH_3CH_2)_4Pb$ burns, it produces not only carbon dioxide and water, but also lead

$$(CH_3CH_2)_4Pb + 13\ O_2 \longrightarrow 8\ CO_2 + 10\ H_2O + Pb$$

This lead can oxidize further to give species such as lead(II) oxide:

$$2\ Pb + O_2 \longrightarrow 2\ PbO$$

The Pb and PbO would quickly accumulate and destroy an engine. For this reason, the lead scavengers 1,2-dibromoethane and 1,2-dichloroethane are used in conjunction with TEL—these agents form volatile lead(II) bromide and lead(II) chloride, respectively, which are flushed from the engine and into the air.

TEL was supplied for mixing with raw gasoline in the form of **ethyl fluid**, which was TEL blended together with the lead scavengers 1,2-dibromoethane and 1,2-dichloroethane. Ethyl fluid also contained a reddish dye to distinguish treated from untreated gasoline and discourage the use of leaded gasoline for other purposes such as cleaning. Ethyl fluid was added to gasoline at rate of 1:1260, usually at the refinery. The formula for ethyl fluid is :

- Tetraethyl lead 61.45%
- 1,2-Dibromoethane 17.85%
- 1,2-Dichloroethane 18.80%
- Inerts & dye 1.90%

Dibromoethane and dichloroethane act in a synergistic manner, where a particular mixing ratio provides the best lead scavenging ability.

Alternative Antiknock Agents

Antiknock agents are grouped into "high-percentage" additives, such as alcohol, and "low-percentage" additives based on heavy elements. Since the main problem with TEL is its lead content, many alternative additives that contain less poisonous metals have been examined. A manganese-carrying additive, **methylcyclopentadienyl manganese tricarbonyl** (MMT), is used as an antiknock agent in but its use as a fuel additive. **Ferrocene**, an organometallic compound of iron, has also been reported as an effective antiknock agent.

High-percentage additives are organic compounds that do not contain metals, but they require much higher blending ratios, such as 20-30% for benzene and ethanol. Oxygenates, mainly methanol-derived MTBE and ethanol-derived ETBE have largely substituted the need for TEL. MTBE has environmental risks of its own and there are also bans on its use. ETBE, on the other hand, requires more expensive ethanol as a starting material.

Improvements of the gasoline itself decrease the need for separate antiknock agents. Synthetic iso-octane and alkylate are examples of such blending stocks. Benzene and other high-octane aromatics can be also blended to raise the octane number, but they are disfavored today because of toxicity and carcinogenity.

Methylcyclopentadienyl Manganese Tricarbonyl (MMT) is an organomanganese compound with the formula $(CH_3C_5H_4)Mn(CO)_3$. Marketed initially as a supplement to the gasoline additive tetraethyl lead to increase the fuel's octane rating, MMT was later used in unleaded gasoline It is sold under the tradenames HiTec 3000 and AK-33X. It is also marketed as Ecotane.

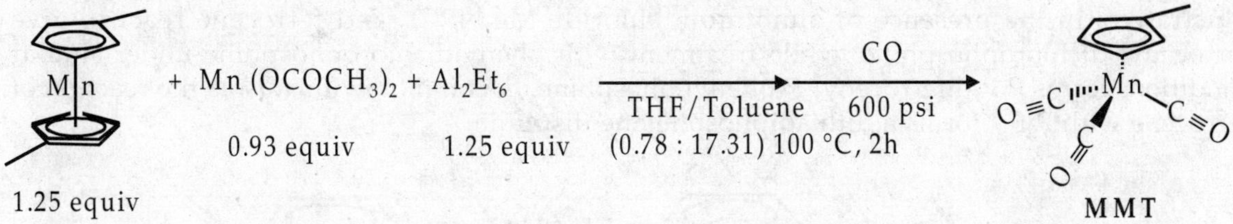

The dermal absorption from accidental gasoline spills, misuse of gasoline as a solvent cleaner, and deliberate gasoline fume inhalation were the main sources of potential MMT exposure. Chronic exposure to manganese has been known to cause manganism. The symptoms of manganism are similar to those of Parkinson's disease. MMT has been shown to be cytotoxic Short-term symptoms resulting from MMT poisoning include giddiness, headache, nausea, and difficulties in breathing. Animal studies have shown that long-term exposure to MMT can result in damage to the liver and kidneys.

Ferrocene : Ferrocene is an organometallic compound with the formula $Fe(C_5H_5)_2$. It is the prototypical metallocene, a type of organometallic chemical compound consisting of two cyclopentadienyl rings bound on opposite sides of a central metal atom. Such organometallic compounds are also known as sandwich compounds.

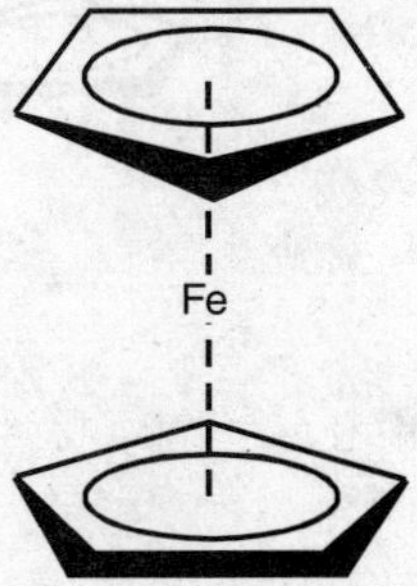

Ferrocene and its derivatives are antiknock agents used in the fuel for petrol engines; they are safer than tetraethyl lead The iron containing deposits formed from ferrocene can form a conductive coating on the spark plug surfaces. In diesel-fuelled engines, ferrocene reduces the production of soot.

Ferrocene is an air-stable orange solid that readily sublimes, especially upon heating in a vacuum. As expected for a symmetric and uncharged species, ferrocene is soluble in normal organic solvents, such as benzene, but is insoluble in water. It is stable to temperatures as high as 400 °C.

Table 7.12 gives typical values of vapor pressure of ferrocene at different temperatures

Table 7.12

Pressure (Pa)	1	10	100
Temperature (K)	298	323	353

Ferrocene is efficiently prepared by the reaction of sodium cyclopentadienide with anhydrous ferrous chloride in ethereal solvents.

$$2\ NaC_5H_5 + FeCl_2 \longrightarrow Fe(C_5H_5)_2 + 2\ NaCl$$

Ferrocene undergoes many reactions electrophilic reaction characteristic of aromatic compounds (Fig. 7.14), enabling the preparation of substituted derivatives. The most common is the Friedel-Crafts reaction of ferrocene with acetic anhydride (or acetyl chloride) in the presence of phosphoric acid as a catalyst. The preparation of phosphorus derivatives of ferrocenes are illustrative. In the presence of aluminium chloride Me_2NPCl_2 and ferrocene react to give ferrocenyl dichlorophosphine, while treatment with phenyldichlorophosphine under similar conditions forms *P,P*-diferrocenyl-*P*-phenyl phosphine. In common with anisole the reaction of ferrocene with P_4S_{10} forms a dithiadiphosphetane disulfide.

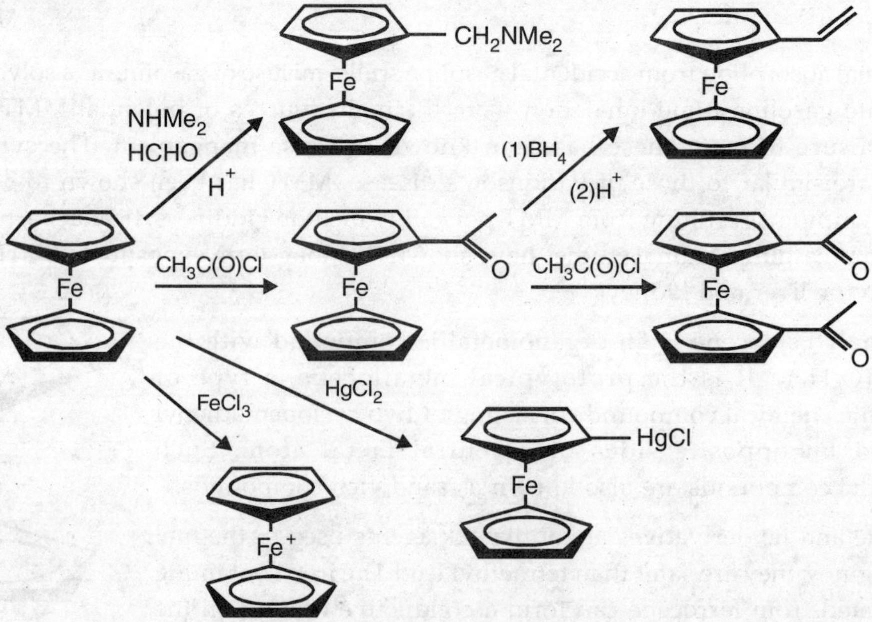

FIG. 7.14. IMPORTANT REACTIONS OF FERROCENE WITH ELECTROPHILES AND OTHER REAGENTS.

Ferrocene reacts readily with butyl lithium (Fig. 7.15) to give 1,1'-dilithioferrocene, which in turn is a versatile nucleophile. It has been reported that the reaction of 1,1'-dilithioferrocene with selenium diethyldithiocarbamate forms a strained ferrocenophane where the two cyclopentadienyl ligands are linked by the selenium atom. This ferrocenophane can be converted to a polymer by a thermal ring-opening polymerization (ROP) to form poly(ferrocenyl selenide). Likewise by the reaction of silicon and phosphorus linked ferrocenophanes the poly(ferrocenylsilane)s and poly(ferrocenylphosphines)s can be obtained.

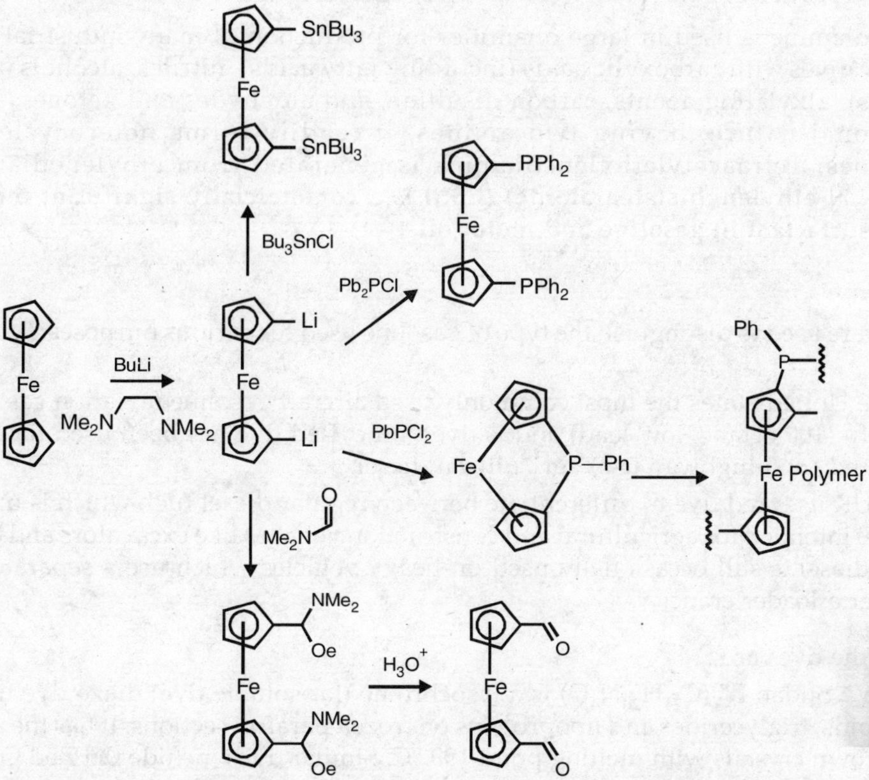

FIG. 7.15. A DIAGRAM SHOWING THE LITHIATION OF FERROCENE WITH BULI, AND THEN THE REACTIONS OF THE DILITHIOFERROCENE WITH A SERIES OF ELECTROPHILES.

A. Butylated Hydroxytoluene

Butylated hydroxytoluene (BHT), also known as **butylhydroxytoluene**, is a lipophilic (fat-soluble) organic compound that is primarily used as an antioxidant food additive as well as in cosmetics, pharmaceuticals, **jet fuels**, rubber, **petroleum products**, electrical transformer oil, and embalming fluid.

BHT is prepared by the reaction of *p*-cresol (4-methylphenol) with isobutylene (2-methylpropene) catalysed by sulfuric acid :

$$CH_3(C_6H_4)OH + 2\ CH_2{=}C(CH_3)_2 \longrightarrow ((CH_3)_3C)_2CH_3C_6H_2OH$$

Alternatively, BHT is prepared from 2,6-di-*tert*-butylphenol by hydroxymethylation or aminomethylation followed by hydrogenolysis.

B. Ethylenediamine

Ethylenediamine is the organic compound with the formula $C_2H_4(NH_2)_2$. This colorless liquid with an ammonia-like odor is a strongly basic amine. It is a widely used building block in chemical synthesis.

Ethylenediamine is manufactured by reacting ammonia and 1,2-dichloroethane. The reaction yields the mixture of ethylenediamine and the linear polyamines. A simplified equation would be:

$$ClCH_2CH_2Cl + 4\ NH_3 \longrightarrow H_2NCH_2CH_2NH_2 + 2\ NH_4Cl$$

Ethylenediamine is used in large quantities for production of many industrial chemicals. It forms derivatives with carboxylic acids (including fatty acids), nitriles, alcohols (at elevated temperatures), alkylating agents, carbon disulfide, and aldehydes and ketones. Because of its bifunctional nature, having two amines, it readily forms heterocycles such as imidazolidines. Tetraacetylethylenediamine is generated from ethylenediamine. **The derivative N,N-ethylenebis(stearamide) (EBS) is a commercially significant mold-release agent and a surfactant in gasoline and motor oil.**

7.4.5. Dye

The dyes are used to distinguish the type of gasoline used for various purposes. Few examples are :

- In the United States the most commonly used aircraft gasoline (aviation gas, known as 100LL) (100 octane, low lead) and is dyed blue. Red dye has been used for identifying untaxed (non-highway use) agricultural diesel.
- The UK uses red dye to differentiate between regular diesel fuel, which is undyed, and diesel intended for agricultural and construction vehicles like excavators and bulldozers. Red diesel is still occasionally used on heavy vehicles which use a separate engine to power a loader crane.

Some of the dyes are :

Sudan IV : Sudan IV ($C_{24}H_{20}N_4O$) is a lysochrome (fat-soluble dye) diazo dye used for the staining of lipids, triglycerides and lipoproteins on frozen paraffin sections. It has the appearance of reddish brown crystals with melting point 199 °C. Similar dyes include Oil Red O, Sudan III, and Sudan Black B. For staining purposes Sudan IV can be made up in propylene glycol. Alternatively, authors have reported using the dye saturated in isopropyl acohol, 95% ethanol, or 0.05% by weight in acetone:ethanol:water (50:35:15) The idea is to use a moderately apolar solvent to solubilize the dye allowing it to partition into the highly apolar fat without the solvent solubilizing the fat to be stained. Sudan I, Sudan III, and Sudan IV have been classified as category 3 carcinogens by the International Agency for Research on Cancer. In its purified form it is called **Biebrich scarlet** R, which should not be confused with the water-soluble Biebrich scarlet. In industry, it is used to color nonpolar substances like oils, fats, waxes, greases, various hydrocarbon products, and acrylic emulsions. Sudan IV is used in United Kingdom as a fuel dye to dye lower-taxed heating oil; because of that it is also known as Oil Tax Red.

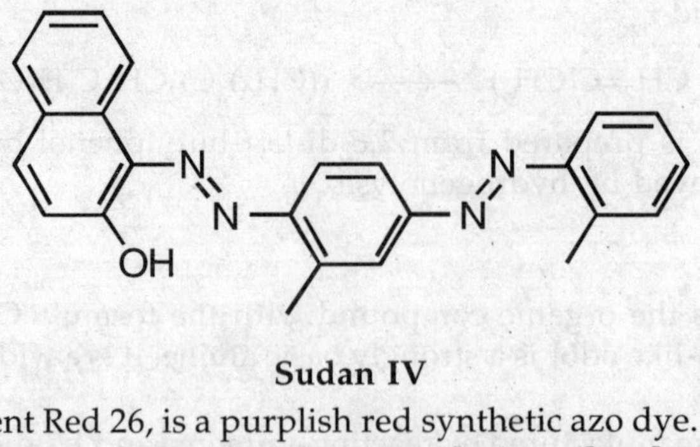

Sudan IV

Solvent Red 26 : Solvent Red 26, is a purplish red synthetic azo dye. Its chemical formula is $C_{25}H_{22}N_4O$, or 1-[[2,5-dimethyl-4-[(2-methylphenyl)azo]-phenyl]azo]-2-naphthol. It is soluble in oils and insoluble in water.

It is used by the EPA (Environmental Protection agency) to mark fuels with higher sulfur content. It is used as a standard fuel dye in the United States of America to distinguish low-taxed heating oil from automotive diesel fuel, it is however increasingly replaced with Solvent Red 164, a similar dye with longer alkyl chains, which is better soluble in hydrocarbons.

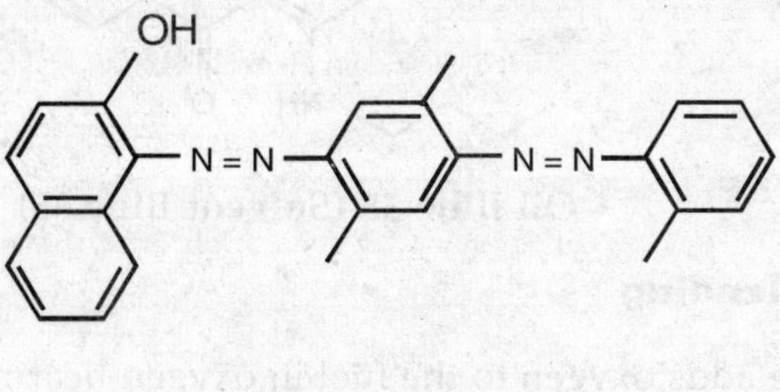

Solvent Red 26

Solvent Yellow 124 : Solvent Yellow 124 is a yellow azo dye used in European Union as a fuel dye. It is used to distinguish diesel fuel intended for heating from a higher-taxed motor diesel fuel. It is added to fuels not intended for motor vehicles and sold under the name **Euromarker**.

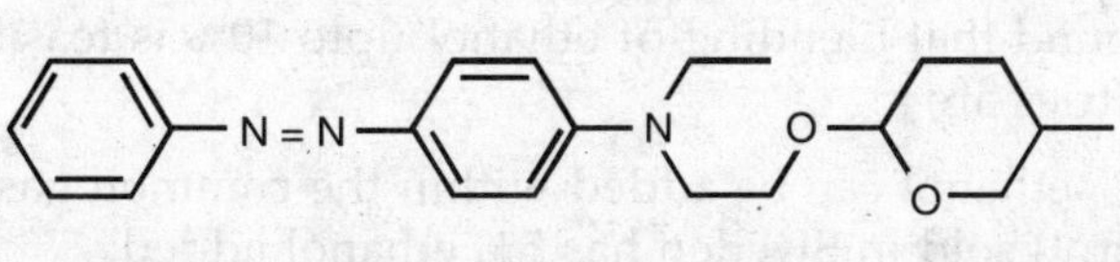

Solvent Yellow 124 is a dye with structure similar to Solvent Yellow 56.

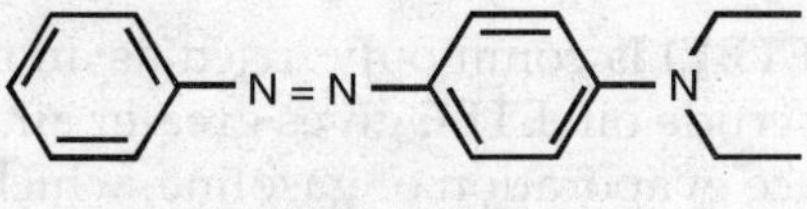

Solvent Yellow 56.

Chemically it is N,N-diethyl-p-(phenylazo)aniline, $C_{16}H_{19}N_3$. Its structure is similar to Solvent Yellow 124, used as a fuel dye, and to Aniline Yellow. It is used to dye hydrocarbon solvents, oils, fats and waxes, notably petrol, mineral oil and shoe polishes, and polystyrene resins. In pyrotechnics, it is used in some yellow colored smokes. This dye can be easily hydrolyzed with acids, splitting off the acetal group responsible for its solubility in nonpolar solvents, and yielding a water-soluble form which is easy to extract to water. Like a similar methyl orange dye, it changes color to red in acidic pH. It can be easily detected in the fuel at levels as low as 0.3 ppm by extraction to a diluted hydrochloric acid, allowing detection of the red diesel added into motor diesel in amounts as low as 2-3%.

Oil Blue 35 : Oil Blue 35, also called Solvent Blue 35, Blue 2N, Blue B, Oil Blue B, 1,4-bis(butylamino) anthraquinone, is a blue dye derived from anthraquinone. It has the appearance of a dark bluish-black powder, soluble in benzene and toluene and insoluble in water, with melting point 104-105 °C. When exposed to 5% hydrochloric acid solution, it becomes dirty green. It is used as fuel and dye for alcoholic, ester, and hydrocarbon based solvents, oils, fats, and waxes. It is used also in lacquers and inks.

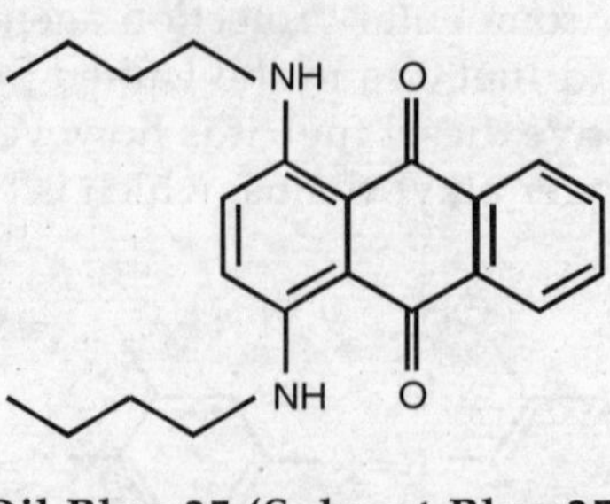

Oil Blue 35 (Solvent Blue 35)

7.4.6. Oxygenate Blending

Oxygenate blending adds oxygen to the fuel in oxygen-bearing compounds such as MTBE, ETBE and ethanol which reduces the amount of carbon monoxide and unburned fuel in the exhaust gas, thereby reducing smog. Oxygenate blending is mandated by EPA regulations to reduce smog and other airborne pollutants. For example, in Southern California, fuel must contain 2% oxygen by weight, resulting in a mixture of 5.6% ethanol in gasoline. The resulting fuel is often known **as reformulated gasoline (RFG) or oxygenated gasoline.**

Ethanol and Ethyl tert-butyl ether (ETBE) : Ethanol blended with gasoline is a common practice. It is been found that blending of ethanol upto 10% is feasible. Blending of ethanol in gasoline in few countries are :

- In the EU, 5% ethanol can be added within the common gasoline spec (EN 228).. Most gasoline (petrol) sold in Sweden has 5% ethanol added.
- In Brazil, the Brazilian National Agency of Petroleum, Natural Gas and Biofuels (ANP) requires that gasoline for automobile use has 23% of ethanol added to its composition.

Ethyl tert-butyl ether (**ETBE**) is commonly used as an oxygenate gasoline additive in the production of gasoline from crude oil. ETBE gives greater air quality benefits as ethanol. Unlike ethanol, ETBE does not induce evaporation of gasoline, which is one of the causes of smog, and does not absorb moisture from the atmosphere.

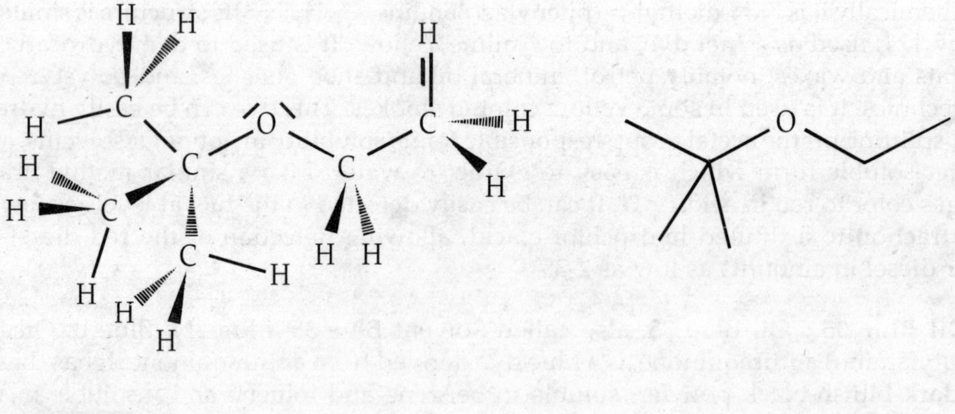

Structure of ETBE

It is synthesized by mixing ethanol and isobutylene and reacting them with heat over a catalyst.

Ethanol, produced by fermentation and distillation, is more expensive than methanol, which is derived from natural gas. Therefore (bio)ethanol makes ETBE partially a biofuel. Especially since ethanol derived from biomatter such as corn, sugar cane or grain is frequent, this will often be referred to as *bio*-ethanol. A common ethanol-gasoline mix of 10% ethanol mixed with gasoline is called gasohol or E10, and an ethanol-gasoline mix of 85% ethanol mixed with gasoline is called E85. The most extensive use of ethanol takes place in Brazil, where the ethanol is derived from sugarcane. E85 is available in much of the United States as much of ethanol is being produced for fuel from corn. The use of bioethanol, either directly or indirectly by conversion of such ethanol to *bio*-ETBE, is encouraged by the European Union Directive on the Promotion of the use of biofuels and other renewable fuels for transport. s

Methyl tert-butyl ether : Methyl *tert*-butyl ether, also known as methyl tertiary butyl ether and MTBE, is a chemical compound with molecular formula $C_5H_{12}O$. MTBE is a volatile, flammable and colorless liquid that is immiscible with water. MTBE is a gasoline additive, used as an oxygenate and to raise the octane number, although its use has declined in response to environmental and health concerns and also due to the alternative ethanol-derived ether ETBE being given more favorable tax treatment. In other parts of the world, the use of MTBE is still a common practice . It has been found to easily pollute large quantities of groundwater when gasoline with MTBE is spilled or leaked at gas stations. MTBE is also used in organic chemistry as a relatively inexpensive solvent with properties comparable to diethyl ether but with a higher boiling point and lower solubility in water. It is also used medically to dissolve gallstones.

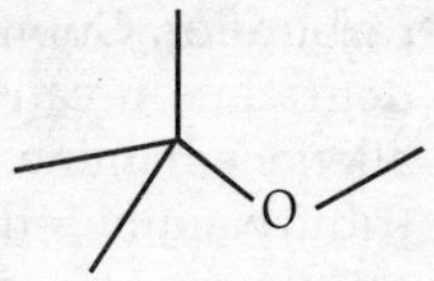

MTBE is manufactured via the chemical reaction of methanol and isobutylene. Methanol is derived from natural gas, and isobutylene is derived from butane obtained from crude oil or natural gas, thus MTBE is a fossil fuel.

It has been used in gasoline to replace tetra-ethyl lead and to increase its octane rating helping prevent engine knocking. Oxygen helps gasoline burn more completely, reducing tailpipe emissions from motor vehicles; dilutes or displaces gasoline components such as aromatics (e.g., benzene) and sulfur; and optimizes the oxidation during combustion. Most refiners chose MTBE over other oxygenates primarily for its blending characteristics and for economic reasons.

Other compounds, which are considered as alternatives are available as oxygenate additives for gasoline including ethanol and related ethers, e.g. tert-amyl methyl ether (TAME). Reasons for using MTBE include economic considerations, as some of the production is obtained by adding methanol to isobutylene produced as a by-product of other processes.

Many of the non-aliphatic hydrocarbons naturally present in gasoline (especially aromatic ones like benzene), as well as many anti-knocking additives, are carcinogenic. Because of this, any large-scale or ongoing leaks of gasoline pose a threat to the public's health and the environment. The chief risks of such leaks come not from vehicles, but from gasoline delivery truck accidents and leaks from storage tanks. Because of this risk, most (underground) storage tanks now have extensive measures in place to detect and prevent any such leaks, such as sacrificial anodes. Gasoline is rather volatile (meaning it readily evaporates), requiring that storage tanks on land and in vehicles be properly sealed. The high volatility also means that it will easily ignite in hot weather conditions, unlike diesel for example. Appropriate venting is needed to ensure the level of pressure is similar on the inside and outside. Gasoline also reacts dangerously with certain common chemicals. Gasoline is also one of the sources of pollutant gases. Even gasoline which does not contain lead or sulfur compounds produces carbon dioxide, nitrogen oxides, and carbon monoxide in the exhaust of the engine which is running on it. Furthermore, unburnt gasoline and evaporation from the tank, when in the atmosphere, react in sunlight to produce photochemical smog. Addition of ethanol increases the volatility of gasoline.

When gasoline is left for a period of time, gums and varnishes may build up and precipitate in the gasoline, causing "**stale fuel**". This will cause gums to build up in the fuel tank, lines, and carburetor or fuel injection components making it harder to start the engine. Motor gasoline may be stored up to 60 days in an approved container. If it is to be stored for a longer period of time, a fuel stabilizer may be used. This will extend the life of the fuel to about 1–2 years, and keep it fresh for the next uses. Fuel stabilizer is commonly used for small engines such as lawnmower and tractor engines to promote quicker and more reliable starting. The gasoline containers and tanks more than half full and properly capped to reduce air exposure, to avoid storage at high temperatures, to run an engine for ten minutes to circulate the stabilizer through all components prior to storage, and to run the engine at intervals to purge stale fuel from the carburetor. Gummy, sticky resin deposits result from oxidative degradation of gasoline. This degradation can be prevented through the use of antioxidants such as phenylenediamines, alkylenediamines (diethylenetriamine, triethylenetetramine, etc), and alkylamines (diethylamine, tributylamine, ethylamine). Other useful additives include gum inhibitors such as N-substituted alkylaminophenols and colour stabilizers such as N-(2-aminoethyl)piperazine, N,N-diethylhydroxylamine, and triethylenetetramine. Improvements in refinery techniques have generally reduced the reliance on the catalytically or thermally cracked stocks most susceptible to oxidation. Gasoline containing acidic contaminants such as naphthenic acids can be addressed with additives including strongly basic organo-amines such as N,N-diethylhydroxylamine, preventing metal corrosion and breakdown of other antioxidant additives due to acidity. Hydrocarbons with a bromine number of 10 or above can be protected with the combination of unhindered or partially hindered phenols and oil soluble strong amine bases such as monoethanolamine, N-(2-aminoethyl)piperazine, cyclohexylamine, 1,3-cyclohexane-bis(methylamine), 2,5-dimethylaniline, 2,6-dimethylaniline, diethylenetriamine and triethylenetetramine.

7.4.7. Synthetic Petrol

The increase in the demand of petrol has created aneed to synthesize petrol from other sources. These include polymerization and liquification of coal.

Polymerization

It aims at the production of superior gasoline and is similar to catalytic cracking. During cracking large amount of olefins and unsaturated compound which includes ethene, propene and butenes.The compounds or gases when subjected to high pressure and high temperaturein presence or absence of catalyst undergo polymerization to form higher hydrocarbons similar to gasoline

$$CH_3\,CH = CH_2 + CH_3\,CH_2\,CH = CH_2 \longrightarrow CH_2 = CH\ CH_2\,CH\,(CH_3)_2 \text{(5-methyl hexene)}$$

Such type of gasoline is known as polymer gasoline and has octane number from 78-83. In the similar way, when cracked gases are passed in 70% sulphuric acid at 10-20°C, two molecules of isobutene polymerize to form di-isobutylene which can be catalytically hydrogenated under pressure over nickel as catalyst at 160 °C to give iso-octane which has octane number of 100

$$\underset{\text{Isobutene}}{CH_3\,C(CH_3){=}\,CH_2} \longrightarrow \underset{\text{Di-isobutylene}}{C(CH_3)_3\,CH_2\,C(CH_3){=}\,CH_2} \longrightarrow \underset{\text{Iso-octane}}{C(CH_3)_3\,CH_2\,CH(CH_3)_2}$$

Polymerization can take place by two process :

- **Thermal Polymerization** carried out by subjecting cracked gases to a temperature of 500-600 °C and a pressue of 70-350 kg/cm^2. The product formed are gasoline and gas oil which are separated by fractional distillation
- **Catalytic Polymerization** is carried out in presence of catalyst such as phosphoric acid, sulphuric acid at the temperature of 150-200 °C

Liquefation of Coal

The process which are commonly used are **Bergius Process and Fischer Tropsch Synthesis** for the synthesis of liquid fuels. The process is already described in Chapter 5.

Conversion of Coal Into Liquid Fuels

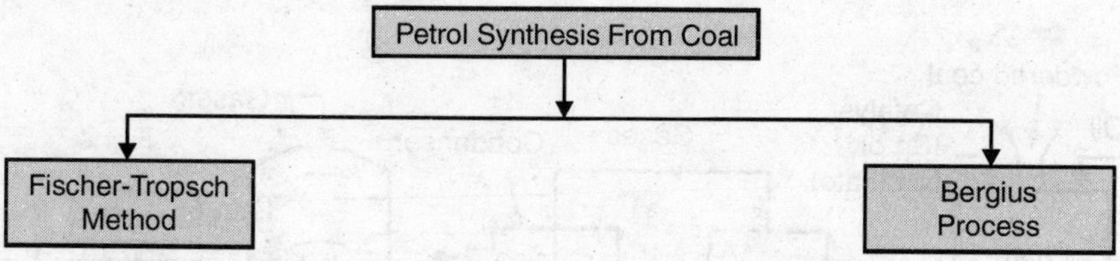

Petrol is synthesized by any one of the following methods :

Fischer-Tropsch Method

The raw material is the hard coke which is converted into water gas (CO + H_2) by passing steam over red hot coke.

$$CO + H_2O \xrightarrow{1200°C} \underset{\text{(Water Gas)}}{CO + H_2}$$

A mixture of hydrogen and water gas (CO + H_2) is first purified by passing through Fe_2O_3 (to remove H_2S) and then into a mixture of Fe_2O_3 + Na_2CO_3 (to remove organic sulphur compounds).

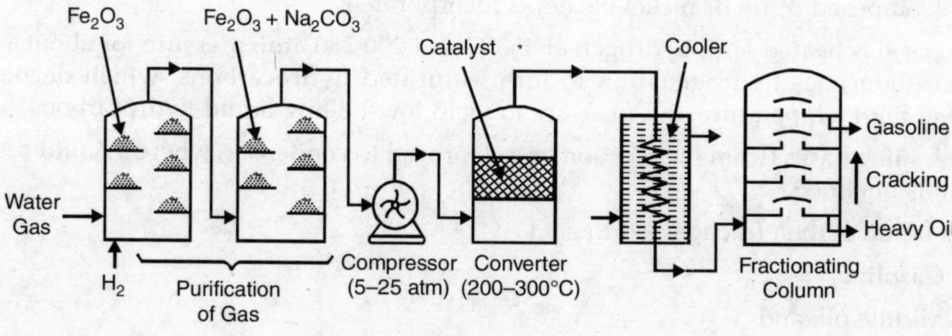

FIG. 7.16. FISCHER-TROPSCH METHOD.

The purified gas is compressed to 5 to 25 atm and then passed through a convertor maintained at about 200–300°C. The convertor is packed with a catalyst, consisting of a mixture of

Cobalt (CO),	Thoria (Th),	Magnesia (MgO)	and Keiselguhr earth
100 parts	5 parts	8 parts	200 parts

A mixture of saturated and unsaturated hydrocarbon results :

$$nCO + 2nH_2 \longrightarrow C_nH_{2n} + nH_2O$$

$$nCO + (2n + 1)H_2 \longrightarrow C_nH_{2n+2} + nH_2O$$

The outcoming hot gaseous mixtures from the exothermic reaction are led to a cooler, where a liquid resembling crude oil is obtained.

The crude oil thus obtained is then fractionated to yield :

(*i*) Gasoline and

(*ii*) High-boiling heavy oil.

The heavy oil can be reused for cracking to get more gasoline.

Bergius Process

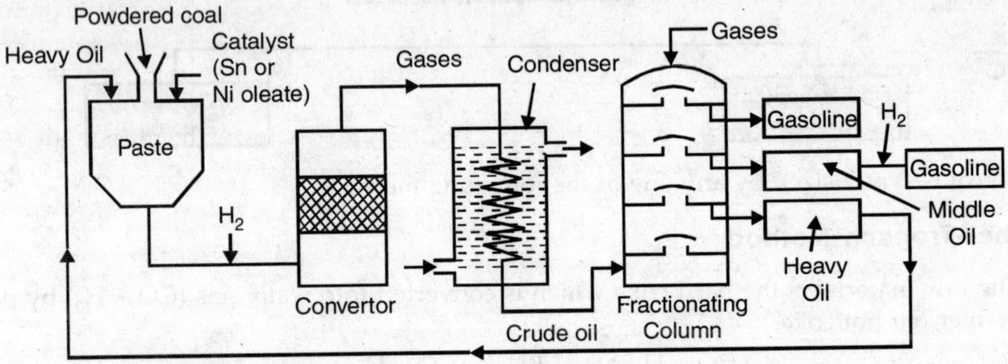

FIG. 7.17. BERGIUS PROCESS OF HYDROGENATION OF COAL TO GASOLINE.

Coal (which is a mixture of high molecular weight, complex organic compounds deficient in hydrogen) is a raw material in Bergius process.

In this process, the low ash coal is powdered well and made into a paste with heavy oil and catalyst (composed of tin or nickel oleate) is incorporated.

The paste is heated with hydrogen at 450°C and 200-250 atm. pressure for about 1.5 hours. The coal undergoes hydrogenation to form saturated hydrocarbons, which decompose at prevailing high temperature and pressure to yield low-boiling liquid hydrocarbons.

The issuing gases (from the reaction vessel) are led to condenser, where a liquid resembling crude oil is obtained.

This liquid is then fractionated to get :

(*i*) Gasoline,

(*ii*) Middle oil, and

(*iii*) Heavy oil

Heavy oil is used again for making paste with fresh coal dust. The middle oil is hydrogenated in vapour-phase in presence of a solid catalyst to yield more gasoline.

The yield of gasoline is about 60% of the coal dust used.

7.5. KEROSENE

Kerosene, sometimes also known as **paraffin**, is a combustible hydrocarbon liquid. The name is derived from Greek *keros*. The word *Kerosene* was registered as a trademark by Abraham Gesner.

- It is usually called **paraffin** (sometimes **paraffin oil**) in the UK, South East Asia and South Africa.
- The term *kerosene* is usual in much of Canada, the United States, Australia and New Zealand.

Kerosene is widely used to power jet-engined aircraft (jet fuel) and some rockets, but is also commonly used as a heating fuel and for fire toys.

Kerosene is a thin, clear liquid formed from hydrocarbons, with density of 0.78-0.81g/cm^3. Kerosene is obtained from the fractional distillation of petroleum between 150 °C and 275 °C, resulting in a mixture of carbon chains that typically contain between 6 and 16 carbon atoms. The flash point of kerosene is between 37 and 65 °C (100–150 °F) and its autoignition temperature is 220 °C (428 °F). Kerosene is insoluble in water (cold or hot), but miscible in petroleum solvents. They are treated mainly for reducing aromatic content to increase their smoke point and hydrofining to reduce sulphur content and to improve odour, colour & burning qualities . The heat of combustion of Kerosene is similar to that of diesel; its lower heating value is around 18,500 Btu/lb, or 43.1 MJ/kg, and its higher heating value is 46.2MJ/kg.

Kerosene is used as heating and lighting, as fuel and as cooking medium.

Heating and Lighting

As a heating fuel, it is often used in portable stoves, and is sold in some filling stations. It is sometimes used as a heat source during power failures. The use of portable kerosene heaters is not recommended for closed indoor areas without a chimney due to the danger of build-up of carbon monoxide gas.

- In India, kerosene is used for lightening of lamp in rural areas.
- Kerosene is widely used in Japan as a home heating fuel for portable and installed kerosene heaters and kerosene can be readily bought at any filling station or be delivered to homes.
- In the United Kingdom and Ireland kerosene is often used as a heating fuel in areas that are unconnected to the gas pipeline network. It is used less for cooking, which has more commonly been LPG for some decades now, owing to its easier lighting.
- The Amish, who forbid use of electric appliances for religious reasons, rely on kerosene for lighting at night.

Transportation

Today kerosene is mainly used in fuel for jet engines (more technically *Avtur*, *Jet A*, *Jet A-1*, *Jet B*, *JP-4*, *JP-5*, *JP-7* or *JP-8*). One form of the fuel known as RP-1 is burned with liquid oxygen as rocket fuel. These fuel grade kerosenes meet specifications for smoke points and freeze points. The combustion reaction can be approximated as follows, with the molecular formula $C_{12}H_{26}$:

$$C_{12}H_{26}(l) + {}^{37}/_{2}\, O_2(g) \longrightarrow 12\, CO_2(g) + 13\, H_2O(g);\ \Delta H° = -7513 \text{ kJ}$$

In the initial phase of liftoff the Saturn V launch vehicle was powered by the reaction of liquid oxygen with RP-1. For the five ~6.4 million newton sea-level thrust F-1 rocket engines of the Saturn V, burning together, the reaction generated roughly 1.62×10^{11} watts (J/s) or 217 million horsepower. Kerosene is sometimes used as an additive in diesel fuel to prevent gelling or waxing in cold temperatures.

Ultra-low sulfur kerosene is a custom-blended fuel used to power its bus fleet.

Cooking

In countries such as India, kerosene is the main fuel used for cooking, especially by the poor. Kerosene stoves have replaced traditional wood-based cooking appliances. The price of kerosene can be a major political and environmental issue; the Indian government subsidizes the fuel to keep the price very low.

The other uses of Kerosene are:

- Liquid pesticides have traditionally used kerosene or some other petroleum distillate as a carrier, though water has recently begun to replace kerosene
- Kerosene has also been found effective in killing bed bugs upon direct spray.
- Kerosene has been used to treat pools of standing water to prevent mosquitoes from breeding, notably in the yellow fever outbreak of 1905 in New Orleans.
- It can be used to remove lice from hair, but this practice is painful and potentially very dangerous. Also, this practice removes all natural oils and fats from the scalp.
- Since kerosene is chemically stable, it is used to store substances with redox tendencies within to prevent unwanted reactions, such as alkali metals.
- It is used in the packaging and storing of white phosphorus to prevent contact with oxygen, which would lead to immediate combustion.
- Kerosene can be used to store crystals. When a hydrated crystal is left in air, dehydration may occur slowly. This makes the colour of the crystal become dull. Kerosene can keep air from the crystal.
- It is used as a solvent.
  - Kerosene can be applied topically to hard-to-remove mucilage or adhesive left by stickers on a glass surface (such as in show windows of stores).
  - Kerosene can be used to remove candle wax that has dripped onto a glass surface; it is recommended that the excess wax be scraped off prior to applying kerosene via a soaked cloth or tissue paper.
  - Kerosene can be used to clean bicycle and motorcycle chains of old lubricant before relubrication.
- It can be used in conjunction with cutting oil as a thread cutting and lubricant. When machining aluminium and its alloys, kerosene on its own is an excellent cutting lubricant.
- Military Applications—used to make "napalm-like" incendiary devices—as reported by the San Diego Union-Tribune.
- Kerosene-based diluent is commonly used as a component of the organic solvent in SX/EW copper refining.
- Hydrotreated kerosene can be used as a starting material to produce high purity linear paraffins which are subsequently dehydrogenated to linear olefins, and when the latter are reacted with benzene in the presence of a catalyst result in the production of linear alkyl benzene.

Kerosene

Normal paraffins in the range of C10 – C13 are extracted from kerosene employing selective adsorption into synthetic zeolites (molecular sieves). This process rely upon the difference in cross-sectional area between the n-paraffins (diameter of cross-section of about 4.9 Å) and the isoparaffins, naphthenes and aromatics (all greater than 5.6 Å). In sieve process the adsorbent is a crystalline alumino silicate which in the sodium/calcium form consists of large internal cavities, access to which is via ports of diameter 5 Å. Thus, only n-paraffins in a straight-run kerosene fraction can enter the internal voids of the sieve, the remaining hydrocarbons not being adsorbed.

The extraction of n-paraffins from kerosene is carries out on a conventionally-desulphurised feedstock. Extraction can occur from the liquid phase Molex process or vapour phase (BP, Linde, Iso Siv, or Texco processes). Desorption of the n-paraffins can be achieved by displacement with a lower n-paraffin (BP, Molex and Texaco process) or by lowering the pressure (Linde, Iso Siv). For several decades solvent extraction of aromatics from kerosine has been practiced in petroleum refining to produce superior kerosene of lower aromatic content and hence of improved quality.

The extracted aromatics are suitable raw materials for the manufacture of aromatic solvents and aromatic process oils. Rigorous extraction or aromatics from kerosene by Edeleanu process results in almost aromatic free raffinates which are used for the production of odourless kerosene (low-odour kerosine). A range of highly aromatic white spirits and kerosine solvents may be manufactured from the kerosine extracts.

Other extraction processes for the separation of aromatics and non-aromatics use solvents with much higher boiling points than that of sulphur dioxide. Such processes are very versatile in that raffinates with very low, as well as extracts with very high aromatics contents can be made. Of the processes available, the Shell sulfolane extraction process is widely used, especially for the extraction of pure aromatics from hydrocarbon mixtures under/or the production of low-aromatic solvents.

7.6. AVIATION TURBINE FUELS

A potential fuel for the gas turbine engines used in aircraft requires high thermal stability, high heat content, low vapour pressure, good combustion, characteristics, good viscosity temperture relationship, high density, high specific heat, uniformity and good handling characteristics. The combustion properties of aviation turbine fuel (ATF) are generally controlled using several of the following five tests : smoke point, luminometer number, aromatics content, aniline point and gravity.

The heat of combustion is normally calculated from the aniline point and gravity. Only two basic types of ATFs, the kerosine type and the wide cut gasoline type, are in general used worldwide. The kerosine type of ATF is a much modified development of the illuminating kerosine orginally used in gas turbine engines, whilst the wide cut gasoline type of ATF is a wider boiling range fuel including some gasoline fraction. In addition, a number of specifalised fuel grades are required for limited military use, as in supersonic aircraft.

The properties that control ATF quality are listed in Table 7.13. Combustion quality relates those properties which directly affect the ability of a fuel to ignite properly. With gas turbine engines, the volatility factor assures delivery of fuel to the turbine nozzle with satisfactory ignition and flame proppagation characteristics. In turbine engines, the fuel should not allow deposits to form in fuel nozzles, fuel control systems, turbine blades, coatings and seals, all of which will degrade performance.

High heat content per unit volume is required in high speed aircraft while weight is more important in low speed aircraft. Naphthenes have high heat content per unit volume and paraffins per unit weight. Low vapour pressure is needed to minimize evaporation losses. A fuel with a dense molecular structure premits more energy to be packed in a given tank space. Fuels capacity as a coolant depends on specific heat and heat of vaporization. Fuels should possess high smoke points, high luminometer number to minimize the temperature rise in combustion zone from flame radiation. Normal paraffins are the best from luminosity and smoke point of view. Specific heat as related to density is the highest for paraffin, lowest for aromatics and intermediate for naphtenes.

TABLE 7.13. Fuel Properties Needed for Acceptable Performance

Fuel Performance Required	Property Controlled
Combustion	Luminosity
	Hydrocarbon composition
	Thermal stability
	Heat of combustion
Handling and storgage	Flash point
	Viscosity
	Contamination (water/surfactant)
	Particulates
	Microbiological growth
Cleanliness during use	Trace metals
	Trace metals
	Distillation
	Sulphur
	Existent gum
	Stability

The larger demand of ATFs cannot be met economically by straight-run ATF alone. Extensive hydrocracking and hydrotreating yield a product of the desired type. Hydrocracking has proven to be a flexible process which can produce additional ATF. Isomax process produces ATF from residual stocks that were previously used as heavy fuel. Arofing is the process to convert aromatics into naphthenes using a highly active sulphur resistant catalyst giving the ATF having high energy content per unit volume, and improved smoke point and luminometer number. This process is suitable for petroleum distillates between 60 – 349°C containing 5-30% aromatics. Previously SO_2 extraction was used to remove aromatics but it reduces the yield of ATF and it is responsible for silver-strip corrosion. Merox process is used to remove mercaptans.

Some fuels having high thermal stability do not retain thermal stability in storage; part of this degradation is caused by trace materials that can catalyze oxidation. Hence to improve stability metal deactivator, anit-oxidant and dispersant are to be added.

7.6.1. Composition of ATFs

Aviation turbine fuels and distillate fractions of crude oil in the boiling range of 150–270°C, and consist predominantly of hydrocarbons. The minor constituents of ATFs consist of sulphur, nitrogen and oxygen compounds and traces of particulate matter and water as contaminants.

Paraffins and naphtheness are the major compounds in ATFs and their ratio varies with crude source. On a weight basis, the heat of combustion is the highest for paraffins, lowest for aromatics and intermediate for naphthenes. However, when heat of combustion is considered on a volume basis, the order is reversed.

Aromatics are present in ATFs principally as monocyclic hydrocarbons with small quantities of two and three ring compounds. These have a more luminous flame than paraffins and cause smokiness and carbon deposition. They cause many elastomers present in fuel hose lines to swell. For these reasons an upper limit to the amount of aromatics is set in specifications which is 20 volume %. The presence of the polynuclear aromatics results is highly luminous flames giving high radiant heat in the combustion chamber. The polynuclear aromatics are good lubricity agents for control valves and fuel pumps in the fuel handling systems.

The hydrotreating of ATFs tends to reduce aromatics. The most critical portion of the fuel system to lubricity is the high pressure fuel pump which is entirely fuel lubricated. This tends to increase the wear on the pump and so reduces its life. A small quantity (10-12 ppm) of an approved corrosion inhibitor added to the fuel will protect the original systems. The most reasonable approach is to design fuel systems, high pressure pumps in particular which can run on low-lubricity fuels.

Olefins are limited to 5 volume % on account of their instability, chemical reactivity and ease with which they form gums. They have good burning qualities and are intermediate between paraffins and naphthenes as regards heat of combustion.

The non hydrocarbons present in ATFs are compounds containing sulphur, nitrogen and oxygen. The amount of sulphur and nitrogen present is controlled by the crude source composition and the processing scheme. The total sulphur content is limited to 0.2 wt.%. Mercaptan is controlled specifcally for its corrosiveness to copper and cadmium and for its disagreeable odour. The specifications limit the amount of mercaptan to less than 0.001%. Mercaptan and/or other complex corrosive compounds are limited by a copper-strip corrosion test. As a further protection against sulphide corrosion on silver parts, present in some fuel pumps, a silver-strip corrosion test has been introduced. Nitrogen compounds occur mainly in the heavy end of the fuel and are detected in gums and sediments. The amount of nitrogen is countrolled by thermal stability tests. Oxygen is controlled by acidity test.

Gums contain a large amount of sulphur.They are, in fact, high molecular weight, non-volatile organic complexes of carbon hydrogen, sulphur, oxygen and sometimes nitrogen. Gum formation can take place in fuels when stored for a long time or under thermal stresses at high temperature in modern high speed aircraft. This latter effect can adversely affect engine performance by blocking burner jets or fouling up heat exchangers.

7.6.2. Properties of ATFs

The proper ATF quality is maintained by limiting and controlling specification characteristics. Indian specifications of ATF are listed in Table 7.14.

Physical properties such as density, viscosity, volatility, auto ignition temperature are of considerable importance with regard to engine and fuel system design. These properties also affect storage and handling facilities.

Density is an important parameter in the design of fuel tanks in aircraft. Fuels of a considerable density range can readily satisfy the essential quality requirement of ATF. However, it is usually necessary to limit the density of a particular type of fuel, e.g. kerosine type to get satisfactory engine control. The specific energy of the fuel on a volume basis increases with increasing density, and it decreases on a weight basis. Hence, the density of the fuel is important in respect of total heat (energy) content. Depending on route structures traffic density and a balance between payload and fuel weight, one can decide the aircraft to be weight limited or volume limited. ATF specifications set minimum values for the net (or lower) calorific value per unit weight. For maximum energy, therefore, in a volume-limited situation, a high density fuel is required.

Viscosity is specified to maintain fuel line pressure losses at low-temperature operation at acceptable levels and to ensure the fuel injection nozzles and fuel system controls at or near to design conditions. For ATFs (kerosine type). Indian specification requires a maximum viscosity of 15 cSt at –34.4°C and 6 cSt at –17.8°C.

TABLE 7.14. Requirements for Aviation Turbine Fuels-Kerosine Type K-50 Grade (IS : 1571 -1976)

Sl. No.	Characteristics	Requirement	Test method IS : 1488
1.	Appearance	Clear bright and free from solid matter and undissolved water at ambient temperature	
2.	Aromatics, vol.%, Max	20	P : 23
3.	Olefins, vol.%, Max	5	P : 23 or P : 48
4.	Net calorific value, cal/g, Min.	10225	P : 6
5.	Product of API gravity and aniline point, Min.	5250	P : 3
6.	Copper content, µg/kg, Max	150	—
7.	Copper-strip corrostion for 2 h at 100°C	Not worse than No. 1	P : 15
8.	Silver-strip corrosion at 45±1°C for 16 h	Not exceeding (0)	IP : 227
9.	Distillation :		P : 18
	(a) Initial boiling point, °C	Not limited but to be reported	
	(b) Recovery up to 200°C, vol. % Min.	20	
	(c) Final boiling point, °C, Max.	288	
	(d) Residue, vol. % Max.	1.5	
	(e) Loss, vol.%, Max.	1.5	
10.	Electrical conductivity, pS/M		
	Min.	50	
	Max.	300	
11.	Existent gum, mg/100 ml, Max	7	P : 29 (Steam jet)
12.	Oxidation stability (16 h ageing) :		P : 28 (Method B)
	(a) Total gum content, mg/100 ml, Max	I4	
	(b) Precipitate mg/100 ml, Max.	To be reported separately	
13.	Flash point, (Abel), °C	38	P : 20 (Method B)
14.	Freezing point, °C, Max	–50	P : 11
15.	Smoke point, mm Min.	20	P : 31
16.	Specific gravity at 15.6/15.6°C		P : 16
	Min.	0.775	
	Max.	0.830	

Sl. No.	Characteristics	Requirement	Test method IS : 1488
17.	Sulphur, mercaption, wt. %, Max.	0.001	P : 36 (Method B)
18.	Doctor test, Max	Negative	P : 19
19.	Total sulphur, wt. %, Max	0.20	P : 34
20.	Thermal stability : (JFTOT)		
	(*a*) Filter pressure (differential), mm Hg, Max.	25	D3241/323
	(*b*) Tube rating visual	Less than 3 : No 'Peacock' or 'abnormal' colour deposits	DERD issue 7 amendment No. 3.
	or		
	(*c*) TDR, SPUN rating (Note 10), Max.	17. No. 'Peacok' or 'Abnormal' colour deposits	
21.	Total acidity, mg KOH/g, Max.	0.012	
22.	Kinematic viscosity, cSt, Max		P : 25
	(*a*) at – 17.8°C	6	
	(*b*) at – 34.4°C	15	
24.	Water reaction :		P : 42 (Method A)
	(*a*) Interface rating, Max	1 (b)	
	(*b*) Separation	Sharp separation, no emulsion or precipitate within or upon either layer.	
25.	Water separometer index modified (WSIM), Min.	85*	D2550-74(ASTM/IP)
26.	Anti-oxidants, mg/100 ml, Min.	8.6	
	Max.	24.0	

*If anti-static additive is present in the fuel, the water separometer index, modified shall be 70 minimum.

The volatility of ATFs is controlled to minimize the loss of fuel from upressurised aircraft tanks due to boiling-off at high altitudes. For this reason the low volatility of kerosine type ATF is desirable. Distillation specification requires at least 20 vol.% recovery up to 200°C. The 20 percent point controls front-end volatility, and ensures that the fuel will readily vaporize for cold starting and limit evaporation losses at high altitudes.

Auto-ignition of ATF can happen with a slight leak of fuel from the engine shortly after shutdown, or from a fuel line facture during operation. The temperature for auto-ignition increases with increase in vapour pressure. As the atmospheric pressure is reduced the vapour pressure of the fuel increases, and so the auto-ignition temperature increases.

Thermal Properties : Jet engines require the use of fuels which have low freezing points and maintain their flow properties. A high degree of thermal stability of fuels is required to avoid the formation of deposits and lacquers in the heat exchangers and in the fuel tanks. The radiant heat emitted in the combustion chamber of the engine of the engine and its effect on the life of the hot parts of the engine is also important. This is controlled by the luminometer number and partially by smoke point.

Aircraft fly at high altitudes with very low outside air temperature. This necessitates good low temperature properties of the fuels. The specification of freezing point limit for ATF is based on the ability of the fuel to be pumped, and to flow effectively through the fuel handling system. At low temperature the formation of wax crystals and ice crystals may clog and block filters preventing fuel flow. It is a safe practice to use a fuel that has a freezing point below filters preventing fuel flow. It is a safe practice to use a fuel that has a freezing point below its expected operating temperature. ATFs have a freezing point specification of –50°C maximum for kerosine type and –60°C maximum for a wide cut gasoline type. The freezing point parameter affects the yield of ATF from a particular crude oil because it controls the final boiling point which decreases with decrease in freezing point of ATF.

Fuels are now used as a coolant for engine oil, hydraulic and electronic equipments. On high speed military aircraft and on supersonic transports the fuel is also used as a heat sink for the air-conditioning systems and in the control of cabin air temperature. The heat absorption at supersonic speeds results in considerable kinetic heating also. Hence, the fuel should not oxidise and form gums, varnishes and deposits which will affect the efficiency of heat exchangers and also be liable to clog fuel filters, fuel meters, or fuel injection nozzles into the combustion chambers.

The thermal stability of an ATF is controlled by the Jet Fuel Thermal Oxidation Tester (JFTOT). The use of an anti-oxidant and a metal deactivator improves thermal stability.

Combustion in a jet engine converts the chemical energy of the fuel into useful work. The quality of combustion depends on the carbon/hydrogen ratio of the fuel (fixed by crude source and type of processing) and the flame speed. The flame speed is also a function of fuel composition, as well as temperature and oxygen concentration. For ideal burning conditions the lowest possible C/H ratio is required. The aromatic compound present in the fuel are the most difficult to burn and can result in the production of carbon, which forms smoke and hence does not release all of its available energy. They also emit a high radiant heat from a luminous flame, For this reason ATF specification contain a maximum limit on aromatics content. A specification of smoke point is a direct control on combustion.

The emissivity of the flame is measured to control the radiant heat in the primary zone of the combustion chamber. The apparatus for this test is a luminometer and the fuel under test is rated on an arbitrary scale known as the luminometer number of the fuel. A high number denotes a fuel of good combustion quality. In Indian specification a smoke point of 20 mm (minimum) and a total aromatics limit of 20 percent by volume give adequate control.

For the wide-cut gasoline type of ATF, the best correlation with combustion deposits is derived from a function of smoke point and fuel volatility known as the smoke volatility index (SVI). It is defined as

$$\text{SVI} = \text{Smoke Point} + 0.2\ (\text{percent of fuel boiling below } 204.4°\text{C})$$

The fuel related problems associated with gas turbine engine operation are listed in Table 7.15. As is shown, the control of combustion characteristics minimizes carbon deposition in burner liners and on stationary turbine hardware, and also minimizes increase in burner liner temperatures caused by incandescent carbon particle radiation. Luminious flames radiate more heat to these metal parts than do non-luminous flames, resulting in higher metal temperatures and reduction in component life. Relative to contamination, gas turbines operate best on clean fuel because of their sensitivity to particulate debris, water, and metallic fuel contaminants. Since these continuous combustion engine system operate at higher average metal temperature and pressures than in other types of power plants, they are more prone to corrosion and erosion problems. The type of contaminants that affect engine performance are trace metals (i.e. sodium, calcium, potassium, vanadium, lead and copper), residual fuel contamination, water suspended dirt/debris, other fuels, surfactants, microorganisms and refinery treating solution carry-overs.

7.6.3. ATF Additives

Various ATF additives and their functions are listed in Table 7.16. These are described below.

Oxidation inhibitors (anti-oxidants) are used in ATFs to prevent gum formation and to retard peroxide formation. The presence of peroxides in the fuel will attack nitrile rubber hoses used in the fuel system of aircraft engines.

As a result of refinery processing copper may be present in ATFs. Copper is undesirable in the fuel as it catalyses oxidation reactions resulting in gum formation. A metal deactivator is added in the fuel to remove any dissolved copper by chelating.

TABLE 7.15. Fuel Effects on Turbine Performance

Performance Problem	Probable Fuel Related Causes
Poor combustion	Low luminometer number Low smoke point High aromatics content High carbon residue values Heavier fuel contamination
Excess linear/blade deposits	High fuel viscosity Low hydrogen content High sulphur/hetero-atom content High aromatics content Soluble metal contaminants
Nozzle plugging/wear	High particulate contamination Soluble metal contaminants Heavy end impurities Marginal thermal stability High sulphur content
Fuel control system malfunction	High sulphur/hetero atom content Heavy ends contamination Thermally reactive hydrocarbons Low fuel viscosity Marginal lubricity
Filter plugging	Water contamination Surfactant contamination Microbiological growth Fuel impurities Improper freezing point Thermally reactive hydrocarbons

TABLE 7.16. ATF Additives

Additive	Functions
Oxidation inhibiters	To prvent gum formation and to retard peroxide formation in certain hydrotreated fuels.
Metal deactivators	To remove by chelation any dissolved copper to prevent catalysis of oxidation reactions.
Anti-icing additives	To avoid the freezing of dissolved water in the fuel system.
Static dissipator additives	To increase the conductivity of the fuel and so reduce the possibility of an electrostatic discharge causing explosion during high speed refuelling through a micronic filter.
Corrosion inhibitors	Used essentially for distribution pipeline. Exhibit beneficial properties with respect to fuel lubricity.
Anti-smoke additives	To reduce the amount of smoke emitted from jet engines on take-off.
Biocides/biostats	To inhibit microbial and fungoidal growth in aircraft fuel systems.

At low temperatures clogging of filters can occur in jet aircraft. The effective additives for this purpose belong to two classes. (*i*) Those which reduce the freezing point of water and (*ii*) Surfactants. Alcohols such as ethanol or isoproanol are useful in 0.5-2 vol.%. Compounds such as dihyric alcohols like hexylene glycol and dipropylene glycol are effective at 0.05-0.2 vol.% Surfactants are effective at 0.002-0.02 vol.% and are cheaper.

Aviation turbine fuels can develop considerable electric charges in refinery storage tanks, airfield fuel loading vehicles and during mechanised filling operations. Usually a spark can form due to static discharge and fuel ignition can occur if the spark energy exceeds 0.2 mJ. The conductivty of ATFs is the range of 0.02-40 picomho/m. The use of static dissipator eadditives can raise this to 1000 picomho/m allowing rapid dissipation of charges.

Rust contamination of fuel is possible when water enters the distribution system either by condensation or accidentally. Fuel soluble corrosion inhibitors have been used in ATF. Their use is mandatory in military specifications. Most popular types are amine salts of high molecular weight carboxylic, sulphonic or phosphoric acids. Some commercial types are alkyl aminophosphate, ethylene diamine dinonyl naphthalene, dimerized linoleic acid.

Anti-smoke additives can be used to reduce the amount of smoke emitted from jet engines on take-off.

Fuel soluble biocides are used to inhibit microbial and fungoidal growth in aircraft fuel systems. These are used periodically once every three to four months as a shock threatment. Another effective but costly approach is to add a biostat to the fuel on a continual basis.

7.6.4. Storage and Handling Problems

Particulate matter and free water are undesirable contaminants in ATFs. During transportation the fuel is liable to pick up dirt, finely-divided rust and/or free water. Filter separators are used in the distribution system to ensure the delivery of clean fuel into aircraft. These remove solid particulate matter by filtration and separate out free water by a coalescing action. These are designed to reduce free water to acceptable levels (30 ppm) and to reduce the dirt level to less than 1 mg/l fuel. The most generally accepted system of measuring particulate contamination is supplied by the Millpore Filter Corporation.

A typical aviation kerosene has a conductivity of about 1 picosiemen/merter (pS/m), known as conductivity unit (CU). The conductivity of a fuel in a tank may be measured directly by portable meters manufactured by Maihak or Emcee Electronics Inc. For comparison, demineralised water has a conductivity of 107 CU. Thus ATF has a very low conductivity. The conductivity of the fuel affects the rate at which the charge in the fuel to fall from the fuel, that is the charge relaxation. The time taken for the charge in the fuel to fall from a given value to half that value (the fuel's half value time = $t_{1/2}$ = 12/conductivity) (CU), sec) gives an indication of whether static charges are likely to be hazardous or not. Charges can accumulate to an appreciable extent only if the relaxation time and the half-value time are high.

The generation of static electrical charges in handling of ATFs has long been recognized as a potential safety hazard. A number of rules covering pumping rates, line velocities, bottom loading (no splash filling) and settling times have been generally adopted by the major oil companies as standard practices for safe handling of these products. As an additional safety precaution a static dissipator additive may be added to ATFs. The additive permitted for use in DERD 2494 specification is called ASA-3 and its addition rate to the fuel is 1 ppm (maximum). The amount of additive required depends upon the crude source from which the fuel was produced, since the additive response is affected by the variation in the chemical composition of the fuel. It is usual to dilute a static dissipator additive with fuel before making an addition to bulk fuel. This is done to ensure adequate dispersion of the additive in the fuel.

Static dissipator additive (at the concentration used) have no effect on the thermal stability or oxidation stability of the fuel or its combustion quality. The presence of surfactant in the additive may reduce the water separometer index (modified (WSIM) by about 5 points.

7.7. FUEL OIL

Fuel oil is a fraction obtained from petroleum distillation, either as a distillate or a residue. It is said that fuel oil is any liquid petroleum product that is burned in a furnace or boiler for the generation of heat or used in an engine for the generation of power, except oils having a flash point of approximately 40 °C (104 °F) and oils burned in cotton or wool-wick burners. In this sense, diesel is a type of fuel oil. Fuel oil is made of long hydrocarbon chains, particularly alkanes, cycloalkanes and aromatics. The term *fuel oil* is also used in a stricter sense to refer only to the heaviest commercial fuel that can be obtained from crude oil, heavier than gasoline and naphtha.

Fuel oil is classified into six classes, numbered 1 through 6, according to its boiling point, composition and purpose. The boiling point, ranging from 175 to 600 °C, and carbon chain length, 20 to 70 atoms, of the fuel increases with fuel oil number. Viscosity also increases with number, and the heaviest oil has to be heated to get it to flow.

No. 1 fuel oil, No. 2 fuel oil and **No. 3 fuel oil** are variously referred to as **distillate fuel oils, diesel fuel oils, light fuel oils, gasoil** or just **distillate**. For example, No. 2 fuel oil, No. 2 distillate and No. 2 diesel fuel oil are almost the same thing. Distillate fuel oils are distilled from crude oil.

Gas oil refers to the process of distillation. The oil is heated, becomes a gas and then condenses.

- No. 1 is similar to kerosene and is the fraction that boils off right after gasoline.
- No. 2 is the diesel that trucks and some cars run on, leading to the name "road diesel". It is the same thing as heating oil
- No. 3 is a distillate fuel oil and is rarely used.
- **No. 4 fuel oil** is usually a blend of distillate and residual fuel oils, such as No. 2 and 6; however, sometimes it is just a heavy distillate. No. 4 may be classified as diesel, distillate or residual fuel oil.
- **No. 5 fuel oil** and **No. 6 fuel oil** are called **residual fuel oils** (**RFO**) or **heavy fuel oils**. Residual fuel oils are sometimes called **light** when they have been mixed with distillate fuel oil, while distillate fuel oils are called **heavy** when they have been mixed with

residual fuel oil. **Heavy gas oil**, for example, is a distillate that contains residual fuel oil. The ready availability of very heavy grades of fuel oil is often due to the success of catalytic cracking of fuel to release more valuable fractions and leave heavy residue.

Bunker Fuel

Bunker fuel is technically any type of fuel oil used aboard ships. It gets its name from the containers on ships and in ports that it is stored in; when steam engine was used they were coal bunkers but now they are bunker-fuel tanks. Bunker A is No. 2 fuel oil, bunker B is No. 4 or No. 5 and bunker C is No. 6. Since No. 6 is the most common, "bunker fuel" is often used as a synonym for No. 6. No. 5 fuel oil is also called **navy special fuel oil.** No. 6 or 5 are also called **furnace fuel oil (FFO)**; the high viscosity of the oil requires heating, usually by a recirculated low pressure steam system, before the oil can be pumped from a bunker tank. Below is the table of fuel oils (Table 7.17)

TABLE 7.17. Types of Fuel Oil.

Table of fuel oils				
Name			Type	Chain Length
No. 1 fuel oil	No. 1 distillate	No. 1 diesel fuel	Distillate	9-16
No. 2 fuel oil	No. 2 distillate	No. 2 diesel fuel	Distillate	10-20
No. 3 fuel oil	No. 3 distillate	No. 3 diesel fuel	Distillate	
No. 4 fuel oil	No. 4 distillate	No. 4 residual fuel oil	Distillate/Residual	12-70
No. 5 fuel oil	No. 5 residual fuel oil	Heavy fuel oil	Residual	12-70
No. 6 fuel oil	No. 6 residual fuel oil	Heavy fuel oil	Residual	20-70

The chief drawback to residual fuel oil is its high initial viscosity, particularly in the case of No. 6 oil, which requires a correctly engineered system for storage, pumping, and burning. Though it is still usually lighter than water (with a specific gravity usually ranging from 0.95 to 1.03 it is much heavier and more viscous than No. 2 oil, kerosene, or gasoline. No. 6 oil must, in fact, be stored at around 100 °F (37.8 °C) heated to 150° (65.6 °C)–250 °F (121.1 °C) before it can be easily pumped, and in cooler temperatures it can congeal into a tarry semisolid. The flash point of most blends of No. 6 oil is, incidentally, about 150 °F (65.6°C). Attempting to pump high-viscosity oil at low temperatures was a frequent cause of damage to fuel lines, furnaces, and related equipment which were often designed with lighter fuels in mind.

Residual fuel oil is less useful because it is so viscous that it has to be heated with a special heating system before use and it contains relatively high amounts of pollutants, particularly sulfur, which forms sulfur dioxide upon combustion. However, its undesirable properties make it very cheap. In fact, it is the cheapest liquid fuel available. Since it requires heating before use, residual fuel oil cannot be used in road vehicles, boats or small ships, as the heating equipment takes up valuable space and makes the vehicle heavier. Heating the oil is also a delicate procedure, which is inappropriate to do on small, fast moving vehicles. However, power plants and large ships are able to use residual fuel oil. The high sulfur content of No. 6 oil—up to 3% by weight in some extreme cases—had a corrosive effect on many heating systems (which were usually designed without adequate corrosion protection in mind), shortening their lifespans and increasing the polluting effects. This was particularly the case in furnaces that were regularly shut down and allowed to go cold; the internal condensation produced sulfuric acid.

Environmental cleanups at such facilities are frequently complicated by the use of asbestos insulation on the fuel feed lines. No. 6 oil is very persistent, and does not degrade rapidly. Its viscosity and stickiness also make remediation of underground contamination very difficult, since it reduces the effectiveness of methods such as air-stripping.

When released into water, such as a river or ocean, residual oil tends to break up into patches or tarballs— mixtures of oil and particulate matter such as silt and floating organic matter- rather than form a single slick. An average of about 5-10% of the material will evaporate within hours of the release, primarily the lighter hydrocarbon fractions. The remainder will then often sink to the bottom of the water column.

Marine Fuel Oil : In marine system another type of classification is used for fuel oils:

- **MGO (Marine gas oil)** - roughly equivalent to No. 2 fuel oil, made from distillate only
- **MDO (Marine diesel oil)** - A blend of gas oil and heavy fuel oil
- **IFO (Intermediate fuel oil)** A blend of gas oil and heavy fuel oil, with less gasoil than marine diesel oil.
- **MFO (Medium fuel oil)** - A blend of gas oil and heavy fuel oil, with less gas oil than intermediate fuel oil.
- **HFO (Heavy fuel oil)** - Pure or nearly pure residual oil, roughly equivalent to No. 6 fuel oil.

Marine diesel oil contains some heavy fuel oil, unlike regular diesels. Also, marine fuel oils sometimes contain waste products such as used motor oil.

Marine fuels were traditionally classified after their kinematic viscosity. This is a mostly valid criteria for the quality of the oil as long as the oil is made only from atmospheric distillation. Today, almost all marine fuels are based on fractions from other more advanced refinery processes and the viscosity itself says little about the quality as fuel. CCAI and CII are two indices which describe the ignition quality of residual fuel oil, and CCAI is especially often calculated for marine fuels.

(The **calculated carbon aromaticity index** (CCAI) is an index of the ignition quality of residual fuel oil.

The running of all internal combustion engines is dependent on the ignition quality of the fuel. For spark-ignition engines the fuel has an octane rating. For diesel engines it depends on the type of fuel, for distillate fuels the cetane numbers are used. Cetane numbers are tested using a special test engine and the existing engine was not made for residual fuels. For residual fuel oil two other empirical indexes are used: CCAI and Calculated Ignition Index (CII). Both CCAI and CII are calculated from the density and kinematic viscosity of the fuel.

Formula for CCAI :

$$\text{CCIA} = D - 140.7 \log (\log(V + 0.85)) - 80.6 - 210 \ln \left(\frac{t + 273}{323}\right)$$

which is equivalent to

$$\text{CCIA} = D - 140.7 \log(\log(V + 0.85)) - 80.6 - 483.5 \log\left(\frac{t + 273}{323}\right)$$

where: D = Density at 15°C (kg/m^3)
V = Viscosity (cST)
t = viscosity temperature (°C)

This will normally give a value somewhere between 800 and 880. The lower the value is the better the ignition quality. Fuels with a CCAI higher than 880 are often problematic or even unusable in a diesel engine.

The **Calculated Ignition Index** (CII) is an index of the ignition quality of residual fuel oil.

Formula for CII:

$$\text{CII} = (270.795 + 0.1038T) - 0.254565D + 23.708 \log\log(V + 0.7)$$

Where: D = Density at 15°C (kg/m^3)
V = Viscosity (cSt)
T = Viscosity temperature (°C)

CII was designed to give out numbers in the same order as the cetane index for distillate fuels)

Despite this marine fuels are still quoted on the international bunker markets with their maximum viscosity (which is set by the ISO 8217 standard) due to the fact that marine engines are designed to use different viscosities of fuel. The unit of viscosity used is the **Centistoke** and the fuels most frequently quoted are listed below in order of cost, the least expensive first-

- **IFO 380** - Intermediate fuel oil with a maximum viscosity of 380 Centistokes
- **IFO 180** - Intermediate fuel oil with a maximum viscosity of 180 Centistokes
- **LS 380** - Low-sulphur (<1.5%) intermediate fuel oil with a maximum viscosity of 380 Centistokes
- **LS 180** - Low-sulphur (<1.5%) intermediate fuel oil with a maximum viscosity of 180 Centistokes
- **MDO** - Marine diesel oil.
- **MGO** - Marine gasoil.

The density is also an important parameter for fuel oils since marine fuels are purified before use to remove water and dirt from the oil. Since the purifiers use centrifugal force, the oil must have a density which is sufficiently different from water. Older purifiers had a maximum of 991 kg/m^3; with modern purifiers it is also possible to purify oil with a density of 1010 kg/m^3.

The first British standard for fuel oil came in 1982. The latest standard is ISO 8217 from 2005. The ISO standard describe four qualities of distillate fuels and 10 qualities of residual fuels. Over the years the standards have become stricter on environmentally important parameters such as sulfur content. The latest standard also banned the adding of used lubricating oil (ULO).

Some parameters of marine fuel oils according to ISO 8217 as shown below in Table 7.18

TABLE 7.18. Parameters of Marine Fuel Oils.

Marine Distillate Fuels (DM)						
Parameter	**Unit**	**Limit**	**DMX**	**DMA**	**DMB**	**DMC**
Density at 15°C	kg/m^3	Max	-	890.0	900.0	920.0
Viscosity at 40°C	mm^2/s	Max	5.5	6.0	11.0	14.0
Viscosity at 40°C	mm^2/s	Min	1.4	1.5	-	-
Water	% V/V	Max	-	-	0.3	0.3
Sulfur[1]	% (m/m)	Max	1.0	1.5	2.0	2.0
Aluminium + Silicon[2]	mg/kg	Max	-	-	-	25
Flash point[3]	°C	Min	43	60	60	60
Pour point, Summer	°C	Max	-	0	6	6
Pour point, Winter	°C	Max	-	-6	0	0
Cloud point	°C	Max	-16	-	-	-
Calculated Cetane Index		Min	45	40	35	-

1. Max sulfur content is 1.5% in designated areas.
2. The aluminium+silicon value is used to check for remains of the catalyst after catalytic cracking. Most catalysts contains aluminium or silicon and remains of catalyst can cause damage to the engine.
3. The flash point of all fuels used in the engine room should be at least 60°C (DMX is used for things like emergency generators and not normally used in the engine room).

The parameters for Marine residual fuels are summarized below in Table 7.19.

Fuel oil is transported worldwide by fleets of supertankers making deliveries to suitably sized strategic ports such as Houston, Singapore, and Rotterdam. Where a convenient seaport does not exist, inland transport may be achieved with the use of barges. The lighter fuel oils can also be transported through pipelines.

7.8. DIESEL FUELS

High Speed Diesel Oil (HSD) and Light Diesel Oil (LDO) are produced in our country. HSD is widely used in diesel engines for automotive purposes, as in buses, lorries, diesel locomotives. Stationary and marine diesel engines as installed in ships or used for electricity generation consume LDO. Compared to the rest of world, India's demand for diesel fuelf is roughly six times that of the gageline. Since straight-run diesel fraction for a given crude oil is fixed, varying amounts of selected cracked distillates from conversion processes—such as fluid catalytic cracking, hydroeracking, coking are usedto increase the volume available for meeting the growing demand of diesel fuels.

The primary quality factors of concern for compression-ignition engines are combustion, volatility and cleanliness (of both the engines and their emission). The properties of diesel fuels that control product quality are listed in Table 7.20 Combustion quality of diesel fuels relates

TABLE 7.19. Parameters of Marine Residual Fuels.

			Marine Residual Fuels (RM)									
Parameter	*Unit*	*Limit*	*RMA 30*	*RMB 30*	*RMD 80*	*RME 180*	*RMF 180*	*RMG 380*	*RMH 380*	*RMK 380*	*RMH 700*	*RMK 700*
Density at 15°C	kg/m^3	Max	960.0	975.0	980.0	991.0	991.0	991.0	991.0	1010.0	991.0	1010.0
Viscosity at 50°C	mm^2/s	Max	30.0	30.0	80.0	180.0	180.0	380.0	380.0	380.0	700.0	700.0
Water	% V/V	Max	0.5	0.5	0.5	0.5	0.5	0.5	0.5	0.5	0.5	0.5
Sulfur[1]	% (m/m)	Max	3.5	3.5	4.0	4.5	4.5	4.5	4.5	4.5	4.5	4.5
Aluminium + Silicon[2]	mg/kg	Max	80	80	80	80	80	80	80	80	80	80
Flash point[3]	°C	Min	60	60	60	60	60	60	60	60	60	60
Pour point, Summer	°C	Max	6	24	30	30	30	30	30	30	30	30
Pour point, Winter	°C	Max	0	24	30	30	30	30	30	30	30	30

1. Max sulfur content is 1.5% in designated areas.
2. The aluminium+silicon value is used to check for remains of the catalyst after catalytic cracking. Most catalysts contains aluminium or silicon and remains of catalyst can cause damage to the engine.
3. The flash point of all fuels used in the engine room should be at least 60°C.

those properties which directly affect the ability of a fuel to ignite properly. Diesel fuels must be suitable for handling by the injection equipment. The handling and storage characteristic is a function of volatility, fluidity, contamination during refinement and product shipment or movement. The characteristic of cleanliness during use is related to composition, trace contamination, and the degree of inherent stability of the particular fuel.

Diesel fuels for automotive use need good ignition quality to ensure easy starting and *smooth* running. Sulphur content must be critically controlled to minimize environmental pollution, corrosion, wear and tear.

TABLE 7.20. Fuel Properties Needed for Acceptable Performance.

Fuel performance required	Property controlled for diesel fuels
Combustion	Cetane number Distillation range Gravity Heat of combustion
Handling and storage	Flashpoint Viscosity Contamination (Water/sediment) Copper corrosion Cloud and pour joint
Cleanliness during use	Carbon residue Ash Sulphur Stability

7.8.1. Compression-Ignition (Diesel) Engine

The compression-ignition (diesel) engine is a reciprocating internal combustion engine. In dicsel engine the air intake to the engine cylinder is unthrottled and not premixed with fuel. Ignition is achieved spontaneously without the aid of a sparking plug. Air induced into the cylinder at atmospheric pressure is compressed to a volume ratio somewhere near to 1:16, determined based on the engine speed and application and the fuel expected to be used. At the end of compression fuel is injected into the cylinder.

Combustion performance of the diesel engine is a function of the compression ratio, fuel injection timing, rapid homogeneous mixing of fuel and air and resulting ignition delay time. The fuel combustion process is divided into separate stages. In the first stage, air is compressed to the required volume. This increases the pressure many times and the temperature to 500°C or more. As fuel is injected into this heated and turbulent air for the second stage, mixing takes place and some precombustion oxidation of the fuel occurs. Combustion begins. Finally, continuous burning of the incoming fuel from the injector proceeds as the third stage. The sooner ignition takes place after injection starts, the smoother and quieter will the engine operate. The quality of the fuel is a major factor influencing this delay.

Relative to combustion Fig. 7.18 is a diagram is a time pressure plot for combustion occurring in a compression-ignition engine. The time interval between injection and the onset of the pressure rise is the ignition delay period. Distillate fuels having a high cetane number will show a short ignition delay period while those fuels with poor cetane numbers will give long

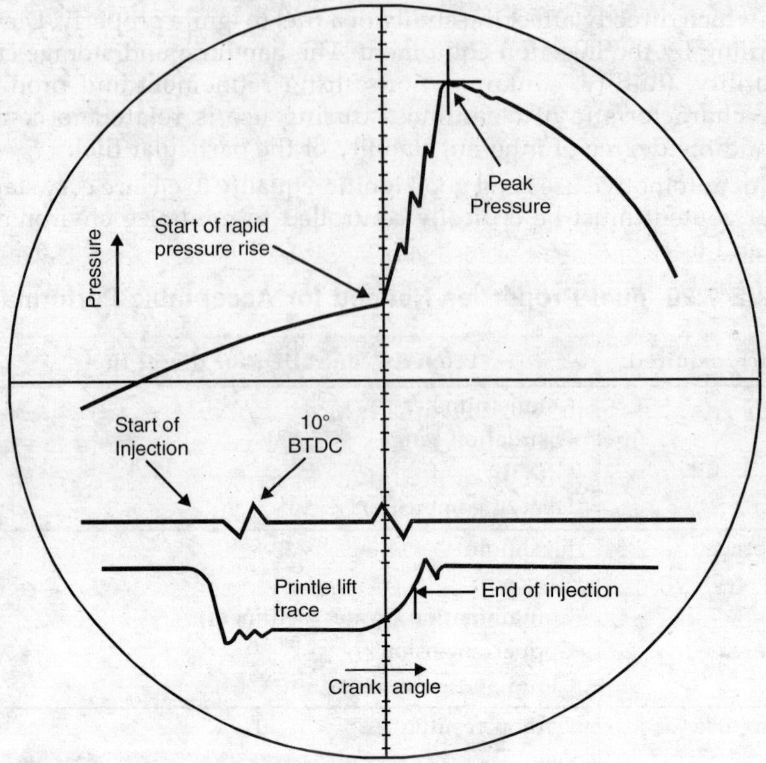

FIG. 7.18. TYPICAL OSCILLOSCOPE TRACES FOR COMPRESSION IGNITION.

ignition delay periods. Hydrocarbon composition directly affects the fuel's physical combustion characteristics. The proper consistent combustion is influenced by physical fuel characteristics, such as viscosity, API gravity, initial boiling point, 50% distillation temperature and final boiling point. A rapid pressure rise results from the accumulation of fuel in the combustion chamber during the ignition delay period and the large number of ignition points occurring throughout the fuel/air mixture. With a long ignition delay time, the rapid pressure rise causes uncontrolled inefficient engine smoking. It is this rapid pressure rise that can cause undesirable audible knocks, high stresses and severe engine vibration. To avoid these unwanted, minute explosions causing knock, the pressure rise must be kept even and consistent. Thus, it is desirable to minimize ignition delay (measured by high cetane number). The fuel injection time can be reduced mechanically by selecting a proper nozzle spray pattern, injection pressure and a combustion chamber designed to create better fuel/air turbulence. A period of controlled burning results from even combustion with continual temperature rises throughout the fuel's distillation range. For a given diesel engine, a fuel of higher cetane number causes a shorter ignition delay period. Consequently, higher cetane fuels, generally cause lower pressure rise rates and lower peak pressures. Both tend to lessen combustion noise and allow improved combustion control, resulting in increased engine efficiency and power output. In addition to these advantages, higher cetane fuels tend to result in easier starting, particularly in cold weather and faster engine warm-up. Reduced exhaust smoke and odour are also associated with higher cetane number fuels.

7.8.2. Composition of Diesel Fuels

It is a general property of hydrocarbons that the auto-ignition temperature is higher for more volatile hydrocarbons. Therefore, the less volatile middle distillate fractions of crude oil boiling in the range of 150-400°C are suitable as diesel fuels. The hydrocarbons present in the diesel fuels include paraffins, naphthenes, olefins, and aromatics. Carbon number of these fuels ranges from 12 to 18. Diesel fuels can contain both saturated and straight or unbranched chain unsaturated hydrocarbons, but the latter are not present in large enough amounts to make fuel oxidation a problem. The aromatics that are present generally are oxidation resistant.

7.8.3. Properties of Diesel Fuels

The important characteristics of diesel fuels are listed in Table 7.21.

Ignition quality is important in high speed automotive diesel engines. It is less relevant in low speed engines used in large stationary and marine applications. The most universally accepted measure of the ignition quality of diesel fuels is cetane number. Cetane index which provides a measure of ignition quality very close to cetane number, is commonly used. It is calculated based on 10%, 50%, 90% distillation temperatures and specific gravity. Diesel index which is determined from aniline point and API gravity, provides a measure of ignition quality less, close to cetane number than cetane index. When ignition quality improvers are added to fuels, the ignition quality cannot be indicated either by cetane index or diesel index. Fuels of poor ignition quality can lead to extended ignition delay and result in the phenomenon of 'diesel knock'. This leads to decrease in efficiency, smoky exhaust, crankcase oil dilution and carbonaceous deposits in piston ring grooves.

In general, the cenetan number seems to be associated with the thermal stability of the fuel hydrocarbons. Thus, higher molecular weight normal alkanes have high cetane numbers. Generally, high performance diesel oils are required to have a cetane number higher than 45. In India, high speed diesel engines are supplied with diesel oil having a cetane number more than 42. Cetane number to be written, Diesel index and CCI to be written viscosity.

Volatility is the major determinant of the tendency of a hydrocarbon to produce potentially explosive vapours. It also is critically important to an engine's start and warm-up. The presence of high-boiling point components in fuels can significantly affect the degree of formation of solid combustion deposits.

Proper operation of an engine depends on the proper viscosity of the liquid fuel. The viscosity of the liquid fuel is important to its flow through pipelines, injector nozzles, and orifices and for atmization of the fuel in the cylinder.

Diesel fuels should not creates any operational problems during its use. For example, diesel fuels should not cause deposits to occur in fuel injector systems or critical piston ring areas. The carbon residue value of diesel fuel correlates with the amount of carbonaceous deposits which are formed in the combustion chamber of the engine. The higher carbon deposits in the combustion chamber are expected with the diesel fuels of higher carbon residues.

TABLE 7.21. Desirable Fuel Properties of Diesel Fuels.

Fuel characteristics	Comments
1. Cetane number	(*a*) A measure of ignition quality of diesel fuels (*b*) Diesel fuels having high cetane number show a short ignition delay (*c*) High performance diesel fuels have cetane number higher than 45 (*d*) Higher molecular weight normal alkanes have high cetane numbers (*e*) Influences both gaseous and particulate emissions (*f*) Has an influence on cold starting and on subsequent white smoke and noise emissions (*g*) Cetane index which is very close to cetane number is calculated based on 10%, 50%, 90% distillation temperatures and specific gravity (*h*) Fuels with high autoignition temperatures are more likely to cause diesel knock
2. Distillation range	(*a*) Affects fuel performance and safety (*b*) Important to an engine's start and warm up (*c*) Presence of high-boiling components affects the degree of formation of solid combustion deposits (*d*) Needed in the estimation of cetane index
3. Specific gravity	(*a*) Required for the conversion of measured volumes to volumes at standard temperature of 15°C (*b*) Used in the calculation of cetane index
4. Heat of combustion	(*a*) A measure of energy available in a fuel (*b*) A enticed property of fuel intended for use in weight-limited vehicles
5. Flash point	(*a*) Indicates the presence of highly volatile and flammable materials (*b*) Measures the tendency of oil to form a flammable mixture with air (*c*) Used to assess the overall flammability hazard of a material
6. Viscosity	(*a*) Proper viscosity of fuel required for proper operation of an engine (*b*) Important for flow of oil through pipelines, injector nozzles and orifices (*c*) Effective atomization of fuel in the cylinder requires limited range of viscosity of the fuel to avoid excessive pumping pressures
7. Contamination (Water/sediment)	(*a*) Causes corrosion of equipments (*b*) Causes problems in processing (*c*) Required to accurately measure net volumes of actual fuel in sales, taxation, exchanges and custody transfer
8. Copper-strip corrosion	(*a*) A measure to assess relative degree of corrosivity (*b*) Indicates the presence of sulphur compounds
9. Cloud point, pour point, cold-filter plugging point	(*a*) A measure of the performance of fuels under cold temperature conditions. (*b*) Used as quality control sepcification or low temperature handling indicators for large storage tanks and pipelines at refineries and terminals.
10. Carbon residue	(*a*) Correlates with the amount of carbonaceous deposits in the combustion chamber (*b*) Greater carbon deposits expected for higher values of carbon residue

Fuel characteristics	Comments
11. Particulate matter	(*a*) Indicates the potential of emission of particulate matter (*b*) Contains primarily carbon particles (*c*) Soot (carbonaceous particulates formed from gas-phase processes) particles absorb and carry carcinogenic materials into environment as emission and can cause an ill effect on human health. Excessive soot particles might clog the exhaust valves.
12. Ash	(*a*) Results from oil, water-soluble metallic compounds or extraneous solids, such as dirt and rust. (*b*) Can be used to decide product's suitability for a given application
13. Sulphur	(*a*) Controlled to minimize corrosion, wear and tear (*b*) Causes environmental pollution from their combustion products (*c*) Corrosive in nature and causes physical problems to engine parts

Particulate matter contains primarily carbon particles. Carbonaceous particulates that form from gas-phase processes are generally referred to as soot, and those that develop from pyrolysis of liquid hydrocarbon fuels are generally referred to as coke. Because soot particulates can absorb and carry carcinogenic materials into the environment as emission, they can cause an ill effect on human health.

Sulphur content must be critically controlled to minimize corrosion, wear and tear. Sulphur compounds pose a dual problem: they not only cause environmental pollution from their combustion products, but these products are also naturally corrosive and cause severe physical problems to engine parts. Sulphur pollution and corrosion were problems Umg before the NO_x were known to affect the atmosphere. The general availability of low-sulphur fuels, however, diminished the concern with respect to the sulphur.

The performance problems versus fuel-related causes for compression-ignition (diesel) engines are summarized in Table 7.21. The more critical problems are those related to combustion, injector nozzle/pump fouling and filter plugging. Indian specifications for diesel fuels are listed in Tables 7.22 to 7.24.

TABLE 7.22. Fuel Effects on Compression-Ignition Engine Performance

Performance problem	Probable fuel-related cause
Poor combustion, smoking	Low cetane number; Water contamination; Improper cloud point; Lighter/heavier fuel contamination
Excess cylinder wear	Fuel dilution ; High sulphur content ; Dirt (silicon) contamination
Injector nozzle/plugging wear	Soluble metal contaminants ; Heavy end impurities ; Preformed gum impurities
Injector pump fouling sticking	High sulphur/hetero-atom content ; Heavy ends contamination ; Gasoline contamination Low fuel viscosity
Filter plugging	Water contamination ; Fuel impurities ; Improper cloud point ; Thermally-reactive hydrocarbons
Excess engine deposits	Heavy ends contamination ; Low cetane number ; High sulphur/hetero-atom content

TABLE 7.23. Requirement for Diesel Fuels (IS : 1460-1974).

Sl. No.	Characteristics	Requirement	Method of test IS : 1448
1.	Acidity, inorganic	Nil	P : 2
2.	Acidity, total, mg KOH/g, Max.	0.50	P : 2
3.	Ash, wt. %, Max.	0.01	P : 4
4.	Carbon residue (Ramsbottom), wt.%, Max.	0.20	P : 8
5.	Cetane number, Min.	42**	P : 9
6.	Pour point, °C, Max.	6*	P : 10
7.	Copper-strip corrosion for 3 h at 100°C	Not worse than No. 1	P : 15
8.	Distillation-Recovery at 366°C, vol%, Min.	90	P : 18
9.	Flash point (Abel), °C, Min.	38	P : 20
10.	Kinematic viscosity at 38°C, cSt	2.0-7.5	P : 25
11.	Sediment, wt. %, Max.	0.05	P : 30
12.	Total sulphur, wt. %, Max.	0.5	P : 33 or P : 35
13.	Water content, vol. %, Max.	0.05	P : 40
14.	Total sediments, mg/100 ml, Max.	1.0	—

*Subject to seasonal changes, pour point varies from 0 to 18°C

**45 (for Naval applications)

TABLE 7.24. Requirements for LDO Fuels (IKS : 1460-1974)

Sl. No.	Characteristics	Requirement	Method of test (IS : 1448)
1.	Acidity, inorganic	Nil	P : 2
2.	Ash, wt.%, Max.	0.02	P : 4
3.	Carbon residue, wt.%, Max.	1.50	P : 8
		Min.	0.25
4.	Pour point, °C, Max.	12 for winter* 18 for summer	P : 10
5.	Copper-strip corrosion for 3 h at 100°C	Not worse than No. 2	P : 15
6.	Flash point, Pensky, Martens (closed), °C, Min.	66	P : 21
7.	Kinematic viscosity at 38°C, cSt	2.0-15.7	P : 25
8.	Sediment, wt.%, Max.	0.10	P : 30
9.	Total sulphur, wt.%, Max.	1.8	P : 33 or P : 35
10.	Water content, vol.%, Max.	0.25	P : 40
11.	Flame height, mm, Max.	1.0	P : 31

* Winter-November to February, both months inclusive.

7.8.4. Additives for Diesel Fuels

Diesel fuel additives are used to enhance fuel quality and are often cost-effective.

Cetane Improvers

These are used to increase cetane number of diesel fuels. Straight-chain paraffins ignite readily under compression, but branched-chain paraffins and aromatics react more slowly. Since diesel fuel's ignition delay characteristic, and subsequent complete versus incomplete fuel burning, directly influence overall engine performance. Cetane number of a diesel fuel reflects the ease with which it undergoes compression indicator of performance related properties which include cold during warm-up, noise, power, exhaust emissions, fuel consumption and engine durability.

Cetane improvers are compounds which readily decompose to give free radicals and thus enhance the rate of chain initiation for hydrocarbon oxidation in diesel combustion. They promote fast oxidation of fuels and thus improve their ignition characteristics. Cetane improvers are mainly alkyl nitrates. Depending on the base fuel's cetane numbers and additive treat rate, it is possible to obtain very cost-effective improvements between three and six cetane numbers with the iso-octyl nitrate. One cetane number increase requires about 100 pounds per thousand barrels (ptb) cetane improver.

Cetane improvers provide refineries with a simple and effective means of operational flexibility on a day to day basis. They are typically used to :

- Upgrade HSD quality to meet specification requirements in conversion refineries where economic benefits from the use of cracked components in HSD production exists.
- Give the flexibility required to process significant quantities of low cetane thenic crude.

Detergents

These are used as cleaners to remove carbonaceous and gummy deposits from fuel injectors and to prevent further formations. Fuel gums cause moving parts in the fuel system to stick, especially close-tolerance components like injector needles and plunger pistons. Lacquer and carbon deposited on injector needles restrict fuel flow, distort the fuel spray and often totally block one or more holes in a multihole injector. Modern detergent additives are composed mainly of ashless amines and amides. Older-style imidazonesare still used. However, some of these older-style detergents under the high temperatures (71°C) and high pressures (1100 kPa) experienced in today's engines, react with dimer acids to form a black, filter-plugging residue. These residues can block fuel filters and wear close-tolerance injection system parts in less than 1500 miles (or 50 hours) of operation. The detergent must be added to the base diesel fuel at a minimum dose of 40 to 50 ptb to effect optimum engine cleanup and prevent further deposits.

Stabilizers

These retard the natural degradation of diesel fuels during long-term storage. Because diesel fuels are organic in nature, they will degrade from the time of manufacture until the time of use. The harmful effects of fuel oxidation include free-carbon particulate fallout and sedimentation sludge buildup in the fuel delivery system and fuel storage tanks. The following stabilizer additives are needed :

- De-emulsifier to improve anti-haze and minimize fuel/water suspension.
- Sludge retardant to limit deposit buildup.
- Dispersant to break, up fuel droplets, reduce surface tension and provide buffer barrier to prevent further deposits.

- Rust-corrosion inhibitor to prevent pitting and erosion of metallic surfaces.
- Metal deactivator to inhibit deterioration by trace metals.

Dispersant restricts the size of fuel particles and reduce the fuel's surface tension. By lowering the fuel's surface tension, dispersant modifier additives such as polyamides, polyisobutane and succinimides cause the fuel droplets to break up and atomize more easily. Because they are surfactants, they also are adsorbed onto metallic surfaces and provide a buffer barrier to prevent future deposits and keep the surfaces of the combustion chambers and injection nozzles clean. Dispersants provide adequate performance when added to diesel fuels at doses of only 20 to 25 ptb.

Anti-corrosion and rust-inhibitor additives are used to protect pipelines through which diesel fuels are transported. These form oily coating over metal surfaces as protection against corrosive attack. Usually the protection of fuel system from corrosion and rust requires only 3 to 5 ptb.

Metal deactivators inhibit catalytic decomposition and deterioration of diesel fuels by copper and other metals. Typical dosage levels of metal deactivators are 2 to 3 ptb.

Anti-oxidants retard the formation of fuel gums, sludges, sedimentation deposits and oxidation byproducts. One notable effect of fuel degradation during long-term storage of hydrotreated or hydrocracked low-sulphur diesels is peroxides formation. The anti-oxidant additive is to be added to the diesel fuel as soon after the manufacture as possible. The peroxide inhibitors are cost-effective at a treat rate of 3 to 5 ptb.

Cold Weather Additives

Paraffins exist in all diesels and vary in amount depending on the crude oil regined. In cold temperatures, paraffins crystallize into wax. Without flow modifier additives, this wax generally creates large, flat platelets crystals that tend to clump together and clog fuel filters by adhering to the screen grids. The resulting engine starvation can cause poor starting and engine shutdown.

Pour point depressants, flow improvers and cloud point reducers are three basic types of cold-flow modifiers. Wax (paraffin) content and the type of fuel will affect each of these parameters as well as an additive's response when treated. Wax content in a finished fuel is mainly a function of boiling range of the diesel fuel and the composition of the crude oil. Fuels with a wide boiling range usually form somewhat smaller, more consistently shaped wax crystals that are more responsive to additive treatments. The narrow boiling range diesel fuels tend to form larger, more dense wax crystals that do not respond well to traditional cold-flow additives. Pour point depressants are chemicals that lower the pour point temperature. Today's pour point depressants work by changing the attraction of wax crystals and modifying the wax crystal's shape and size as they form. These additives reduce the crystal formation rate effectively.

Flow improvers primarily improve cold filterability, but they can also lower the pour point of the fuel. These chemicals change wax crystal formation by forming nuclei and arresting growth. Instead of the larger, flat platelets, a formation of small, multi-axial needle crystals develop. As the fuel's temperature drops below the cloud point', small wax crystals begin to grow on the nuclei. A flow improver's molecules attach to crystal surfaces and block further growth. This results in smaller crystals. By virtue of their small size and multi-axial needle arrangement, these wax crystals pass more easily through fuel filters. Flow improver's responses are measured by cold-filter plugging point (CFPP). The best performing CFPP additive requires dose rates between 65 and 125 ptb.

Anti-wax settling additives (AWSA) are flow improvers that disperse wax crystals uniformly. Normally, small, compact wax crystals tend to settle in tank bottoms, especially in unagitated storage tanks. These compounds create more porous (less dense) wax crystals with repelling charges that tend to prevent wax agglomerations into the larger/heavier wax crystal matrix. These additives work more efficiently in lower wax-content fuels and total wax content will affect performance of all AWSA. Most distillate fuels can be effectively treated with AWSAs at doses between 40 to 60 ptb.

Cloud point reducers lower the temperature at which fuel paraffins present in, the diesel fuels begin to haze or cloud. Generally, a reduction of approximately 2 to 3°C in the cloud point can be achieved by the use of olefin-ester copolymer. Typical levels of cloud point depressant additives are 50 to 170 ptb.

De-emulsifier and de-icer agents break any fuel/water mixtures and prevent water from freezing in cold temperatures. De-emulsifier additives can be of different chemical makeups. The only effective de-icer additives are chemicals of glycol-ether type. These additives chemically combine with water droplets (and dissolved water accumulations), break the surface connection with the fuel and lower the freezing point temperature in direct relation to the amount of de-emulsifier and de-icer present. These additives should be used in combination at dose rates of 10 to 20 ptb.

Lubricity Additives

Rotary or distributor-type injection pumps are used in the majority of light-duty diesel fuel injection equipment that rely on the fuel for lubrication. This contrasts with in-line pumps, most commonly found in medium- and heavy-duty applications, where some components are lubricated by engine oil. A fuel's lubricity can be loosely defined as its ability to reduce friction between moving metal parts and prevent wear. A more rigorous definition is as follows: For two liquids of the same viscosity, if liquid *A*, gives lower friction, wear or scuffing than liquid B, A is said to have better (or higher) lubricity. Generally, the new lower-sulphur diesel fuels contain less aromatics (diaromatic and polyaroroatic) because of the severe hydrotreating necessary to lower the total sulphur content. These low-aromatic fuels negatively impact the fuel's natural lubrication tendency and increase the need for supplemental lubricity enhancement. For low-sulphur diesels, lubricity is improved by using an additive which impart additional lubricity properties. Most effective lubricity additives require 3 to 8 ptb.

7.8.5. Alternative Diesel Fuels

Efforts are under way in many countries including India to search for suitable alternative diesel fuels that are environment friendly. The need to search for an alternative diesel fuels arises mainly from the standpoint of preserving the global environment and the concern about long-term supplies of conventional hydrocarbon-based diesel fuels. The alternative diesel fuels must be technically acceptable, economically competitive, environmentally acceptable, and easily available. From the view point of these requirements, the following may be considered as viable alternatives for diesel fuels :

- Vegetable oils and their derivatives
- Petrocrops

- Coal
- Agricultural wastes

Vegetable Oils and Their Derivatives

The basic constituent of vegetable oils is triglyceride. Figure 7.19 shows a typical triglyceride molecule. Vegetable oils comprise of 90-98 percent triglycerides and small amounts of mono- and diglycerides. These also contain free fatty acids (generally 1-5 percent), phospholipids, phosphatides, carotenes, tocopherols, sulphur compounds and traces of water. Triglycerides are esters of three fatty acids and one glycerol. Fatty acids vary in their carbon chain length and in number of double bonds. The fatty acids which are commonly found in vegetable oils are stearie, palmitic, oleic, linoleic and linolenic.

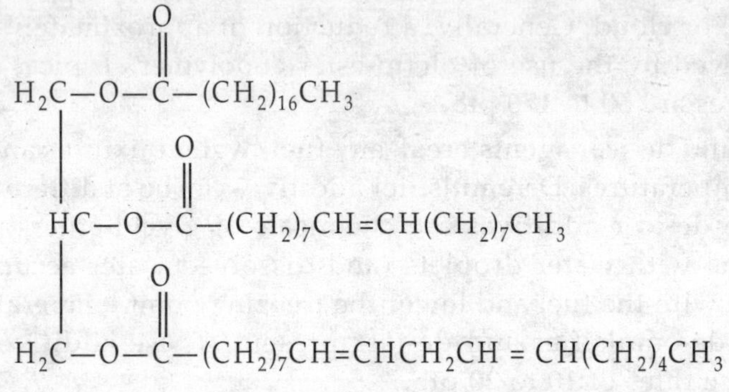

FIG. 7.19. STRUCTURE OF A TYPICAL TRIGLYCERIDE MOLECULE.

The properties of vegetable oils are listed in Table 7.25. The kinematic viscosity of vegetable oils varies in the range of 30-40 cSt at 38°C. High viscosity of these oils is due to large molecular mass and chemical structure. Vegetable; oils have high molecular weights in the range of 600-900, which are three or more times higher than diesel fuels. The flash point of vegetable oils are very high (above 200°C). The volumetric heating values of these oils are in the range of 39-40 MJ/kg which are low compared to diesel (feels (about 45 MJ/kg). The cetane numbers are in the range of 32-40. The iodine value ranges from 0 to 200 depending upon unsaturation. The cloud and pour points of vegetable oils are higher than that of diesel fuels.

It has been found that the use of vegetable oils as diesel fuels in conventional diesel engines leads to a number of problems which are related to the type and grade of oil and local climatic conditions. The high viscosity of vegetable oils interferes with the injection process and leads to poor fuel atomization. The inefficient mixing of oil with air contributes to incomplete combustion. The high flash point attributes to its lower volatility characteristics. This leads to more deposit formation, carbonization of injector tips, ring sticking and lubricating oil dilution and degradation. The combination of high viscosity and low volatility of vegetable oils causes poor cold engine start up, misfire and ignition delay. Oxidative and thermal polymerization of vegetable oils cause a deposition on the injector, forming a film that will continue to trap fuel and interfere with combustion. The injection, atomization, and combustion characteristics of vegetable oils in diesel engines are significantly different from those of hydrocarbon-based diesel fuels. In long-term operation, vegetable oils normally develop gumming, injector coking and ring sticking. Another problem is incompatibility with conventional diesel fuels. Therefore, the engine must be more or less modified according to the conditions of use and the oil involved.

TABLE 7.25. Fuel Properties of Vegetable Oils.

Vegetable oil	Kinematic viscosity @ 38°C mm^2/s	Cetane number MJ/kg	Lower heating value °C	Cloud point °C	Pour point °C	Flash point kg/l	Density vol.%	Water and sediment wt.%	Carbon residue wt.%	Ash wt.%	Sulphur	Copper strip corrosion
Castor	297	-	37.3	-	–31.7	260	0.9537	Tr	0.22	< 0.01	0.01	1a
Corn	34.9	37.6	39.5	–1.1	–40.0	277	0.9095	Tr	0.24	0.01	0.01	1a
Cottonseed	33.5	41.8	39.5	1.7	–15.0	234	0.9148	0.04	0.24	0.01	0.01	1a
Crambe	53.6	44.6	40.5	10.0	–12.2	274	0.9044	0.2	0.23	0.05	0.01	1a
Linseed	27.2	34.6	39.3	1.7	–15.0	241	0.9236	Tr	0.22	< 0.01	0.01	1a
Peanut	39.6	41.8	39.8	12.8	–6.7	271	0.9026	Tr	0.24	0.005	0.01	1a
Rapeseed	37.0	37.6	39.7	–3.9	–31.7	246	0.9115	Tr	0.30	0.054	0.01	1a
Safflower	31.3	41.3	39.5	18.3	–6.7	260	0.9144	Tr	0.25	0.006	0.01	1a
H.O.Safflower	41.2	49.1	39.5	–12.2	–20.6	293	0.9021	Tr	0.24	<0.001	0.02	1a
Sesame	35.5	40.2	39.3	–3.9	–9.4	260	0.9133	Tr	0.25	0.01	<0.01	1a
Soyabean	32.6	37.9	39.6	–3.9	–12.2	254	0.9138	Tr	0.27	<0.01	0.01	1a
Sunflower	33.9	37.1	39.6	7.2	–15.0	274	0.9161	Tr	0.23	<0.01	0.01	1a
Palm	39.6	42	-	31	-	267	0.9180	-	-	-	-	-
Babassu	30.3	38	-	20	-	150	0.9460	-	-	-	-	-
Tallow		40.0	-	-	201	-	-	6.21	-	-	-	

Tr = traces

The use of vegetable oils as diesel fuels in conventional diesel engines requires modification in their properties through physico-chemical processes. Considerable efforts have been made to develop vegetable oil derivatives that would approximate the properties and performance of the hydrocarbon-based diesel fuels. The problems with substituting pure vegetable oils for diesel fuels are mostly associated with their high viscosities. The process of transesterification is widely used to reduce the viscosity of vegetable oils. Transesterification also called al-coholysis, is the displacement of alcohol from an ester by another alcohol. The transesterification reaction of triglyceride using methanol is represented by the general equation given below :

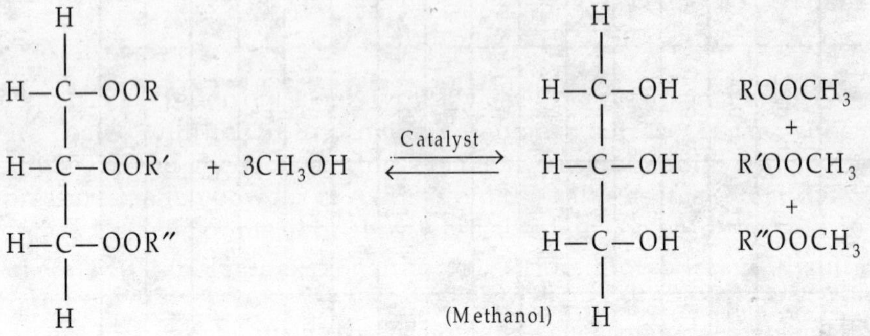

(Triglyceride) (Methanol) (Glycerol) (Fatty acid methyl esters)

The fatty acid methyl esters are referred to as biodiesel. The properties of biodiesel are given in Table 7.26. The characteristics of biodiesel are close to diesel fuels, and therefore biodiesel becomes a strong candidate to replace the diesel fuels if the need arises. The conversion of triglycerides into methyl esters through transesterification process reduces the molecular weight to one-third that of the triglyceride, reduces the viscosity by a factor of about eight-and increases the volatility marginally. Biodiesel has viscosity close to diesel fuels. These esters contain 10 to 11 percent oxygen by weight, which may encourage more combustion than hydrocarbon-based diesel fuels in an engine. Since the volatility increases marginally, the starting problem persists in cold conditions. Methyl esters have lower volumetric heating values (about 12 percent) than diesel fuels but has high cetane number and flash point. These esters have cloud point and pour point that are 15-25°C higher than those of diesel fuels.

The main advantages of using biodiesel are its renewability, better quality exhaust gas emissions; and biodegraedability. The organic carbon present in the biodiesel is photosynthetic in origin, and it does not contribute to a rise in the level of carbon dioxide in the atmosphere and consequently to the greenhouse effect.

Different technologies are currently available and used in the industrial production of biodiesel, which is sold under different trademarks. For example, there are the Italian processes Novamont and Ballestra, the French IFF, the German Henkel and ATT. Triglycerides can readily be transesterified batch wise at atmospheric pressure and at slightly elevated temperature of approximately 60-70°C with an excess of methanol and in the presence of alkaline catalyst. The mixture at the end of the reaction is allowed to settle. The lower glycerine layer is drawn off while tibe upper methyl ester layer is washed to remove entrained glycerine and is then processed further. The excess methanol is recovered in the

TABLE 7.26. Fuel Properties of Biodiesel.

Vegetable Oil methyl ester	*Kinematic viscosity @ 37.8°C*	*Cetane Number*	*Lower heating value*	*Cloud point*	*Pour point*	*Flash point*	*Density*	*Water and sediment*	*Carbon residue*	*Ash*	*Sulphur*	*Copper-strip corrosion*
	mm^2/s		*MJ/L*	°C	°C	°C	*kg/l*	*vol.%*	*wt.%*	*wt.%*	*wt.%*	
Peanut	4.9	54	33.6	+0.5	—	176	0.883	—	—	—	—	—
Soyabean	4.5	45	33.5	+0.1	−7	178	0.885	< 0.05	1.74	—	—	1a
Babassu	3.6	63	31.8	+04	—	127	0.879	—	—	—	—	—
Palm	5.7	62	33.5	+13	—	164	0.880	—	—	—	—	—
Sunflower	4.6	49	33.5	+01	—	183	0.860	—	—	—	—	—
Tallow	—	—	—	+12	9	96	—	<0.05	1.83	—	—	1a

condenser, sent to a rectifying column for purification and recycled. The Henkel process operates at 9000 kPa and 240°C and uses unrefined oil as feedstock. Unrefined oil, methanol in excess and catalyst are metered and heated to 240°C before feeding into the reactor. The bulk of the excess methanol is flashed on as it leaves the reactor and is fed to a bubble tray column for purification. The recovered methanol is recycled into the system. The mixture from the reactor enters *a* separator where the glycerine in excess of 90 percent concentration is removed. The methyl ester is subsequently fed to a distillation column for purification. Further fractionation into special cuts may follow if desired. The Lurgi process operates at normal pressure. This process requires the use of a degummed and deacidified feedstock. The refined vegetable oil and methanol are reacted in a two-stage mixer-settler arrangement in the presence of a catalyst. The glycerine produced in the reaction, dissolved in the surplus methanol, is recovered in the rectification column. Most of the entrained methanol and glycerine are recovered from the methyl ester in the countercurrent scrubber. The methyl ester can be further purified by distillation.

Depending upon climate and soil conditions, different nation can look into different vegetable oils for diesel fuels. For example, soyabean oil in United States, rapeseed and sunflower oil in Europe, palm oil in southeast Asia (mainly Malaysia and Indonesia), and coconut oil in The Philippines are being considered as substitute for diesel fuels. The production of these oilseeds can be stepped up if it is decided to use vegetable oils as sources of diesel fuels. It is also necessary to develop new and more productive plant sources whose seeds should have high oil content.

Petrocrops

Several plants are known whose latex contain petroleum like hydrocarbons which can be extracted by organic solvents as a biocrude (petroleum type crude), and can be hydrocracked to get oil. These plants are known as petrocrops. These are laticiferous plants occurring in arid, semi-arid and uncultivated lands. These plants occur all over India. Arid and semi-arid lands can be upgraded by growing certain plants of cactus type which can grow in these areas without much agricultural management and inputs. Since these plants will not be used as food, therefore even an organic nitrogenous waste or humates may be tried as manure or fertilizer for growing these hydrocarbon crops. In fact, these plants grow wildly and their growth is an environmental menace. Even animals do not use these plants as fodder. Nonpolluting solid and liquid fuels can be obtained from petrocrops by solvent extraction followed by pyrolysis techniques. Hydrocracking or liquefaction of latex or of petrocrop plants as a whole can yield petroleum oil.

The potential petrocrops are those plants that are perennial, have high hydrocarbon and protein contents and yield large biocrude. Biocrude obtained from different plants belonging to the families of Euphorbiaceae, Asclepiadaceae and Apocynaceae contains oils which are similar to crude oil. Plants belonging to Euphorbia species have high latex flow and therefore are promising as petrocrops, the important being Jatropha curcas, E. lathyris, E. hirta and E. microphylla. The oil extract from Jatropha plant generates a wide range of hydrocarbons as an effective fuel substitute. It can be converted into petroleum products using metal oxide catalyst. Thus the use of petrocrops for producing diesel fuels may not be an immediate prospects but studies and research in the subsequent years may turn this impossible fact into reality.

7.8.6. Coal

A scheme for producing diesel fuels from coal is given in Fig. 7.20. A company, Sasol, in South Africa is producing diesel fuels from coal. The first step is the complete gasification of coal with steam and oxygen. After purification of the gas, a synthesis gas results that consists mainly of hydrogen, carbon monoxide and some methane. The next major step is the synthesis of hydrocarbons from the hydrogen and carbon monoxide over an iron catalyst. This is the Fischer-Tropsch synthesis. It gives a wide range of hydrocarbons of different chain lengths and simultaneously a number of oxygenates such as alcohols and ketones. After separation and recovery of these synthesis products from the reactor tail gas, the hydrocarbons boiling above 180°C are used as refinery feedstock for diesel production. The characteristics of diesel fuels from coal are given in Table 7.27. The hydrocarbons in this fraction stiow the higher degree of branching in the synthol processi and-high degree of paraffinieity in the fixed bed product. From these properties it is clelar that fixed bed material with some hydrocracking of the heavy stocks will produce excellent diesel.

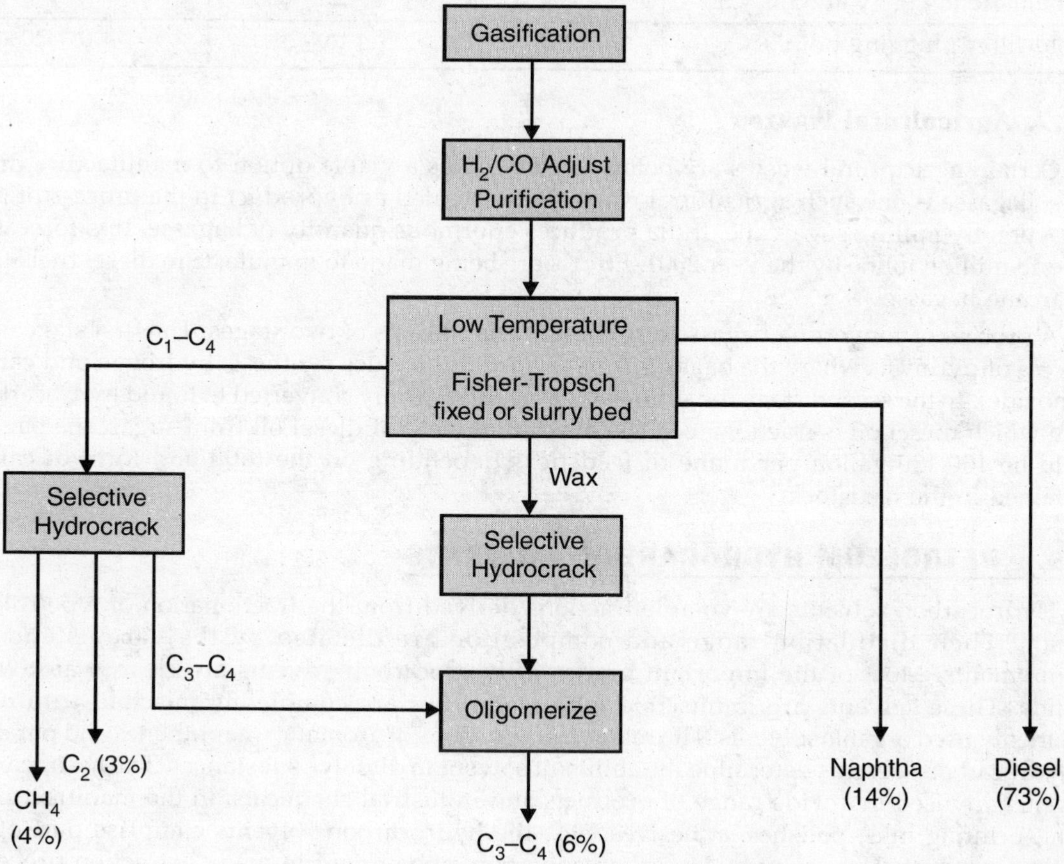

FIG. 7.20. SCHEME FOR PRODUCING DIESEL FUELS.

TABLE 7.27. Characteristics of Diesel Fuels from Coal.

	Sasol II (As marketed)	Synthol achievable (Gasoline mode)	Fixed bed achievable (Diesel mode)
Diesel fraction of liquid products	35-45	20	75
Components :			
Paraffins, wt%	96-98	96-98	65-75
Olefins, wt%	—	—	20-30
Aromatics, wt%	2-4	2-4	5-10
Characteristics :			
Cetane Number	47	50	55-65
Total sulphur, wt.%	0.0002	0.0001	0.0001
Density, kg/l	0.80	0.82	0.8
Kinematic viscosity at 38°C, cSt	1.85	2.3	2-3
Cold-filter plugging point, °C	< -12	–4 to –20	–4 to –20

7.8.7. Agricultural Wastes

Certain agricultural wastes are being considered as a viable option to manufacture diesel fuels. Bagasse is one such agricultural waste. It is obtained as byproduct in the process of juice extraction by milling sugarcane. India produces enormous quantity of bagasse. It is forecasted to be 18 million tonne by the year 2001. Efforts are being made to manufacture diesel fuels from sugarcane bagasse.

A process of converting bagasse into diesel fuels consists of two stages. The first stage is the process of pyrolysis where the bagasse is pyrolysed to produce synthesis hydrogen and carbon monoxide). In the second stage the synthesis gas is catalytically converted to liquid hydrocarbons from which diesel oil is fractionated. The maximum yield of diesel oil from sugarcane bagasse could be 100-120 gallon per tonne of feedsticks depending on the total and form of carbon contained in the feedstock.

7.9. PETROLEUM HYDROCARBON SOLVENTS

Hydrocarbon solvents are volatile fractions derived from the fractionation of a petroleum stream. Their distillation range and composition are dictated by the various end-use requirements. Most of the important grades of hydrocarbon advents are clear, water-white liquids. These solvents are immiscible with water, but are completely miscible with other commonly used organic solvents. The relative proportions of aromatics, naphthenes and paraffins in hydrocarbon solvents determine the ability of solvent to dissolve substances. The hydrocarbon solvents are used in a wide range of products and industrial chemicals. In the manufacture of paint, printing inks, polishes, adhesives, etc., the hydrocarbon solvents comprise part of the finished product. The use of hydrocarbon solvent may be found in many extraction processes such as the extraction of vegetable oils from oil seeds, pharmaceuticals etc. In chemical processes these solvents are used as an inert reaction medium or catalyst carrier. These may be used simply for clean up and maintenance operations.

7.9.1. Composition of Hydrocarbon Solvents

Almost all hydrocarbon solvents are mixtures of chemically different hydrocarbon molecules. The composition of a particular solvent is usually quoted in terms of type of hydrocarbons present, i.e. paraffins, naphthenes, aromatics and olefins. The proportions of these groups of hydrocarbon types have an effect on physical properties such as solvent power, viscosity, density and colour.

Small amounts of non-hydrocarbon compounds such as sulphur and oxygenated com pounds can be present. However, these compounds are usually removed to very low levels during the manufacturing process. The olefin content, characterized by the bromine index is usually low.

7.9.2. Classification of Hydrocarbon Solvents

Different types of hydrocarbon solvents are mainly classified according to the distillation range. The hydrocarbon solvents which evaporate after doing its job, e.g. in paint are usually classified as slow, medium or fast evaporating. Various classes of hydrocarbon solvents are given below.

SBPs (Special Boiling Point Spirits). These represent a range of products, often characterized by a number, e.g. SBP1, SBP2 etc. These solvents are essentially aliphatic solvents within the temperature range 45 - 160°C. Both narrow and wide boiling grades of SBPs are available. There is no universally accepted nomenclature. Petroleum spirit is also used for SBPs. Names such as hexane, isohexane, heptane etc. for SBPs which may consist mainly (but not entirely) of the named hydrocarbon are also used. Petroleum ether is used for SBPs below 100°C.

Food-grade hexane is a hydrocarbon solvent having a boiling range 63-70°C . It consists of n-hexane and its isomers, methyfcyclopentane, cyclohexane and small concentration of other C_5 - C_7 hydrocarbons. These hydrocarbons are mainly present in natural-gas liquids and straight-run naptha obtained from crude oils. Freedom from carcinogenic impurities is the prime quality requirement of the product as it is used for the extraction of vegetable oils and pharmaceutical intermediates. Table 7.28 lists BIS specifications (IS : 3470-1966) of this product.

Isoparaffinic solvents. These solvents consist mainly of isoparaffins with negligible amounts of other types of hydrocarbons. These are manufactured from lower molecular weight hydrocarbons by alkylation. For all practical purposes these are considered odourless. These have a lower solvent power than conventional grades.

Aromatic solvents. These solvents are available in a full range. The faster evaporating types are essentially toluene and xylenes. The composition of slower evaporating types of aromatic solvents is complex.

White spirits. These solvents boil within the range 150 - 220°C, but rarely within less than 20°C. The other names used for white spirits include mineral turpentine, mineral spirits, stoddard solvent. White spirits are not de-aromatized. Crude oil is usually selected to give approximately 17 vol% aromatics in white spirits.

De-aromatized advents. These solvents have much lower odour than conventional regular grades. For both performance and toxicological reasons there is an increasing demand for solvents that are essentially benzene-free, and have very low aromatics contents. Examples of applications, where solvents with very low benzene contents are required, are oil seed extraction, printing inks, adhesives and for performance reasons in polyolefin manufacture. The removal of aromatics from a solvent fraction gives a marked improvement to the solvent odour. For this reason de-aromatised solvents in the 150 - 250°C distillation range are gaining popularity.

TABLE 7.28. Specification of Food-Grade Hexane (IS: 3470-1966)

Sl. No.	Characteristics	Requirements
1.	Distillation	
	(a) IBP, °C, Min.	63
	(b) Dry point, *C, Max.	70
	(c) Temperature, range of final 10 vol.%, °C, Max.	2
2.	Composition	
	(a) Aromatics, vol.%, Max.	1
	(b) Saturates, vol.%, Min.	98.5
3.	Density at 25*C, g/m1, Max.	0.687
4.	Colour (Saybolt), Max.	+30
5.	Sulphur, ppm, Max.	75
6.	Copper-strip corrosion for 3 h at 50°C, Max.	1
7.	Doctor test	Negative
8.	Bromine number. Max.	1
9.	Non-volatile residue, g/100 ml, Max.	0.001
10.	Lead, g/1, Max.	0.0005
11.	Chloride, ppm, Max.	20
12.	Reaction of non-volatile residue	To pass the test

Indian specifications of petroleum hydrocarbon solvents are listed in Table 7.29.

7.9.3. Manufacture of Hydrocarbon Solvents

There are three main steps involved in the manufacture of hydrocarbon solvents. These are :

1. Distillation
2. Chemical conversion and/or separation
3. Treatment

Blending is sometimes done for very specific requirements. For example, blending of aliphatic fractions with pure aromatics is done to achieve the required solvent power.

Distillation. Volatility of the hydrocarbon solvents is controlled by distillation. Most hydrocarbon solvents are manufactured in solvent distillation column from hydrotreated/hydrodesulphurised feedstocks. The increased demand for narrow boiling range products requires the solvent distillation columns consisting of more than 70 trays for a reflux ratio of 3:1. Depending on the quality requirements of solvents many refinery streams are used as feedstocks.

Certain high boiling solvents such as ink solvents require vacuum distillation to avoid an adverse effect of high distillation temperature on product quality.

Chemical conversion and/or separation. For many applications solvents are essentially benzene-free and have very low aromatic contents. Hydrocarbon solvents with very low benzene contents are required in oil seed extraction* printing inks, adhesives and polyolfenic manufacture. The removal of aromatics from a solvent fraction improves the solvent power.

De-aromatized solvents are manufactured by catalytic hydrogenation and solvent extraction processes. Catalytic hydrogenation converts aromatics into naphthenes. This process also converts unsaturated impurities to inert paraffins. This process results in products where the loss in solvent power due to aromatics removal is largely compensated by the increased naphthenes content. This process is used in locations where hydrogen is readily available.

Solvent extraction processes produce solvents with very low aromatics contents, but the unsaturates remain unchanged. Solvent extraction of aromatics from kerosine produces kerosine of lower aromatics content and hence of improved quality. For the extraction of aromatics from kerosine Edeleanu process, which uses liquid sulphur dioxide as extraction solvent, is usually employed. A range of highly aromatic white spirits and kerosine solvents may be manufactured from kerosine extracts.

Solvent extraction technology is used to dearomatize naphtha to the desired limits. Different solvents have been used by various process. These include sulfolane(IIP-EIL process, Shell process), Glycol-water mixture (UDEX process, UOP), tetraethylene glycol (Carom process, UOP), N-methyl pyrroh'done (Arosol van process), dimethyl sulfoxide (IFF process), N-fbrmyl morpholine (FORMEX process, Snamprogetti), morpholine derivatives (Aromex process, Krupp-Kroppers), tetraethylene glycol (Union Carbide process). Due to serious sludge handling and disposal problems involved in the oleum treatment of 63 - 69°C cut naphtha for the manufacture of food-grade hexane, solvent extraction technology using sulfolane is being employed by BPCL, Bombay and MKL, Madras.

Catalytic reforming of naphthas is carried out to produce high-octane gasoline components . This process converts naphthenes into aromatics. It is the basis for the manufacture of benxene, toluene and xylenes. Aromatic solvents boiling over 160°C are manufactured from the heavier fractions from the reformer. This heavy fraction is generally separated into three commercial solvents which are widely used in stoving paint formulations.

Alkylation process is employed to manufacture isoparaffinic solvents. In this process low-boiling iso-paraffins, e.g. isobutane reacts with low-boiling olefins, e.g. isobutene to produce isoparaffin of higher molecular weight. A range of solvents are manufactured from the fraction boiling above 100°C by distillation. These are practically odourless due to their high isoparaffin content (more than 90 percent). These solvents have a somewhat low solvent power. These solvents are the best choice in photocopying, household products and odourless paints etc. since these uses require a no-odour product.

Treatment. Depending on the feedstock source and other conditions in a refinery, treatment of hydrocarbon solvents is often required to produce marketable solvents.

Hypochlorite treating process converts mercaptans to less obnoxious compounds. It also extracts part of the sulphur or oxygen containing compounds and provides an oxidative clean-up of the fraction handled. However, the removal of chlorinated hydrocarbons which are formed in the process may pose problems.

Caustic soda solution may be used to remove acidic components such as hydrogen sulphide, carboxylic acids and to a certain extent phenols and mercaptans present in the solvents.

Hydrocarbon solvents may be treated with sulphuric acid for the conversion and/or removal of various types of contaminants present. This treatment process also reduces the total sulphur content. The disposal of arid sludge formed and considerable hydrocarbon losses are the two drawbacks of the process.

Clay treatment is usually required for hydrotreated products to produce products of acceptable odour. Clay/activated carbon treatments can improve arid wash colour and remove undesirable impurities responsible for unacceptable odour of the product.

The production of hydrocarbon solvents requires blending of components to achieve the particular evaporation characteristic or to produce a solvent of specific solvent power. A solvent of required evaporation characteristics may be obtained from a number of narrow-cut fractions.

TABLE 7.29. Specification for Petroleum Hydrocarbon Solvents

Sl. No.	Characteristics	Solvent 60/80	Solvent 50/120	Solvent 90/135	Solvent 125/240	Solvent 145/205 (Low aromatic)	Solvent 146/205 (High aromatic)	Solvent 150/300	Method of test Ref. to (P:) of IS : 1448
1.	Colour (Saybolt), Min.	+25	+25	+25	+21	+20	+20	—	P:14
2.	Colour (ASTM), Max.	—	—	—	—	—	—	3.5	P:12
3.	Density at 15°C, g/ml	Not limited but to be reported							P:16
4.	Flash point (Abel), °C, Min.	—	—	—	30	35	35	35	P:20
5.	Distillation range :								
	(*a*) IBP, Min.	60	50	90	126	146	146	160	P:18
	(*b*) 50 vol.% recovery, °C	Not limited but to be reported							
	(*c*) 90 vol.% recovery, °C	Not limited but to be reported							
	(*d*) FBP, °C, Max.	—	—	—	240	205	205	300	—
	(*e*) Dry point, °C, Max.	80	120	135	—	—	—	—	—
6.	Aromatic content, vol.% Max.	6	35	—	40	40	—	—	P : 23 or
	Min.	—	—	20	—	—	45	50	P:48
7.	Sulphur, total, wt.%, Max.	0.05	—	—	—	—	—	—	P:34
8.	Copper-strip corrosion for 3h at 50°C	Not worse than No. 1							P:15
9.	Residue on evaporation, mg/100 ml, Max.	5	5	5	5	5	5		P:29 (Air-jet)
	Uses	Solvent extraction	Rubber industry	Paint industry	Textile printing paint	Paint industry	Paint industry	Insecticide formulations	

The required solvent power may be achieved by blending two products of approximately the same boiling range, one of which is low in aromatics, the other a high-aromatic solvent or a pure aromatic.

7.9.4. Properties of Hydrocarbon Solvents

A solvent is expected to perform several functions. These include :

(*a*) The preferential removal of one substance from another;

(*b*) To cause a solid or viscous material to become fluid enough for use;

(*c*) To permit its removal by evaporation; and

(*d*) It must be inert to the material it dissolves.

The quality of a petroleum hydrocarbon solvents is judged by the characteristics of solvent power, rate of evaporation and odour. Appearance, odour and chemical stability of the hydrocarbon solvents are of utmost importance. All solvents should be clear and free from suspended matter. Most of the hydrocarbon solvents except a few high boiling are water white. Odour of a solvent is becoming increasingly important. Odour in hydrocarbons is due to impurities such as sulphur or oxidized degradation products. Odour is important in end-uses, such as photocopying, degreasing, low-odour paints and household products. Aromatic hydrocarbons usually impart a sharp unpleasant odour. Naphthenes have a lower and sweeter odour. Isoparaffins are considered odourless. Hydrocarbon solvents have low viscosities which increase with increase in the density and boiling range.

The drying rate in surface coating end-uses is of utmost importance. In such applications the evaporation rate or volatility of a hydrocarbon solvent is the most important characteristic.

In order to minimize dust collection and to allow early handling of the coated products, the drying time should be short. This requires sufficiently fast rate of evaporation. At the same time, the evaporation rate should be slow enough to allow the development of a good film. For defining the evaporation rates of solvents, n-butyl acetate has been assigned a standard evaporation rate of 1.

Solvent power, or solvency, is a measure of the ability of a solvent to dissolve a given product. The available tests for measuring solvent power in industry are kauri-butanol number, aniline point and dilution ratio. The composition of a hydrocarbon solvent fraction also gives an indication of solvent power. For most polymers and resins, solvent power decreases in the order.

Aromatics > Naphthenes > n-Paraffins > Isoparaffins

Aniline point provides a useful means of determining the solvent power of hydrocarbon solvents since aniline is a poor solvent for aliphatic hydrocarbons but an excellent one for aromatics. The aniline point test is usually selected to measure the solvent power of aliphatic solvents or those with low-solvent power and low aromatic content.

The Kauri-butanol (KB) number is the point at which a solvent added as a thinner will cloud a solution of kauri resin dissolved in n-butanol. It is the best known criteria of the hydrocarbon solvent power. This test can be used to measure the relative solvent power. A higher KB number implies higher solvent power, since more of a good solvent can be tolerated than a poor one. Standards are toluene (KB number =105) and 75 vol.% n-heptane plus 25 vol.% toluene (KB number - 40). Commercial solvents have KB values in the range of 25 to 105.

7.9.5. Uses of Hydrocarbon Solvents

The end-uses of hydrocarbon solvents are listed in Table 7.30. The surface coating industries (paints, lacquers and printing inks) are by far the largest users of petroleum hydrocarbon solvents. The word paint is usually applied very broadly to cover pigmented materials. Non-pigmented paints are often termed varnishes. Examples of paints are oil paint, chlorinated rubber paint, bituminous paint, stoving paint. The oil paint, based on a combination of vegetable oil and synthetic resin, is used in house decoration. The requirement of higher chemical or water resistance is met by chlorinated rubber paints. Bituminous paints are based on cut-back bitumins. The resin (binder) in these systems is soluble in the hydrocarbon solvent. In stoving paints, a proportion of an oxygenated organic solvent, usually n-butyl alcohol, is also present. The type of resin and the desired viscosity dictate the chemical type of solvent to be used.

Most paints dry by solvent loss as well as by chemical change in the film-forming materials by polymerization. The volatility and the solvent power affect the appearance of the final coating of the paint. Rapid drying and high viscosities may contribute towards unevenness. Slow drying and low viscosities may cause sagging, *i.e.* the paint runs down vertical surface.

Air drying paints are based on drying oils such as linseed combined with natural or synthetic resins, mainly alkyd resins. White spirit containing 15-20 percent aromatics is widely used in the oil paint formulations. This dissolves a wide range of resins, giving solutions with an acceptable viscosity. The solubility of the resin in the hydrocarbon solvents depends largely on the resin/oil ratio. Low aromatic hydrocarbon solvents are required if the oil content is considerably higher than the resin content. If the oil content is lower than the resin content, high aromatic hydrocarbon solvents are required. Xylene is widelyused in the stoving enamels. Lifting of the paint film cannot occur with stoving paints in which aromatic hydrocarbon solvents are mainly used. For household and similar paints, white spirits with a very low or nil aromatics content are used to give low odoiir or odourless paint.

TABLE 7.30. End-Uses of Hydrocarbon Solvents.

Hydrocarbon solvent	End-uses
1. SBPs (30-160°C)	
Low aromatics	Degreasing, oil extraction, paints, printing inks, rubber solutions, lacquers
High aromatics	Lube oil dewaxing, paints, lacquers, printing inks, rubber solutions
2. Pure aromatics	
Benzene	Lube oil dewaxing, alcohol dehydration
Toluene	Lube oil dewaxing, rubber solutions, paints
Xylenes	Alkyd resin azeotroping, insecticides, rubber solutions, printing inks, paints, lacquers
3. White spirit (150-210°C)	
Low aromatics	Degreasing, dry cleaning, printing inks, paints, lacquers
High aromatics	Degreasing
4. Kerosine solvents	(160-300°C)
Low aromatics	Degreasing, catalyst carrier in polyolefin production, insectides, paints, lacquers, printing inks
High aromatics	Degreasing, insecticides, paints, lacquers, printing inks, rubber reclaiming

Lacquers dry purely by solvent evaporation and hence no chemical change takes place during drying. Nitrocellulose lacquers, a pigmented lacquer are used in repairs to motor cars. Unpigmented, *i.e.* clear lacquers are used in furniture finishing, the solvents for cellulose, vinyl and acrylic lacquers are ketones, esters and ether alcohols in conjuction with hydrocarbon solvents. In lacquer formulation the hydrocarbon solvent is used as a diluent and its diluting power refers to the proportion of diluent that can be added successfully. A correct balance between the different components in the solvent system during evaporation must be maintained. Due to short drying time, quite volatile solvents are chosen. SBPs of suitable boiling ranges with high aromatic content may be used.

In addition to most of the factors mentioned in connection with paints and lacquers, solvent requirements for printing inks are influenced by the method of printing (i.e. lithographic, letterpress, gravure and silk screen), the type of printing press, the speed of operation and the properties of the paper or other material on which printing is carried out.

The function of hydrocarbon solvent in extraction processes is to extract selectively some particular ingredient from a mixed product or raw material, e.g. vegetable oils, pharmaceuticals, chaining and degreasing, etc. Vegetable oils provide a valuable source of food as well as raw materials for soap and paint manufacture. These oils are extracted from oil seeds using hydrocarbon solvents. Vegetable oils of commercial importance include soyabean, sunflower, groundnut, castor, cotton seed, rape seed and rice bran. The solvent for vegetable oil extraction should be fairly low boiling because the solvent from the oil is usually recovered by distillation. To minimize solvent vapour losses, solvent must have a relatively high initial boiling point. Solvent losses during oil see extraction are usually below 1 wt.% of raw material treated. The solvent must be selective for the oil and dissolve the minimum of undesirable pigments, resins, etc. occurring in the seeds. The preferred solvent for the extraction of vegetable oils is the food-grade hezane (SBP 65/70). Vegetable oil industry is the main user of food-grade hexane, consuming around 90% of the total hexane production in India.

Other important extraction processes include perfume extraction and pyrethrum extraction. The perfume and pyrethrum present in flower petals dissolve in the hydrocarbon solvent. These are recovered by carefully evaporating the solvent. These processes use iso-hexane or similar low-boiling SBP.

Solvent is used in polishes to aid spreadability prior to buffing to a fine gloss. The early paste polishes were made by dissolying beeswax in wood turpentine. For general furniture and shoe polishes, etc. turpentine has been replaced by white spirit and beeswax by paraffin waxes. Now many emulsion-based formulations (aerosol products) with combined polishing and cleaning properties are available. The solvent should have sufficient solvent power to dissolve the wax. The boiling range is also important in terms of evaporation which provides ah ideal buffing interval! The concentration of a solvent in a polish varies tremendously, depending on type. It can be as high as 70 percent in a paste to almost zero in polymer/wax water-based emulsions. Oxygenated solvents of glycol ether type are used in some of the newer formulations.

The hydrocarbon solvents are used in the production of adhesives and in the manufacture of tyres. Adhesives play an important role in several manufacturing processes: clothing and footwear, automobiles, the building industry, furniture and other consumer durables, the packaging and labelling industry.

Adhesives based on rubber are a relatively low-cost water-resistant adhesives. The rubber absorbs solvent to give a sticky solution which may be used as an adhesive. The choice of solvent is mainly dependent on the type of adhesive required, *i.e.* whether fast drying or slow drying. For adhesives based on natural rubber, aliphatic SBPs are suitable.

For the manufacture of tyre, a medium- to high-boiling SBP may be used. The natural rubber bales are coated with a fine white powder to prevent the bales from sticking to each other. Bale coating solution is prepared using white spirit of 15 or 45 percent aromaticity.

In the chemical industry, hydrocarbon solvents are widely used as reaction medium during the polymerization of olefins, such as ethylene, propylene to form plastics and elastomers. The function of the hydrocarbon solvents differs from one process to another. A catalyst for the manufacture of polyplefins is usually introduced into the reactor as a suspension or slurry in the hydrocarbon solvent. At the same time, the hydrocarbon solvent acts as a solvent for the monomer. The hydrocarbon solvent helps to reduce the viscosity of the reaction products. It also helps in controlling the reactor temperature by taking up some of the heat of reaction. To prevent catalyst poisoning, the hydrocarbon solvent should be free from polar impurities, in particular water, sulphur compounds and other volatile compounds of oxygen, nitrogen, chlorine etc. It is desirable that the olefins should be absent and aromatic hydrocarbon content must be low. If the polymer is formed as a suspension, a low-boiling SBP, for example, pentane or hexane is preferred. If the polymer is formed in solution, the solvent may be aromatic-free cyclohexane or high-boiling SBP or white spirit. The solvent in the manufacture of low-density polyethylene is used as carrier for the peroxide catalyst used to initiate polymer reaction. It also helps in the removal of heat from the reactor. As a peroxide catalyst carrier, highly refined mineral oils are generally used. In some instances, isoparaffinic, highly refined white spirit or kerosene solvents are preferred.

7.10. LUBRICANT

Petroleum fractions having an average volatility lower than that of gas oil form the base of various types of lubricating oils. These fractions constitute only 2–4 percent of the total crude oil processed.

A **lubricant** is a substance introduced between two moving surfaces to reduce the friction between them, improving efficiency and reducing wear. They may also have the function of dissolving or transporting foreign particles and of distributing heat. One of the single largest applications for lubricants, in the form of motor oil, is to protect the internal combustion engines in motor vehicles and powered equipment. Typically lubricants contain 90% base oil (most often petroleum fractions, called mineral oils) and less than 10% additives.

Lubricants perform the following key functions.

- Keep moving parts apart
- Reduce friction
- Transfer heat
- Carry away contaminants & debris
- Transmit power
- Protect against wear
- Prevent corrosion
- Seal for gasses
- Stop the risk of smoke and fire of objects

Keep moving parts apart: Lubricants are typically used to separate moving parts in a system. This reduces friction and surface fatigue together with reduced heat generation, operating noise and vibrations. Lubricants achieve this by several ways. The most common is by forming a physical barrier i.e. a thin layer of lubricant separates the moving parts. This is termed hydrodynamic lubrication. In cases of high surface pressures or temperatures the fluid film is much thinner and some of the forces are transmitted between the surfaces through the lubricant. This is termed elasto-hydrodynamic lubrication.

Vegetable oils or synthetic liquids such as hydrogenated polyolefins, esters, silicones, fluorocarbons and many others are sometimes used as base oils. Additives deliver reduced friction and wear, increased viscosity, improved viscosity index, resistance to corrosion and oxidation, aging or contamination, etc. Lubricants such as 2-cycle oil are also added to some fuels. Sulfur impurities in fuels also provide some lubrication properties, which has to be taken in account when switching to a low-sulfur diesel; biodiesel is a popular diesel fuel additive providing additional lubricity. Non-liquid lubricants include grease, powders (dry graphite, PTFE, Molybdenum disulfide, tungsten disulfide, etc.), teflon tape used in plumbing, air cushion and others. Dry lubricants such as graphite, molybdenum disulfide and tungsten disulfide also offer lubrication at temperatures (up to 350 °C) higher than liquid and oil-based lubricants are able to operate. Limited interest has been shown in low friction properties of compacted oxide glaze layers formed at several hundred degrees Celsius in metallic sliding systems, however, practical use is still many years away due to their physically unstable nature.

Another approach to reducing friction and wear is to use bearings such as ball bearings, roller bearings or air bearings, which in turn require internal lubrication themselves, or to use sound, in the case of acoustic lubrication.

Reduce Friction

The lubricant-to-surface friction is much less than surface-to-surface friction in a system without any lubrication. Thus use of a lubricant reduces the overall system friction. Reduced friction has the benefit of reducing heat generation and reduced formation of wear particles as well as improved efficiency. Lubricants may contain additives known as friction modifiers that chemically bind to metal surfaces to reduce surface friction even when there is insufficient bulk lubricant present for hydrodynamic lubrication, e.g. protecting the valve train in a car engine at startup.

Transfer Heat

Both gas and liquid lubricants can transfer heat. However, liquid lubricants are much more effective on account of their high specific heat capacity. The liquid lubricant is constantly circulated to and from a cooler part of the system, although lubricants may be used to warm as well as to cool when a regulated temperature is required. This circulating flow also determines the amount of heat that is carried away in any given unit of time. High flow systems can carry away a lot of heat and have the additional benefit of reducing the thermal stress on the lubricant. Thus lower cost liquid lubricants may be used. The primary drawback is that high flows typically require larger lumps and bigger cooling units. A secondary drawback is that a high flow system that relies on the flow rate to protect the lubricant from thermal stress is susceptible to catastrophic failure during sudden system shut downs. An automotive oil-cooled turbocharger is a typical example. Turbochargers get red hot during operation and the oil that is cooling them only survives as its residence time in the system is very short i.e. high flow rate. If the system is shut down suddenly (pulling into a service area after a high speed drive and stopping the engine) the oil that is in the turbo charger immediately oxidizes and will clog the oil ways with deposits. Over time these deposits can completely block the oil ways, reducing the cooling with the result that the turbo charger experiences total failure typically with seized bearings. Non-flowing lubricants such as greases & pastes are not effective at heat transfer although they do contribute by reducing the generation of heat in the first place.

Carry Away Contaminants and Debris

Lubricant circulation systems have the benefit of carrying away internally generated debris and external contaminants that get introduced into the system to a filter where they can be removed. Lubricants for machines that regularly generate debris or contaminants such as automotive engines typically contain detergent and dispersant additives to assist in debris and contaminant transport to the filter and removal. Over time the filter will get clogged and require cleaning or replacement, hence the recommendation to change a car's oil filter at the same time as changing the oil. In closed systems such as gear boxes the filter may be supplemented by a magnet to attract any iron fines that gets created. It is apparent that in a circulatory system the oil will only be as clean as the filter can make it, thus it is unfortunate that there are no industry standards by which consumers can readily assess the filtering ability of various automotive filters. Poor filtration significantly reduces the life of the machine (engine) as well as making the system inefficient.

Protect Against Wear

Lubricants prevent wear by keeping the moving parts apart. Lubricants may also contain anti-wear or extreme pressure additives to boost their performance against wear and fatigue.

Prevent Corrosion

Good quality lubricants are typically formulated with additives that form chemical bonds with surfaces to prevent corrosion and rust.

Seal for gasses: Lubricants will occupy the clearance between moving parts through the capillary force, thus sealing the clearance. This effect can be used to seal pistons and shafts.

Lubricants are generally classified depending on the compositions: (*i*) Paraffinic (*ii*) Naphthenic (*iii*) Aromatic. Lubricants are generally composed of a majority of base oil and a minority of additives to impart desirable characteristics.

7.10.1. Types of Lubricants

- Gas
- ·Liquid including emulsions and suspensions
- Solid
- Greases
- Adhesive

Gaseous lubricants: Gaseous lubricants have a much lower viscosity and higher compressibility compared to liquid lubricants, but the fluid-film principles apply analogous to gases. Some examples for gaseous lubricants are air (used in fluid bearings), technical gases, steam or liquid-metal vapours.

Liquid lubricants: Liquid lubricants may be characterized in many different ways. One of the most common ways is by the type of base oil used. Following are the most common types.

- Lanolin (wool grease, natural water repellant)
- Water

- Mineral oils
- Vegetable (natural oil)
- Synthetic oils
- Other liquids

Lanolin : A natural water repellent, lanolin is derived from sheep wool grease, and is an alternative to the more common petro-chemical based lubricants. This lubricant is also a corrosion inhibitor, protecting against rust, salts, and acids.

Water : Water can be used on its own, or as a major component in combination with one of the other base oils. Commonly used in engineering processes, such as milling and lather turning.

Mineral oil : This term is used for lubricating base oil derived from crude oil. The several types of lubricant base oil identified as:

- Group I - Saturates <90% and/or sulphur >0.03%, and Society of Automotive Engineers (SAE) viscosity index (VI) of 80 to 120
  - These lubricants are manufactured by solvent extraction, solvent or catalytic dewaxing, and hydro-finishing processes. Common Group I base oil are 150SN (solvent neutral), 500SN, and 150BS (brightstok)
- Group II – Saturates >90% and sulfur <0.03%, and SAE viscosity index of 80 to 120
  - Such lubricants are manufactured by hydrocracking and solvent or catalytic dewaxing processes. Group II base oil has superior anti-oxidation properties since virtually all hydrocarbon molecules are saturated. It has water-white color.
- Group III – Saturates > 90%, sulfur <0.03%, and SAE viscosity index >120
  - These groups of lubricants are manufactured by special processes such as isohydromerization. It can be manufactured from base oil or wax from dewaxing process.
- Group IV – Polyalphaolefins (PAO)
- Group V – These group includes lubricants other than mentioned above such as h as naphthenics, PAG, esters, and etc.
- Group IV and V are generally considered as synthetic esters.The lubricant industry commonly extends this group terminology to include:
- Group I with a Viscosity Index of 103 - 108
- Group II with a Viscosity Index of 113 - 119
- Group III with a Viscosity Index of >= 140

While lubricants for use in internal combustion engines may solely consist of one of the abovementioned oil groups, it is not desirable in practice. Additives to reduce oxidation, improve lubrication are added to the final product. The main constituent of such lubricant product is called the base oil, base stock. While it is advantageous to have a high-grade base oil in a lubricant, proper selection of the lubricant additives is equally as important.

Vegetable (Natural) Oils

These are primarily triglyceride esters derived from plants and animals. For lubricant base oil use of vegetable derived materials are preferred. Common ones include high oleic canola oil, castor oil, palm oil, sunflower seed oil and rapeseed oil from vegetable, and Tall oil from animal sources. Many vegetable oils are often hydrolyzed to yield the acids which are subsequently combined selectively to form specialist synthetic esters.

Synthetic Oils

- Polyalpha-olefin (PAO)
- Synthetic esters
- Polyalkylene glycols (PAG)
- Phosphate esters
- Alkylated naphthalenes (AN)
- Silicate esters
- Ionic fluids

Solid Lubricants

Teflon or PTFE : Teflon or PTFE is typically used as a coating material. For example, cooking utensils are coated with teflon to provide a non-stick surface. It can withstand temperature range up to 350°C and chemical inertness make it a useful additive in special greases. Under extreme pressures, teflon powder or solids is of little value as it is soft and flows away from the area of contact. Ceramic or metal or alloy lubricants must be used then.

Non-metals : Graphite, hexagonal boron nitride, molybdenum disulfide and tungsten disulfide are examples of materials that can be used as solid lubricants. These lubricants can withstand very high temperature. The use of some such materials is sometimes restricted by their poor resistance to oxidation (e.g., molybdenum disulfide can only be used up to 350°C in air, but 1100°C in reducing environments).

Metal/Alloy : Metal alloys, composites and pure metals can be used as grease additives or the sole constituents of sliding surfaces and bearings. Cadmium and Gold are used for plating surfaces which gives them good corrosion resistance and sliding properties, Lead, Tin, Zinc alloys and various Bronze alloys are used as sliding bearings, or their powder can be used to lubricate sliding surfaces alone, or as additives to greases.

Glaze' Formation (High Temperature Wear) : A further phenomenon that has undergone investigation in relation to high temperature wear prevention and lubrication, is that of 'glaze' formation. This is the generation of a compacted oxide layer which sinters together to form a crystalline 'glaze' (not the amorphous layer) generally at high temperatures, from metallic surfaces sliding against each other (or a metallic surface against a ceramic surface). Due to the elimination of metallic contact and adhesion by the generation of oxide, friction and wear is reduced. Effectively, such a surface is self-lubricating. As the 'glaze' is already an oxide, it can survive to very high temperatures in air or oxidising environments. However, it is disadvantaged by it being necessary for the base metal (or ceramic) having to undergo some wear first to generate sufficient oxide debris.

A large number additives families which are used to impart performance characteristics to the lubricants include:

- Antioxidants
- Detergents
- Anti-wear
- Metal deactivators
- Corrosion inhibitors, Rust inhibitors
- Friction modifiers
- Extreme Pressure

- Anti-foaming agents
- Viscosity index improvers
- Demulsifying/Emulsifying
- Stickiness improver, provide adhesive property towards tool surface (in metalworking)
- Complexing agent (in case of greases)

Application of Lubricants

The various applications of lubricants are summarized below

- **Automotive**
  - Engine oils
  - Petrol (Gasoline) engine oils
  - Diesel engine oils
  - Automatic transmission fluid
  - Gearbox fluids
  - Brake fluids
  - Hydraulic fluids
- **Industrial**
  - Hydraulic oils
  - Air compressor oils
  - Gas Compressor oils
  - Gear oils
  - Bearing and circulating system oils
  - Refrigerator compressor oils
  - Steam and gas turbine oils
- **Aviation**
  - Gas turbine engine oils
  - Piston engine oils
- **Marine**
  - Crosshead cylinder oils
  - Crosshead Crankcase oils
  - Trunk piston engine oils
  - Stern tube lubricants

Disposal and Environmental Issues

It is estimated that 40% of all lubricants are released into the environment. **Recycling, burning, landfill and discharge into water may achieve disposal of used** lubricant. There are typically strict regulations in most countries regarding disposal in landfill and discharge into water as even small amount of lubricant can contaminate a large amount of water. Most regulations permit a threshold level of lubricant that may be present in waste streams and companies spend huge amount of money annually in treating their waste waters to get to acceptable levels. Burning the lubricant as fuel, to generate electricity, is also governed by regulations mainly on account of the relatively high level of additives present. Burning generates both airborne pollutants and ash rich in toxic materials, mainly heavy metal compounds. Thus lubricant burning takes place in specialized facilities that have incorporated special scrubbers to remove airborne pollutants and have access to landfill sites with permits to handle the toxic ash. Unfortunately, most lubricant that ends up directly in the environment is due to general public discharging it onto the ground, into drains and directly into landfills as trash. Other direct contamination sources include runoff from roadways, accidental spillages, natural or man-made disasters and pipeline leakages. Improvement in filtration technologies and processes has now made recycling a viable option (with rising price of base stock and crude oil). Typically various filtration systems remove particulates, additives and oxidation products and recover the base oil. The oil may get refined during the process. This base oil is then treated much the same as virgin base oil however there is considerable reluctance to use recycled oils as they are generally considered inferior. Basestock fractionally vacuum distilled from used lubricants has superior properties to all natural oils, but cost effectiveness depends on many factors. Used lubricant may also be used as refinery feedstock to become part of crude oil. Again there is considerable reluctance to this use as the additives, soot and wear metals will seriously poison/deactivate the critical catalysts in the process. Cost prohibits carrying out both filtration (soot, additives removal) and re-refining (distilling, isomerisation, hydrocrack, etc.) however the primary hindrance to recycling still remains the collection of fluids as refineries need continuous supply in amounts measured in cisterns, rail tanks. Occasionally, unused lubricant requires disposal. The best course of action in such situations is to return it to the manufacturer where it can be processed as a part of fresh batches.

Lubricants both fresh and used can cause considerable damage to the environment mainly due to their high potential of serious water pollution. Further the additives typically contained in lubricant can be toxic to flora and fauna. In used fluids the oxidation products can be toxic as well. Lubricant persistence in the environment largely depends upon the base fluid, however if very toxic additives are used they may negatively affect the persistence. Lanolin lubricants are non-toxic making them the environmental alternative which is safe for both users and the environment.

Large volumes of hydrocarbon solvents are consumed in metal degreasing, electronic component cleaning, car-dewaxing, industrial dry cleaning, household products, photocopying etc.

7.11. PETROLEUM WAXES

Available references show that petroleum waxes were first recovered in quantity in the U.K. in about 1856-1857 by distilling Rangoon crude, imported in drums. These waxes, were utilized for replacing bees wax and spermaceti in the manufacture of candles. Availability of these cheap waxes immediately brought down the prices of candles in those days which were till then made with costly bees wax and spermaceti. As a result, it was possible even for the poorest to light a candle, which was till that time confined only to the rich and the churches. From this beginning, the trade started to grow and today petroleum waxes have many industrial applications. Now these waxes constitute a range of valuable products from crude oil.

7.11.1. Types of Petroleum Waxes

These are hydrocarbons type waxes and consist of normal and iso-paraffins and alkylated naphthenes in the range of C_{18}-C_{70}. The commercial petroleum waxes can be broadly classified as paraffin wax, microcrystalline wax and petrolatum. Paraffin waxes constitute the major bulk of such waxes. The other two types, produced in comparatively smaller quantities, also command a good market because of their certain specific end uses. The paraffin waxes are produced from waxy distillate of paraffin-base or mixed-base crudes of high pour point. The microcrystalline waxes, on the other hand, are manufactured from the distillation residue of the above crudes, waxy sludge deposited in their storage tanks or from sucker rod wax. The latter material is a deposit generally scraped out of the casings of crude producing wells. Petrolatum, the third category of these waxes, are popularly known as Vaseline. These contain microcrystalline waxes and oil, and can be produced only from certain suitable heavy distillates or residue. The feasibility of manufacturing these waxes from a certain crude, therefore, depends not only on *the* presence of wax in the crude but also on the nature, quantity and the distribution of the wax content in the crude itself. Incidentally, most of crudes currently produced in India are suitable for manufacturing paraffin and microcrystalline waxes.

Paraffin waxes. These consist mainly of normal and slightly branched paraffins and has a relatively pronounced crystal structure. In India, three different types of paraffin waxes are produced. In each of these types, there are a number of melting point grades spreading over a range of 5°F.

Paraffin wax of Type 1 is an extra refined paraffin wax. It contains very low oil (maximum of 0.25 wt.%) and is used mainly in food stuff packaging industry. This wax is also used for some cosmetic preparations. Melting point grades produced are 135/140°F, 140/145°F and 145/150°F.

Paraffin wax of Type 2 is a refined paraffin wax, containing a maximum of 0.5 percent oil by weight. This is mainly used for candle manufacture, coating of paper for general packaging (other than foodstuff packaging), tarpaulins, polishes and hard boards, water proofing formulations, preparing wax emulsions, processing rubber goods etc. Major grades produced in this type are 135/140°F, 140/145°F and 145/150°F.

Paraffin wax of Type 3 is a semi-refined paraffin wax containing a maximum of 3.5 percent oil by weight. Only one grade is produced in this type, i.e. 128/132°F. This is a comparatively soft wax and is used mainly in the impregnation of match sticks, textile and pencil industries water proofing formulations, making cerfain wax emulsions, etc.

The important specifications of waxes are melting point, blocking point, colour, odour, oil content, penetration, viscosity, PDA purity etc. These specifications were derived as per end uses. The requirements for the three types of paraffin waxes are given in Table 7.31. In addition, the requirements for paraffin wax for explosives and pyrotechnic industry (IS: 7401-1974) are given in Table 7.32. This specifies two grades, Soft and Hard, for use in explosive compositions and as a water proofing agent for paper shells.

Microcrystalline waxes. These contain a considerable quantity of branched-chain hydrocarbons, though some straight-chain hydrocarbons are also present. It differs from paraffin wax both in the size and shape of wax crystals. The commonly used microcrystalline waxes in India are graded as follows :

Grade I is a refined microcrystalline wax of flexible type with a melting point of 165°F. The oil content is about 4 percent. This is mostly used in laminations, sealing compounds, petroleum jellies, rust preventives and cosmetics.

Grade II is also a flexible type of microcrystalline wax. This differs from Grade I in that it is an extra refined wax. Its oil content is 4 percent. It is used in applications similar to Grade I.

TABLE 7.31. Requirement for Paraffin Wax (IS : 4654-1974)

Sl. No.	Characteristics	Requirement for Type 1	Type 2	Type 3
1.	Melting point, °C (cooling curve)	45 to 75		
2.	Ash, wt.%, Max.	0.03	0.03	0.03
3.	Acidity :			
	(*a*) Organic, mg KOH/g, Max.	0.10	0.10	0.10
	(*b*) Inorganic, mg KOH/g, Max.	0.02	0.02	0.02
4.	Saponification value, Max.	1.0	1.0	1.0
5.	Oil content, wt.%, Max.	0.25	0.5	3.5
6.	Colour			
	(*a*) Lovi bond, in 18" cell, Max.	0.5Y	2.0Y	—
	(*b*)ASTM, Max.	—	—	2.0
7.	Acid discolouration test	To pass the test		
	or			
	Ultra voilet (UV) absorbance per-centimeter path length			
	280-289 millimicrons	0.15 (Max.)	—	—
	290-299 millimicrons	0.12 (Max.)	—	—
	300-359 millimicrons	0.08 (Max.)	—	—
	360-400 millimicrons	0.02 (Max.)	—	—
8.	Odour, Max.	1	—	—

TABLE 7.32. Requirement for Paraffin Wax for Explosive and Pyrotechnic Industry (IS: 7401-1974)

SI. No.	Characteristics	Requirement far Soft grade	Hard grade
1.	Melting point (cooling curve) °C	52-56	60-64
2.	Acidity, mg KOH/g, Max.	0.10	0.10
3.	Loss on drying at 90°C, wt.%, Max.	0.5	0.5
4.	Saponification value, Max.	1.0	1.0
5.	Penetration (by vica+ needle), mm, Max.	3	1
6.	Ash, wt.%, Max.	0.03	0.03

Grade III is a hard, higher melting type of wax. Its melting point is 190°F and oil content is about 1.5 percent. It is used in applications where low oil content is the major consideration. Main applications are polishes, carbon paper, printing inks, paper coating and electrical insulation.

The requirements for microcrystalline waxes are listed in Table 7.33.

7.11.2. Properties of Petroleum Waxes

Paraffin waxes are snow-white solid crystalline material having melting points between 45 to 75°C. These tasteless, odourless, chemically stable and translucent solid substances are a mixture of saturated paraffinic hydrocarbons having anything from 18 to 35 carbon atoms per molecule. These waxes contain mainly normal paraffins with or without small proportions of iso-and/or cycloparaffins.

Microcrystalline waxes as the name implies are amorphous in appearance and may have any colour from dark brown through amber to white depending upon the degree of refining. The average molecular weight of these waxes is normally in the range of 600-700. They contain besides straight chain paraffins, iso-paraffins, monocyclopentyl-, monocyclohexyl and dicyclohexyl paraffins, aromatics and polycyclo-paraffms. These waxes have melting points higher than 74°C.

TABLE 7.33. Requirements for Microcrystalline Wax

Sl. No.	Characteristics	Requirement for		
		Grade 1	Grade 2	Grade 3
1.	Melting point, °C, Min.	74	74	88
2.	Needle penetration at 25°C, 100g, 5 sec, 1/100 cm	25-35	25-35	2-7
3.	Kinematic viscosity at 98.9°C			
	cSt	15.52-17.98	15.52-17.98	15.52-17.98
	SUS	80-90	80-90	80-90
4.	Colour:			
	(a) Saybolt, Min.	+16	—	—
	(b) ASTM, Min.	—	2	1
5.	Flash point, °C	274	274	288
6.	FDA	To pass	To pass	To pass

Petrolatum which are also refereed to as petroleum jellies consist of certain types of microcrystalline waxes and oil. All microcrystalline waxes however do not p ive satisfactory petrolatum. Their formation seems to depend mainly upon the structure and i Jecular weight of the wax constituents rather than the nature of the oily constituent.

7.11.3. Manufacture of Petroleum Waxes

Initially cold settling, chilling and pressing methods were practiced for producing petroleum waxes. But these methods had certain limitations in respect of filtration, proper development of wax crystals etc. and were not suitable for processing a wide range of waxy stocks.

The first major breakthrough in the wax production technology came up in 1927 when the first commercial ketone deoiling plant was installed by Indiana Refining Co., USA. In coming years, alternate processes, such as propane dewaxing/deoiling, emulsion deoiling, spray deoiling, electric precipitation and magnetic separation were developed. However, the ketone dewaxing/deoiling continues to be the most widely used process today. The ketone solvents generally used are methyl ethyl ketone (MEK) and methyl iso-butyl ketone (MIBK). In India, the production of petroleum wax started with the commissioning of the wax sweating plant of Assam Oil Company at Digboi Refinery. The wax sweating is a process of fractional melting solution and drainage. The molten wax is solidified in the coils of sweating stove and then heated at a slow and a controlled rate. The oil sweats, drips and finally leaves a wax cake of very low oil content which is then melted and treated with acid and clay for improving its colour. This process cannot be employed for deoiling of all varieties of feedstocks particularly slack waxes containing large percentage of iso-paraffins which must be handled by the solvent deoiling process only. The production of paraffin wax at MRL, Madras and IOC, Barauni is based on MIBK solvent deoiling process.

7.11.4. Uses of Petroleum Waxes

The age old use of paraffin wax for making candles is still continuing as a major outlet for this product. In the meantime, more and more new uses have come into being during the years demanding different melting point grades and qualities of wax. One such use of great significance is in paper conversion, *i.e.* manufacture of waxed papers and cartons. Its compatibility with some of plastics has opened up a new field in the coating technology where properties of paraffin wax and plastics are combined and imparted to base paper. As coating of wax makes paper impervious to air, water and water vapour, and wax being odourless, tasteless, non-toxic and chemically stable, these waxed papers and cartons are ideally suitable for packing foodstuff. Cold drinks, milk, ice cream and even coffee are available now-a-days in waxed paper tumblers, bottles, bowls or cups which are cheap, hygienic and could be discarded after use. Biscuits kept fresh in waxed cartons are a common experience of everybody. These waxed papers are also used as a wrapping material for a variety of products like bread, soap tablets, razor blades, tobacco etc.

In the processing of paper, paraffin wax along with film fanning materials such as starch, protein or polymethyl alcohol etc. is used as sizing material. Wax sizing contributes many improvements to the paper and paper board, including redaction in the tendency to curl, improved fold and pliability and enhanced resistance to water vapour, ink and improved appearance with coating mixtures.

A low melting point grade of paraffin wax is used to impregnate match sticks to give it a smooth burning property. This type of paraffin wax is popularly known as match wax.

These waxes are also used for water proofing fabric and leather, as a modelling composition, and in the manufacture of crayons, glass pencils, coloured pencils, polishes etc. The range of polishes in which petroleum waxes are used, are very wide and include shoe polishes, wood and furniture creams, floor and linoleum polishes, cleansing and polishing fluids, etc.

Another interesting use is in the coating of fresh fruits and vegetables to preserve them from drying up and withering during transport from the producing centre to the consumers. This is done by spraying a solution of wax in a volatile solvent.

Dry cells of various types have a paper lining inside the zinc case which is coated or impregnated with paraffin wax. The carbons used in cells are also impregnated with paraffin wax to reduce gas permeability.

In the electrical industry, waxes are used in cable making as an ingredient of the insulating material in admixture with other substances and as a coating of the covering for certain types of cable. These are also used as dielectric medium in electrical paper condensers.

Paraffin waxes are now being used also as a raw material for the manufacture of chlorinated hydrocarbons, detergents of alkane carboxylate and primary or secondary alkyl sulphate types. Paraffin wax of approximate carbon number range C_{25} - C_{30} is a source of C_6 - C_{20} linear α-olefins via steam cracking.

Microcrystalline waxes are used to dope paraffin wax used in paper conversion and polishes etc. Solution of paraffin waxes doped with these waxes forms a smooth creamy paste instead of a rough solid mass on cooling due to diminished crystal size of the former. These waxes are also used as a substitute for expensive vegetable waxes required for specific uses, as sealing wax and in the electrical industry.

Petrolatum or petroleum jellies are almost universally referred to as vaseline. In fact, vaseline is a registered trade mark of an American company for such materials they produce, and hence should not be used to describe these materials as a class. These are valued as salves and the highly refined white jellies find use in pharmaceutical and cosmetic preparations.

7.11.5. Quality Requirements-Industry-wise/End-Use Wise

Match industry. The Indian standard for paraffin waxes specifies Type 3 as suitable for match industry. This is a semi-refined wax with a maximum oil content of 3.5 wt.%, and maximum ASTM colour of 2.0. Digboi refinery produces paraffin wax of Type 3 with melting point range of 128 - 132°F for this purpose and it is this grade which is mainly consumed by the match industry.

Candle manufacture. As per Indian specification, Type 2, refined paraffin wax is suitable for making candles. The specification for tins particular type stipulates a maximum oil content of 0.5 wt.%. and maximum Lovibond colour of 2.0 Yellow in 18" cell.) t Textiles. This covers the use of wax for sizing by textile mills. Normally, paraffin wax of Type 3 is used for this, purpose. For some coarser varieties of cloth, match wax with an oil content up to 3.5 wt.% is also used.

Tarpaulin. While a manufacturer of tarpaulin can use any suitable wax composition, the sweating test in the specifications for tarpaulin can be met only if a high melting point wax with low oil content is used. The grade of wax preferred by the industry is Type 2, which is a general purpose refined paraffin wax. However, match wax with an oil content up to 3.5 wt.% is also used in admixture with paraffin wax of Type 2.

Paper waxing. The paper industry is by far tiie largest user of petroleum waxes. Absence of colour, odour and taste makes paraffin wax suitable for the entire range of food packaging. Its compatibility with some of the plastics has opened a new field in the coating technology where properties of paraffin wax and plastics are combined and imparted to base paper. In the case of food packaging, where the wax coating comes in direct contact with the edible products, only paraffin wax of Type 1, which is a fully refined wax specially produced to meet the requirements of food packaging, should be exclusively used. IOC, Digboi makes this grade and identifies it as P.I. grade (Paper industry grade). Wax used as ingredient in edible products (e.g. chewing gum) must also conform to Type 1. For waxed paper used in general packaging, i.e. engineering goods, consumer articles, etc. paraffin wax of Type 2 is to be used for such applications depending on the quality of the finished products.

Chlorinated paraffin wax. Chlorinated paraffin wax is mainly used in flexible type of polyvinyl chloride (PVC) plastics as a secondary plasticiser. The primary plasticiser, di-octyl phthalate, is used up to 40 wt.% of the PVC and about 5-15 percent can be replaced by chlorinated paraffin wax. In some uses, like compounding for PVC shoes, up to 40 percent chlorinated paraffin wax can be used.

While a certain minimum amount of chlorinated paraffin wax (about 5 percent) is required purely from technical consideration, larger quantities are often used because of the price differential in favour of chlorinated paraffin wax. As regards the quality of wax required for manufacturing chlorinated paraffin wax for use as secondary plasticiser, it is reported that a high ofl content is not desirable since it affects the heat stability of the product. Heat'stability is one of the important requirements of secondary plasticisers and it is measured by keeping the chlorinated wax sample in an oven at 180°C for one hour and measuring the colour change: A satisfactory product should not show any colour change. Paraffin wax of Type 2 is generally used for the manufacture of chlorinated paraffin wax.

Explosives. The ordnance factories as well as manufacturers of explosives and detonators use paraffin wax. The ordnance factories use paraffin wax partly to prepare protective compositions for small arms.

Chemical products. Wax based emulsions are one of the major products in this field. These emulsions are mainly employed by the food, textile and paper industries. In the food industry, these emulsions are used for prolonging storage life of perishable fruits and vegetable. For the textile emulsions, either paraffin wax of Type 2 or match wax is used. Emulsions made for use as a resin substitute by the paper industry are made from paraffin wax containing a maximum oil content of 0.5 percent (Type 2).

Besides emulsions, chemical products- based on wax find usage in wood preservatives, paints, wax formulations for surface coating industry, impregnation solutions, wire-rope products, rust preventives, etc.

Rubber industry The tyre factories are the main consumers of wax. Mainly paraffin wax of Type 2 is consumed by this industry. Wax is used as a component in specialized formulations with little scope for change in the wax specification. However, the total quantity mvolvedisnotlarge.Waxisusedasacompoundmgagentinotherrubbergoods,likebatteries, shoes, hoses, belts, bands, toys, etc. For this application, paraffin wax ofType 2 is used, though some uses match wax also.

Petrolatum. The three categories of this product are: (1) Yellow commercial grade petrolatum, (2) Pharmaceutical grade petrolatum and (3) Mineral jelly. Petroleum wax used for the manufacture of these jellies is mainly microcrystalline wax, though to some extent paraffin wax is also used in some cases. Conventionally, refined lubricating oils form an important constituent of commercial grade petrolatum and mineral jelly. As in the case of food packaging industry, paraffin wax used in the formulation of IP grade petrolatum should be exclusively paraffin wax ofType 1. For commercial grades and mineral jelly, paraffin wax of Type 2 or 3 can be used.

Hard boards. Under this category, chip, particle and hard boards are covered. The chip boards are made after cutting wood into small chips which are sprayed with wax for waterproofing and then treated with resin which acts as binding agent. The consumption of wax for this industry varies from 0.3 to 1.0 wt.%.

Hard boards are made from wood pulp or wood fibre. For this two processes used are wet process and dry process. Both these processes use wax to make the wood waterproof before making it into boards. It is stated that the wet process consumes 1 to 1.5 wt.% wax, whereas dry process uses slightly more up to 2 wt.%.

Other miscellaneous uses. This category covers a large number of miscellaneous uses which individually require fairly small quantities of wax, but together contribute a sizeable volume of demand.

Crayons of many types are made from paraffin waxes and microcrystafline waxes plus a dye of pigment.

Waxes are used in making polishes. This use includes car, floor, shoe and furniture polishes. These polishes use paraffin wax as well as certain grades of natural/synthetic waxes apart from microcrystalline wax. Paraffin wax of Type 2 is normally used and for certain general purpose polishes, match wax with oil content up to 3.5 wt.% is also acceptable.

The industries mentioned above are the major users of paraffin wax. hi addition to these, wax finds numerous other retail applications. For instance, one can find paraffin wax (often blended with natural waxes) being sold in small shops in all towns and most of the villages for small applications which could be as varied as for Batik printing to sizing of stitching threads.

7.12. ASPHALT

Asphalt is a sticky, black and highly viscous liquid or semi-solid that is present in most crude petroleums and in some natural deposits sometimes termed **asphaltum**. It is most commonly modelled as a colloid, with *asphaltenes* as the dispersed phase and *maltenes* as the continuous phase Asphalt is also called **bitumen** in countries like India. Asphalt or bitumen can sometimes be confused with tar, which is a similar black thermo-plastic material produced by the destructive distillation of coal. During the early- and mid-twentieth century when town gas was produced, tar was a readily available product and extensively used as the binder for road aggregates. The addition of tar to macadam roads led to the word tarmac, which is now used in common parlance to refer to road making materials.

Asphalt can be separated from the other components in crude oil (such as naphtha, gasoline and diesel) by the process of fractional distillation, usually under vacuum conditions. A better separation can be achieved by further processing of the heavier fractions of the crude oil in a de-asphalting unit, which uses either propane or butane in a supercritical phase to dissolve the lighter molecules which are then separated. Further processing is possible by "blowing" the product (reacting it with oxygen). This makes the product harder and more viscous.

Natural deposits of asphalt include lake asphalts , Gilsonite, the Dead Sea, and Tar Sands.

Asphalt is typically stored and transported at temperatures around 300 degrees Fahrenheit (150°C). Sometimes diesel oil or kerosene are mixed in before shipping to retain liquidity; upon delivery, these lighter materials are separated out of the mixture. This mixture is often called **bitumen feedstock**, or BFS. Some dump trucks route the hot engine exhaust through pipes in the dump body to keep the material warm. The backs of tippers carrying asphalt, as well as some handling equipment, are also commonly sprayed with a releasing agent before filling to aid release. Diesel oil is sometimes used as a release agent, although it can mix with and thereby reduce the quality of the asphalt.

Uses

Natural asphalt deposits were used for mortar between bricks and stones, ship caulking, and waterproofing. The Persian word for asphalt is *mumiya,* which is related to the English word mummy. Asphalt was also used by ancient Egyptians to embalm mummies. In the ancient Far East, natural asphalt was slowly boiled to get rid of the higher fractions, leaving a material of higher molecular weight which is thermoplastic and when layered on objects, became quite hard upon cooling. This was used to cover scabbards and other objects that needed water-proofing. Statuettes of household deities were also cast with this type of material in Japan, and probably also in China. In North America, archaeological recovery has indicated that asphaltum was sometimes used to apply stone projectile points to a wooden shaft. Poured bitumen has also been used as a damp-proof course in building. Roads are been paved with asphalt.

Rolled asphalt concrete: The largest use of asphalt is for making asphalt concrete for road surfaces and accounts for maximum use of asphalt. . Due to its highly viscous nature, asphalt is commonly heated for road construction to make it soft and easy to work. Special heated tanker trucks are employed in road building to apply hot asphalt to the road bed. Roofing shingles account for most of the remaining asphalt consumption. Other uses include cattle sprays, fence post treatments, and waterproofing for fabrics. In areas of the world with winter climates, asphalt must be protected from cracking with a procedure called **seal coating**. The freezing and thawing process that occurs causes small cracks in the asphalt. As more freezing water gets into these small cracks, the expansion causes further damage to the asphalt. Sealcoat is applied in an effort to prevent additional damage. The use of sealcoat can lengthen the life of asphalt driveways or parking lots significantly.

Mastic Asphalt

Mastic asphalt is a type of asphalt which differs from dense graded asphalt (asphalt concrete) in that it has a higher bitumen (binder) content, usually around 7–10% of the whole aggregate mix, as opposed to roller asphalt, which has only around 5% added bitumen. Another asphalt which is fast gaining global popularity is stone mastic asphalt (SMA). SMA's advantages over rolled asphalt is its high anti-skid qualities due to its high aggregate density and the lack of void content (air pockets). Another advantage of SMA is its longer durability over alternative road asphalt surfaces, but its manufacture and application, if not controlled closely, can result in slippery road surfaces due to excess bitumen pooling (bleeding) onto the surface. Asphalt is widely used in airports around the world. Due to the sturdiness, it is widely used for runways dedicated to aircraft landing and taking off.

Asphalt Emulsion

A number of technologies allow asphalt to be mixed at much lower temperatures. These involve mixing the asphalt with petroleum solvents to form "cutbacks" with reduced melting point or mixtures with water to turn the asphalt into an emulsion. Asphalt emulsions contain up to 70% asphalt and typically less than 1.5% chemical additives. There are two main types of emulsions with different affinity for aggregates, cationic and anionic. Asphalt emulsions are used in a wide variety of applications. Chipseal involves spraying the road surface with asphalt emulsion followed by a layer of crushed rock or gravel. Slurry seal involves the creation of a mixture of asphalt emulsion and fine crushed aggregate that is spread on the surface of a road. Cold mixed asphalt can also be made from asphalt emulsion to create pavements similar to hot-mixed asphalt, several inches in depth and asphalt emulsions are also blended into recycled hot-mix asphalt to create low cost pavements.

Indian Oil refineries at Panipat, Mathura, Koyali, Haldia and CPCL produce grades Bitumen 80/100; Bitumen 60/70; Bitumen 30/40; Bitumen CRMB and Bitumen Emulsion- 'Indmul'. It is available both in packed and in bulk.

General uses of Bitumen/Asphalt are as follows :

For civil engineering works

- Constructions of roads, runways and platforms.
- Water proofing to prevent water seepage.
- Mastic floorings for factories and godowns.
- Canal lining to prevent eroding.
- Dump-proof courses for masonry.
- Tank foundation.
- Joint filling material for mason

Industries :

- Electrical cables and junction boxes.
- Battery manufacturers as sealing compound.
- Paint industries for manufacturing black paints and anti corrosive paints.
- Ceramic industries.
- Printing inks.

- Water proof papers.
- Electrical capacitors.
- Bituminous felts.
- Bituminous grease for lubricating open gears.

Many roofing fetts today employ a glass or polyester felt. If these are used no saturant is required. Roofing felts made in this way, because of the characteristics of the bitumens used, can be rolled for storage and provide flexible characteristics in the variable temperature conditions experienced in roof structures.

Impregnated felts are also used for damp-proof courses in masonry. Mastic asphalts, which are mixtures of finery-divided limestone or other fine aggregate with a high content (12-17 wt.%) of a hard penetration bitumen (15 to 30 penetration) are used in flooring and roofing.

The electrical properties of bitumens enable them to be used in wrappings and jointing compounds for heavy duty cables. Bitumens also find wide use in sound insulation purposes, e.g. in car bodies and floor mats and in floor mountings for factory machinery.

To protect pipelines for oil, water, etc. coatings of bitumen enamel are applied after a primer onto the cleaned metal surface. The enamel is an oxidized bitumen (e.g. 115/15) with up to 30 percent of an inert filler, such as slate dust. The primer is a cutback of the oxidized bitumen with a volatile solvent (white spirit).

Considerable quantities of bitumen are used as binders to briquette coal fines. Bitumen is used hot, or cold (grounded) or as an emulsion/Bitumen is also used as a pelletizing binder in the recovery of mill scale and "red mud" steel wastes. The most suitable bitumens are hard and of low penetration index, e.g. 85/2 or 90/1, (softening point/penetration at 25°C). Other grades such as 15 penetration or H80/90 may also be satisfactory.

7.13. PETROLEUM COKE

Petroleum coke is considered a valuable solid byproduct of crude oil refining. It is produced in refinery coking units using residual oils, i.e. heavy fractions of the crude oil as feedstocks. Basically, coking operations consist of heating the heavy oil feedstock to high temperatures, thereby cracking the oil and driving off lighter products. The residue remaining after the removal of lighter products under controlled temperature and pressure conditions is petroleum coke.

Qualities required for a petroleum coke are low sulphur, ash and metals contents, as well as certain properties such as low porosity, good conductivity and low coefficient of thermal expansion. Good quality petroleum coke for electrode manufacture can be manufactured from high aromatic, high molecular weight petroleum stocks.

7.13.1. Types of Petroleum Cokes

There are three different types of coking processes-delayed coking, fluid coking and flexicoking. These coking processes are all effective in converting residual oils to a higher value products and concentrating the contaminants (sulphur, metals, etc.) in the coke. The petroleum coke produced from each of the three processes is distinctly different.

Petroleum coke from the more common delayed coking is described as delayed sponge, shot or needle coke, depending on its physical structure. Most of the material produced in delayed cokers is interspaced with irregular voids and is commonly called sponge coke and contains greater surface area. When the delayed coker feed is a waxy residue, high in paraffins and low in cyclic molecules, shot coke can result. Shot coke can also result if the coker is not operated correctly or if there is an operating upset. Shot coke has much less surface area, can be extremely hard, has a high coefficient of thermal expansion, and is difficult to handle. Needle coke has large unidirectional pores, elliptical, largely interconnected and surrounded by thick walls. The naming is also well suited because the coke forms needle-shaped pieces when broken. Needle coke is produced from selected aromatic feedstocks. Its unique structure tends its use for graphitization.

Fluid coking, a continuous process, produces a hard, dense material similar in consistency to coarse sand. This coking process typically yields coke with relatively higher concentrations of sulphur and metals than delayed coking process. Fluid coke, the petroleum coke from the fluid coking process, is spherical in shape, contains less volatile materials and is much harder than sponge coke. Normal size is less than 6 mm, and it does not tend to agglomerate like shot coke. Flexicoking process is an extension of the fluid coking process in that some (over 60 percent) of the coke is gasified to low heat content gas for refinery use. The resultant purge coke is called flexicoke. The flexicoke has relatively small particle size, i.e. 80 percent of the product passes 200 mesh. It is either very wet (cake) or very dry. Both these conditions create difficult problems in subsequent handling, storage and transportation.

The characteristics of the cokes from the three coking processes are illustrated in Table 7.34.

TABLE 7.34. Characteristics of Petroleum Cokes

Process	Type of coke	Characteristics
Delayed coking	Sponge coke	Sponge-like appearance Higher surface area Lower contaminants levels Higher volatiles Higher HGI* (>50)
	Shot coke	Spherical apperarance Lower surface area Lower volatiles Lower HGI (< 50) Tends to agglomerate
	Needle coke (Aromatic feed - thermal tars, decant oils)	Needle-like appearance Low volatile High carbon content
Fluid coking	Fluid coke	Low volatile Higher contaminants levels Low HGI (< 40) 3 mm particle size
Flexicoking	Flexicoke	Highest contaminants levels 80 percent passes 200 mesh

*Hardgrove GrindSibility Index

7.13.2. Properties of Petroleum Cokes

The physical properties of petroleum coke are determined by the type of coking process, feedstock properties and coker operating conditions. Petroleum coke is a dark gray or black infusible solid that can go through aplastic stage at high temperature (1500°C). It is insoluble in water. Depending on the processing, it may contain from a few tenths of a percent to about 15 percent volatile matter and hence may have a slight hydrocarbon odour. These volatile components may be soluble in hydrocarbon solvents. Typical density of petroleum coke is 830kg/m^3.

Carbon is the major chemical constituent of petroleum coke. Depending on the type of coke, petroleum coke can contain about 85-99 wt.% carbon. In addition, petroleum coke contains hydrogen, sulphur, nitrogen and oxygen. The substantial portion of this is sulphur. Cokes can contain up to 1.5 wt.% sulphur. Low sulphur content of coke is desirable for making electrodes. Coke with high sulphur content of approximately 1.5 wt.% may lead to vapour formation (called puffing) during electrode manufacture and can result in anode fractures. Sulphur in the coke is a function of sulphur in feed, coke make and source of the crude oil. In general, a low-sulphur residue containing, say 0.25 wt.% sulphur, produce a coke with twice the sulphur content. Under normal conditions to meet a 1.5 wt.% sulphur in the coke, the sulphur content of the feed should be less than about 0.9wt,%. Larger amount of sulphur exists in sponge cokes than in needle cokes.

Petroleum coke also contains trace amounts of metals such as silicon, iron, vanadium, magnesium, manganese, sodium and calcium. Under normal conditions, petroleum coke is stable and non-reactive and does not polymerize. Combustion of petroleum cokes produces oxides of sulphur and carbon. Incomplete combustion generates carbon monoxide also.

TABLE 7.35. Requirements of Petroleum Coke (IS : 8506-1977)

SI. No.	Characteristics	Requirement far Raw petroleum coke	Calcined petroleum coke
1.	Moisture content (as received) wt.%, Max.	10	—
2.	Moisture content (after initial drying for raw petroleum coke only), wt.%, Max.	2	0.1
3.	Ash, wt.%, Max.	0.45	0.5
4.	Volatile matter, wt.%, Max.	12.0	0.4
5.	Density		
	(*a*) Bulk at 2TC, kg/l	To be reported	
	(*b*) Real* at 27°C, kg/1, Min.	—	2.03
6.	Fixed carbon, wt.%, Min.	85.0	99.0
7.	Total sulphur, wt.%, Max.	1.25	1.25
8.	Trace metals		
	(*a*) Silicon, wt.%, Max.	—	0.04
	(*b*) Iron, wt.%, Max.	—	0.04
	(*c*) Vanadium$^+$, wt.%, Max.	—	0.03

*For graphite industry a higher real density and low vanadium content product is required; the limits for this may be settled between the purchaser and the supplier

Indian specifications of petroleum cokes are listed in Table 7.35. The most important properties of coke to be used as graphite electrode in steel production relate to the electrode's ability to carry a current and to withstand the thermal and mechanical stresses in the arc furnace. Typical coke properties that relate to the properties of the electrode include (1) coefficient of thermal expansion (CTE), (2) bulk density, (3) mechanical strength of the coke grains, (4) particle-size distribution and (5) electrical resistivity of coke particles. Other properties such as sulphur, ash and metal contents are also important. Although many of the same properties are specific for the anode-grade coke used in aluminium production, some critical differences exist.

7.13.3. Storage and Transportation of Petroleum Cokes

Green petroleum coke is normally stored in open piles with a stacker reclaimer. Front-end loaders or grab cranes can be used to move the green coke. Dust from the piles is normally controlled with water sprays. The coke can be transported by truck, rail car, barge, or ship, depending on quality and destination.

Calcined petroleum coke is usually stored in closed containers to keep the material dry and to prevent environmental problems from dust. The coke is loaded via conveyor belts and shipped by truck, rail car, barge, or ship. Calcined coke must be shipped in closed system to prevent its contamination.

7.13.4. Uses of Petroleum Cokes

Typical end-uses of petroleum cokes are listed in Table 7.36. The different types of petroleum coke are used as an industrial carbon or an energy source. The chemical and physical properties of cokes generally dictate the end-use application. As an industrial carbon, petroleum cokes function as a reductant in the manufacture of aluminium, as a carbon source in steel manufacturing and as an inert electrical conductor in électric arc steel manufacturing. The quality requirements for the production of both steel and aluminium differ significantly. In the electric arc steel process, electric current is transferred to scrap metal through graphite electrodes, the major constituent of which is petroleum coke. In the production of aluminium, petroleum coke constitutes the major raw material for the anode in this electrolysis. Approximately 0.4-0.5 kg of calcined petroleum coke is used per kg of alumina produced. Calcined needle coke is used to manufacture graphite products as well as prebacked graphite electrodes used in electro-metallurgical furnaces.

Smaller volumes of petroleum coke are used in a variety of applications, including its use as a carbon source in the production of titanium dioxide and various steel alloys, as a carbon anode in production of direct-current batteries. Many other small-volume applications continue to emerge, making use of the excellent thermal conductivity, and high temperature mechanical properties of this carbon source.

The use petroleum coke as an industrial carbon requires calcination of green coke to increase the purity. Higher quality petroleum coke with low sulphur and metals concentrations and desirable physical characteristics is further processed in calcining unit. In the manufacture of electrodes the suitability of calcined petroleum coke is due to its improved binding capacity and facility for moulding and extrusion. Due to loss of volatile matter it results in higher real density electrode.

In the energy (fuel) applications, the raw/green cokes are used. Many applications of petroleum coke require screening and grinding prior to combustion. Low HGI can reduce crushing/pulverizing capabilities. Petroleum coke has been substituted for steam coal in some instances because it is usually softer and requires less grinding capacity and input energy for grinding. Petroleum coke produced from the residue of high-sulphur and high-metal crude oils can generally be used only as a source of energy. This coke is usually blended in coal-fired furnaces and cement kilns. The coke in this case provides additional heating value and also reduces the ash content of the blend. However, in many cases it slightly increases the sulphur content of the blend. The high sulphur levels in coke can contribute to the SO_2 emissions. Petroleum coke for use in cement/lime industry can tolerate high sulphur level (most SO_2 is absorbed in the cement).

TABLE 7.36. Typical End-Uses of Petroleum Coke

Application	Type of coke	State	End-use
Carbon source	Needle coke	Calcined	Electrodes for electrometallurgical industries
			Synthetic graphite
	Sponge	Calcined	Aluminium anodes
			TiO_2 pigments
			Carbon raiser
	Sponge	Green	Silicon carbide
			Foundaries
			Coke ovens
Fuel use	Sponge	Green lump	Space heating
	Sponge	Green	Industrial boilers
	Shot	Green	Utilities
	Fluid	Green	Co-generation
	Flexicoke	Green	Lime
			Cement

REVIEW QUESTIONS

1. Give composition and properties of natural gas.
2. What are the characteristics of LPG ? Give its application.
3. Discuss briefly about gasoline. What are the additives used to improve its octane rating ?
4. Explain the functioning of a typic 4-stroke petrol engine.
5. What are the alternative antiknock agents used in gasoline ?
6. Explain the functioning of a compression ignition diesel engines.
7. What are the fuel-related problem of using vegetable oils directly in diesel engines ?
8. What are the various process used to reduce the viscosity of vegetable oil ?
9. What are the advantages of using vegetable oil as alternative diesel fuel ?
10. Write short note on synthetic petrol.
11. Explain Bergius and Fischer-Tropsch process for liquefaction of coal.
12. What are the uses of kerosene ?
13. What are synthetic and natural lubricants ?
14. Write short notes on petroleum wax.
15. What are the uses of asphalt ?

❑❑❑

CHAPTER

8

PETROCHEMICALS INDUSTRY

Petrochemicals are chemicals made from petroleum (crude oil) and natural gas. The **origin of India petroleum industry** may be traced back to the mid 1800. During India's independence in 1947, the petroleum industry was owned and managed by foreign companies. With the Industrial Policy Resolution of 1954, however, the Indian government placed the petroleum industry at the forefront as a core sector industry. This led to the formation of the India Petrochemical Industry. The origin of India Petrochemical Industry was about 70 years ago, with the first production of organic chemical compounds from propane. The term 'petrochemical' refers to the chemicals that are formed directly or indirectly from petroleum-based hydrocarbons and natural gases. Petrochemicals production is one the most vital industries in all developed countries and a growth-determining factor as well. The main purpose of the India Petrochemical Industry was to capture the growth in this industry in the future.

The India petrochemical industry has regenerated the use of benzene as a petrochemical raw material. At present, around 6% of the total barrel of crude oil is used for the production of petrochemicals. The three major methods involved in the processing of petrochemicals are BTX (Benzene, Toluene, and Xylene) Synthesis Gas (Syn Gas), and Olefins. They act as basic components for different petrochemical arbitrates and consumer products. Around

80% of the total inventory for petrochemicals is developed from petroleum refinery liquids and gases. The second most important base material is the natural gas.

The main end-use products of the petrochemical industry in India are :

- Plastics and Resins
- Pharmaceuticals
- Automotive Chemicals including Antifreeze Agents
- Detergents, Solvents, Plasticizers, and Paint Varnishes
- Agricultural Chemicals such as Fertilizers, Pesticides, and Herbicides
- Synthetic Elastomers

The Indian **petrochemical products overview** is quite fascinating. The wide spectrum of products that are manufactured in India are at par with world standards. Thus, India shares a good portion of the petrochemical business in the world market. Asian, African, and even Arab countries buy Indian petrochemical products. The demand for Indian petrochemical product is high mainly because of its quality and competitive pricing. India's low cost and high-end petrochemical product manufacturing expertise coupled with developing world-class infrastructure is the main leveraging factor for the rise of this industry. Further, with the consolidation of the infrastructure development, this industry is all set to strengthen itself in the near future. India offers petrochemical products at a substantial discount compared to its western counterparts while delivering the same class of output.

The advantages of manufacturing high class of petrochemical products in India are :

- Friendly government of India policies
- Low cost labour
- Low and world class infrastructure
- Strong technical education
- Large number of science and engineering graduates
- Quality output
- Highly skilled workforce
- Usage of innovative process
- Good client relationships
- Huge scope for innovation
- Expansion of existing relationships
- Huge demand in overseas markets
- Availability of more technical work force
- Increased number and quality of training facilities

The locations of various petrochemicals industries in India are given in the map shown in Fig. 8.1.

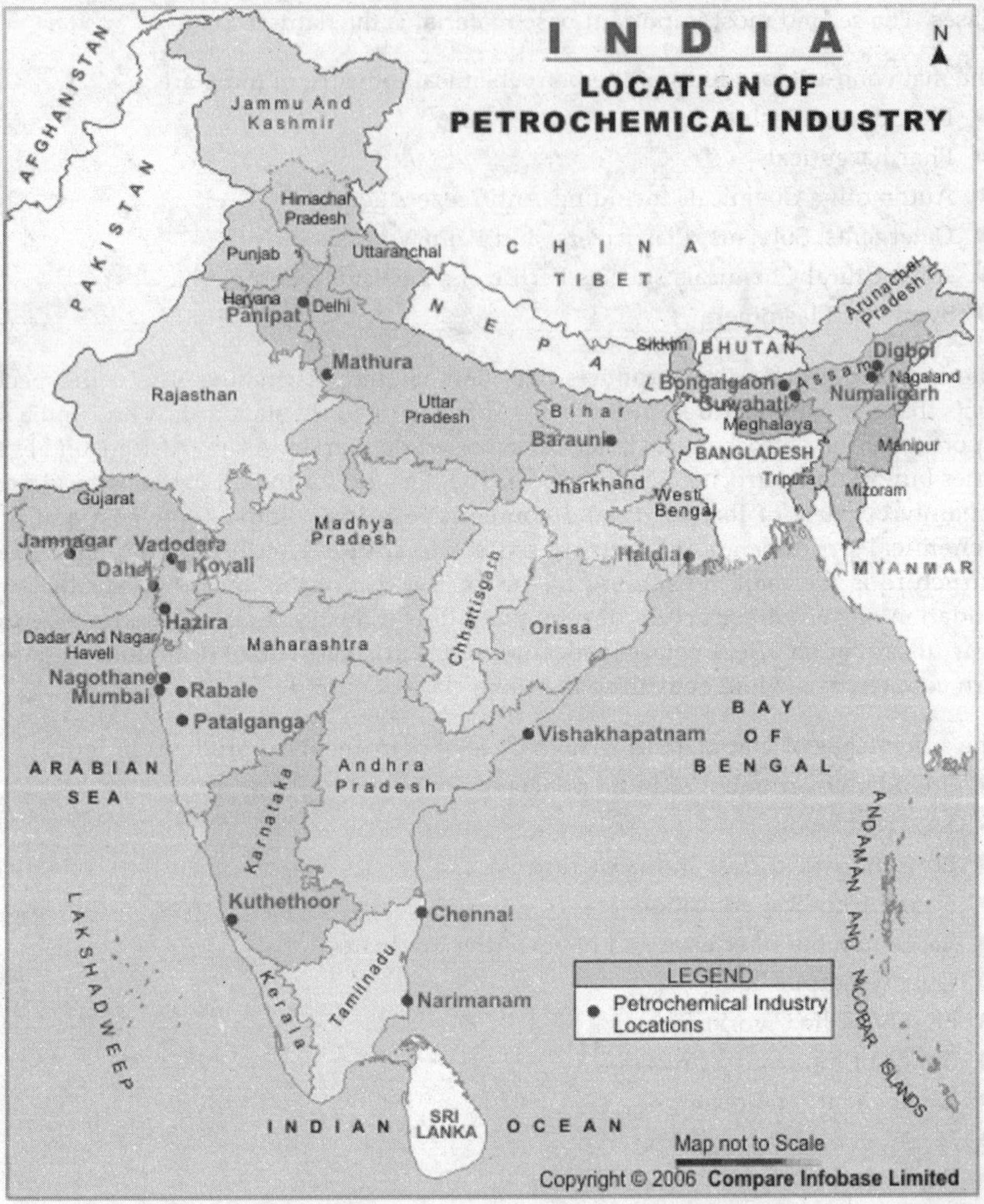

FIG. 8.1. VARIOUS PETROCHEMICAL INDUSTRIES IN INDIA.

8.1. PRIMARY PETROCHEMICALS

"Primary Petrochemicals" include: olefins (ethylene, propylene and butadiene) aromatics (benzene, toluene, and xylenes); and methanol. Olefins are unsaturated molecules of carbon (C) and hydrogen (H) that appear as short chains, of two, three or four carbons in length. Aromatics contain a six carbon ring structure. The oxygen/hydrogen (OH) group in methanol denotes that it is an alcohol. Figure 8.2 shows flow chart for primary petrochemicals

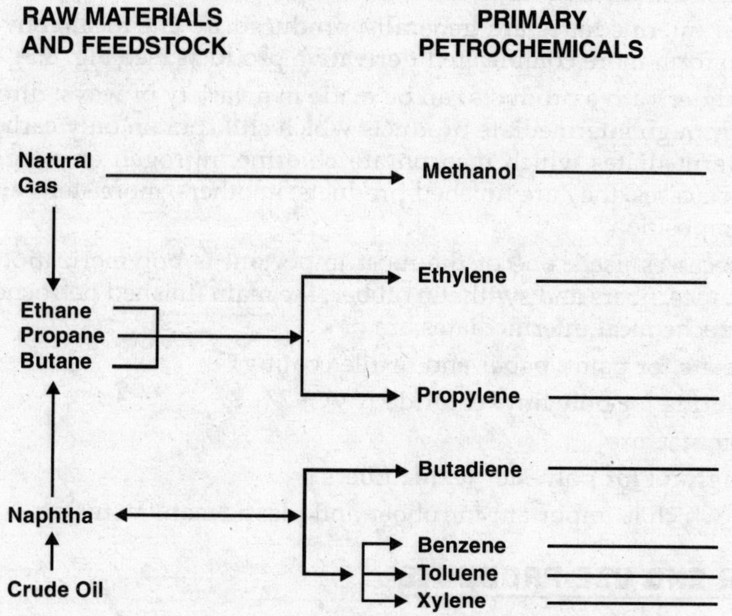

FIG. 8.2. LIST OF PRIMARY PETROCHEMICALS.

8.2. INTERMEDIATES AND DERIVATIVES

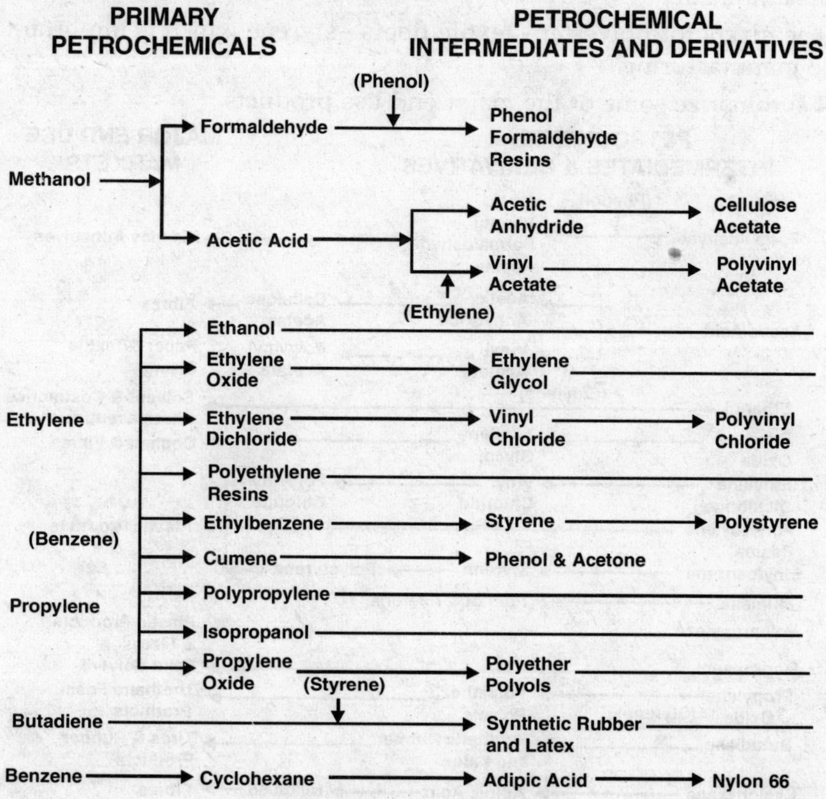

FIG. 8.3. VARIOUS INTERMEDIATES.

Petrochemical intermediates are generally produced by chemical conversion of primary petrochemicals to form more complicated derivative products (see Fig. 8.3).

Petrochemical derivative products can be made in a variety of ways: directly from primary petrochemicals; through intermediate products which still contain only carbon and hydrogen; and, through intermediates which incorporate chlorine, nitrogen or oxygen in the finished derivative. In some cases, they are finished products; in others, more steps are needed to arrive at the desired composition.

Of all the processes used, one of the most important is polymerization. It is used in the production of plastics, fibers and synthetic rubber, the main finished petrochemical derivatives. Some typical petrochemical intermediates are :

- Vinyl acetate for paint, paper and textile coatings
- Vinyl chloride for polyvinyl chloride (PVC)
- Resin manufacture
- Ethylene glycol for polyester textile fibers
- Ethelene which is important in rubber and plastic manufacturing.

8.3. MAJOR END USE PRODUCTS

Some of the major end use petrochemical include

- Vinyl acetate for paint, paper and textile coatings vinyl chloride for
- Polyvinyl chloride PVC)
- Resin manufacture
- Ethylene glycol for polyester - textile fibers - styrene which is important in rubber and plastic manufacturing

Figure 8.4 summarize some of the major end use products.

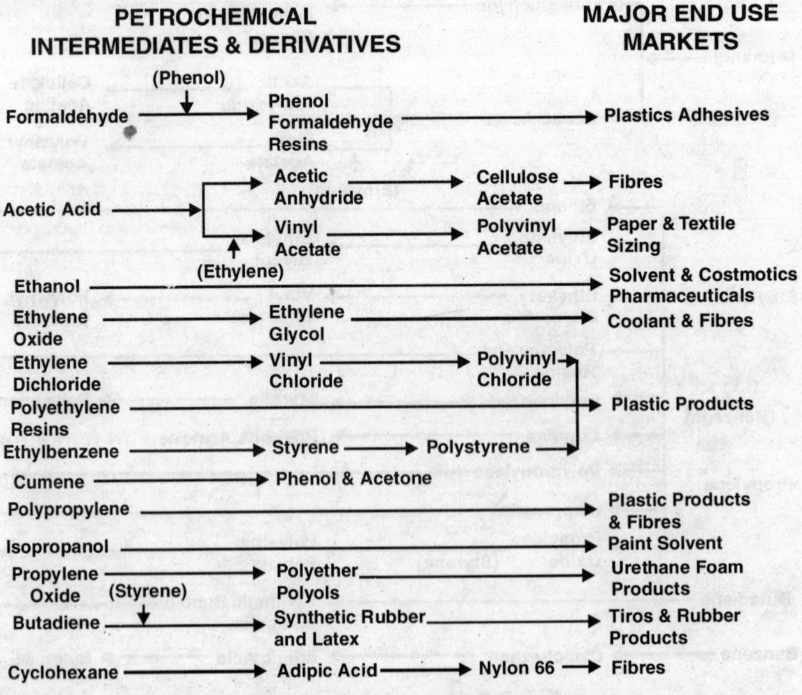

FIG. 8.4. LIST OF MAJOR END PRODUCT USE.

The following is a partial list of the major commercial petrochemicals and their derivatives:

- **Ethylene** - the simplest olefin; used as a ripening hormone, a monomer and a chemical feedstock
  1. Polyethylenes - polymerized ethylene
  2. Ethanol - made by hydration (chemical reaction adding water) of ethylene
  3. Ethylene oxide - sometimes called oxirane; can be made by oxidation of ethylene
     (*a*) ethylene glycol - from hydration of ethylene oxide or oxidation of ethylene
     (*b*) engine coolant - contains ethylene glycol
     (*c*) polyesters - any of several polymers with ester linkages in the backbone chain
     (*d*) glycol ethers - from condensation of glycols
     (*e*) ethoxylates
  4. Vinyl acetate
     (*a*) 1,2-dichloroethane
     (*b*) trichloroethylene
     (*c*) tetrachloroethylene - also called perchloroethylene; used as a dry cleaning solvent and degreaser
     (*d*) vinyl chloride - monomer for polyvinyl chloride
     (*e*) polyvinyl chloride (PVC) - type of plastic used for piping, tubing
- Propylene - used as a monomer and a chemical feedstock
- Isopropyl alcohol - (2-propanol) often used as a solvent or rubbing alcohol
- Acrylonitrile - useful as a monomer in forming Orlon, (acrylonitrile butadiene styrene)
- Polypropylene - polymerized propylene
- Propylene oxide
  - propylene glycol - sometimes used in engine coolant
  - glycol ethers - from condensation of glycols

5. Acrylic acid
   (*a*) acrylic polymers
6. Allyl chloride -
   (*a*) epichlorohydrin - chloro-oxirane; used in epoxy resin formation
   (*b*) epoxy resins - a type of polymerizing glue from bisphenol A, epichlorohydrin, and some amine

C_4 hydrocarbons - a mixture consisting of butanes, butylenes and butadienes

7. Isomers of butylene - useful as monomers or co-monomers
   (*a*) isobutylene - feed for making methyl *tert*-butyl ether (MTBE) or monomer for copolymerization with a low percentage of isoprene to make butyl rubber
8. 1,3-butadiene - a diene often used as a monomer or co-monomer for polymerization to elastomers such as polybutadiene or a plastic such as acrylonitrile-butadiene-styrene (ABS)
   (*a*) synthetic rubbers - synthetic elastomers made of any one or more of several petrochemical (usually) monomers such as 1,3-butadiene, styrene, isobutylene, isoprene, chloroprene; elastomeric polymers are often made with a high percentage of conjugated diene monomers such as 1,3-butadiene, isoprene or chloroprene

- higher olefins

9. Polyolefins such poly-alpha-olefins which are used as lubricants
10. Alpha-olefins - used as monomers, co-monomers, and other chemical precursors. For example, a small amount of 1-hexene can be copolymerized with ethylene into a more flexible form of polyethylene.
11. Other higher olefins
12. Detergent alcohols

- benzene - the simplest aromatic hydrocarbon

13. Ethylbenzene - made from benzene and ethylene
    (*a*) styrene made by dehydrogenation of ethylbenzene; used as a monomer
    (*b*) polystyrenes - polymers with styrene as a monomer
14. Cumene - isopropylbenzene; a feedstock in the cumene process
    (*a*) phenol - hydroxybenzene; of dassdcten made by the cumene process
    (*b*) acetone - dimethyl ketone; also often made by the cumene process
    (*c*) bisphenol A - a type of "double" phenol used in polymerization in epoxy resins and making a common type of polycarbonate
    (*d*) epoxy resins - a type of polymerizing glue from bisphenol A, epichlorohydrin, and some amine
    (*e*) polycarbonate - a plastic polymer made from bisphenol A and phosgene (carbonyl dichloride)
    (*f*) solvents - liquids used for dissolving materials; examples often made from petrochemicals include ethanol, isopropyl alcohol, acetone, benzene, toluene, xylenes
15. Cyclohexane - a 6-carbon aliphatic cyclic hydrocarbon sometimes used as a non-polar solvent
    (*a*) adipic acid - a 6-carbon dicarboxylic acid which can be a precursor used as a co-monomer together with a diamine to form an alternating copolymer form of nylon.
    (*b*) nylons - types of polyamides, some are alternating copolymers formed from copolymerizing dicarboxylic acid or derivatives with diamines

- caprolactam - a 6-carbon cyclic amide
  (*a*) nylons - types of polyamides, some are from polymerizing caprolactam

16. Nitrobenzene - can be made by single nitration of benzene
    (*a*) aniline - aminobenzene
    (*b*) methylene diphenyl diisocyanate (MDI) - used as a co-monomer with diols or polyols to form polyurethanes or with di- or polyamines to form polyureas
    (*c*) polyurethanes
17. Alkylbenzene - a general type of aromatic hydrocarbon which can be used as a presursor for a sulfonate surfactant (detergent)
    (*a*) detergents - often include surfactants types such as alkylbenzenesulfonates and nonylphenol ethoxylates
18. Chlorobenzene
    (*a*) toluene - methylbenzene; can be a solvent or precursor for other chemicals
19. Benzene

20. Toluene diisocyanate (TDI) - used as co-monomers with diols or polyols to form polyurethanes or with di- or polyamines to form polyureas
    (*a*) polyurethanes - a polymer formed from diisocyanates and diols or polyols
21. Benzoic acid - carboxybenzene
    (*a*) caprolactam
    (*b*) nylon

- mixed xylenes - any of three dimethylbenzene isomers, could be a solvent but more often precursor chemicals

22. Ortho-xylene - both methyl groups can be oxidized to form (*ortho-*)phthalic acid
    (*a*) phthalic anhydride
23. Para-xylene - both methyl groups can be oxidized to form terephthalic acid
    (*a*) dimethyl terephthalate - can be copolymerized to form certain polyesters
    (*b*) polyesters - although there can be many types, polyethylene terephthalate is made from petrochemical products and is very widely used.
    (*c*) purified terephthalic acid - often copolymerized to form polyethylene terephthalate
    (*d*) polyesters

Some of the organic compounds are described below:

8.4. 2-ETHYLHEXANOL

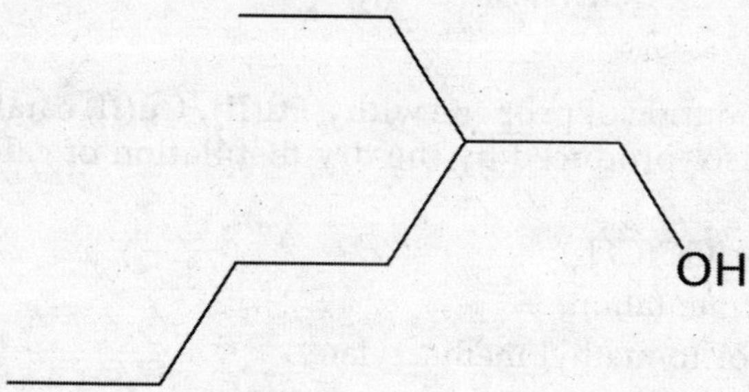

2-Ethylhexanol (2-EH), or isooctanol, is a fatty alcohol, an organic compound used in the manufacture of a variety of products. It is a branched, eight-carbon alcohol. It is a clear, colorless liquid that is nearly insoluble in water, but well soluble in most organic solvents.

2-Ethylhexanol can be readily converted into esters that have a variety of uses. The primary use of 2-ethylhexanol is in the manufacture of the diester bis(2-ethylhexyl) phthalate (DEHP), a plasticizer. Because it is a fatty alcohol, esters of 2-ethylhexanol tend to have emollient properties. For example, the sunscreen octocrylene contains a 2-ethylhexyl ester for this purpose. It is also commonly used as a low volatility solvent.

2-Ethylhexanol is a combustible liquid above 60°C. It is a marine pollutant and it irritates skin, eyes and respiratory tract. It can also cause nausea and headache after inhalation, and diarrhoea and vomiting after ingestion. Long-term or repeated exposure can result in defatting of the skin.

8.5. ACETONE

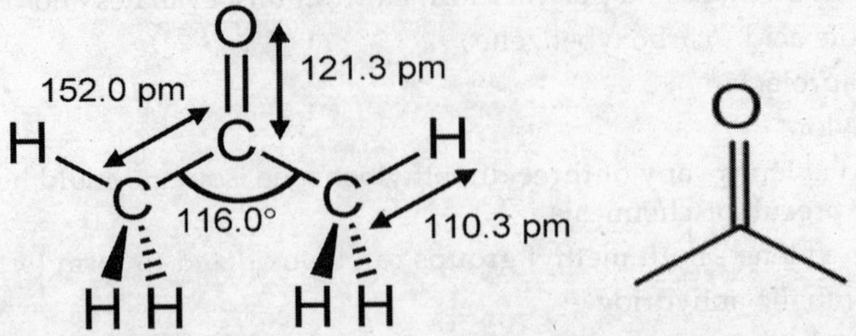

Acetone is the organic compound with the formula $OC(CH_3)_2$. This colorless, mobile, flammable liquid and is the simplest example of the ketones. Owing to the fact that acetone is miscible with water it serves as an important solvent for cleaning purposes in the laboratory. Acetone act as a precursor to polymers. Familiar household uses of acetone are as the active ingredient in nail polish remover and as paint thinner and sanitary cleaner. It is a common building block in organic chemistry. In addition to being manufactured, acetone also occurs naturally, even being biosynthesized in small amounts in the human body.

Acetone is produced directly or indirectly from propene. Most common is the **cumene process.** In this process benzene is alkylated with propene and the resulting cumene (isopropylbenzene) is oxidized to give phenol and acetone:

$$\underset{\text{Cumene (isopropylbenzene)}}{C_6H_5CH(CH_3)_2 + O_2} \longrightarrow C_6H_5OH + \underset{\text{acetone}}{OC(CH_3)_2}$$

Acetone is also produced by the direct oxidation of propene with a Pd(II)/Cu(II) catalysts, similar to the Wacker process. Acetone is also produced by the dry distillation of calcium acetate.

$$(COOCH_3)_2 \longrightarrow OC(CH_3)_2$$

Acetone is also produced by bacterial fermentation.

Uses : Acetone is consumed as a precursor to methyl methacrylate.

- This application begins with the initial conversion of acetone to its cyanohydrin:

$$(CH_3)_2CO + HCN \longrightarrow (CH_3)_2C(OH)CN$$

Nitrile is hydrolyzed to the unsaturated amide, which is further esterified

$$(CH_3)_2C(OH)CN + CH_3OH \longrightarrow CH_2{=}(CH_3)CCO_2CH_3 + NH_3$$

The second process involves acetone in condensation with phenol to give bisphenol A:

$$(CH_3)_2CO + 2\ C_6H_5OH \longrightarrow (CH_3)_2C(C_6H_4OH)_2 + H_2O$$

Bisphenol-A is a component of many polymers such as polycarbonates, polyurethanes, and epoxy resins.

As a solvent : Acetone is a good solvent for most plastics and synthetic fibres including those used in Nalgene bottles made of polystyrene, polycarbonate and some types of polypropylene. It is ideal for thinning fiberglass resin, cleaning fiberglass tools and dissolving epoxies and superglue before hardening. It is used as a volatile component of some paints and varnishes. As a heavy-duty degreaser, it is useful in the preparation of metal prior to painting; it also thins polyester resins, vinyl and adhesives. Acetone is used in the production of the solvents - methyl isobutyl alcohol and methyl isobutyl ketone. These products are formed by aldol condensation to give diacetone alcohol.

$$2\ (CH_3)_2CO \longrightarrow (CH_3)_2C(OH)CH_2C(O)CH_3$$

Acetone is used as a solvent by the pharmaceutical industry and as a denaturation agent in denatured alcohol.

Laboratory uses : In the laboratory, acetone is used as a polar aprotic solvent in a variety of organic reactions, such as S_N2 reactions. The use of acetone solvent is also critical for the Jones oxidation. It is a common solvent for rinsing laboratory glassware because of its low cost, volatility, and ability to for a low boiling temperature azeotrope with water. For similar reasons, acetone is also used as a drying agent. Acetone can be cooled with dry ice to –78 °C without freezing; acetone/dry ice baths are commonly used to conduct reactions at low temperatures. Acetone is fluorescent under ultraviolet light, and its vapor may be used as a fluorescent tracer in fluid flow experiments.

Acetone can be used as an artistic agent; when rubbed on the back of a laser print or photocopy placed face-down on another surface and burnished firmly, the toner of the image is allowed to transfer to the destination surface. Some automotive enthusiasts add acetone at around 1 part in 500 to their fuel, following claims of dramatic improvement in fuel economy and engine life. This practice is controversial as the body of systematic testing shows that acetone has no measurable effect or may in fact reduce engine life by adversely affecting fuel system parts.

Flammability : The most common hazard associated with acetone is its extreme flammability. It auto-ignites at a temperature of 465 °C (869 °F). At temperatures greater than acetone's flash point of –20 °C (–4 °F), air mixtures of between 2.5% and 12.8% acetone, by volume, may explode or cause a flash fire. Vapors can flow along surfaces to distant ignition sources and flash back. Static discharge may also ignite acetone vapors.

Acetone peroxide : When oxidized, acetone forms acetone peroxide as a byproduct, which is a highly unstable compound. It may be formed accidentally, e.g. when waste hydrogen peroxide is poured into waste solvent containing acetone. Acetone peroxide is more than ten times as friction and shock sensitive as nitroglycerin. Due to its instability, it is rarely used, despite its easy chemical synthesis.

Toxicology : Acetone is believed to exhibit only slight toxicity in normal use, and there is no strong evidence of chronic health effects if basic precautions are followed. At very high vapor concentrations, acetone is irritating and, like many other solvents, may depress the central nervous system. It is also a severe irritant on contact with eyes, and a potential pulmonary aspiration risk. It has been hypothesized that the high-fat low-carbohydrate ketogenic diet used clinically to control drug-resistant epilepsy in children works by elevating acetone in the brain.

Environmental effects : Acetone evaporates rapidly, even from water and soil. Once in the atmosphere, it is degraded by UV light with a 22-day half-life. Acetone dissipates slowly in soil, animals, or waterways since it is sometimes consumed by microorganisms; however, it is a significant issue with respect to groundwater contamination due to its high solubility in water.. Acetone may pose a significant risk of oxygen depletion in aquatic systems due to the microbial activity consuming it.

8.6. ACETYLENE

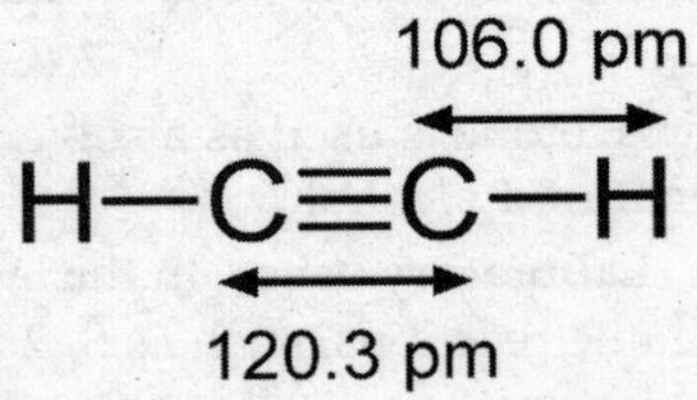

Acetylene (IUPAC name: ethyne) is the chemical compound with the formula HC_2H. It is a hydrocarbon and the simplest alkyne. This colourless gas is widely used as a fuel and a chemical building block. It is unstable in pure form and thus is usually handled as a solution. As an alkyne, acetylene is unsaturated because its two carbon atoms are bonded together in a triple bond. The carbon-carbon triple bond places all four atoms in the same straight line, with CCH bond angles of 180°.

Acetylene is mainly manufactured by the partial combustion of methane or appears as a side product in the ethylene stream from cracking of hydrocarbons. Its presence in ethylene is usually undesirable because of its explosive character and its ability to poison Ziegler-Natta catalysts when ethylene is used for the synthesis of polyethylene. It is selectively hydrogenated into ethylene, usually using Pd-Ag catalysts. It was prepared by the hydrolysis of calcium carbide, a reaction discovered by Friedrich Wöhler

$$CaC_2 + 2H_2O \longrightarrow Ca(OH)_2 + C_2H_2$$

Reactions : Important application is the conversion of acetylene to ethylene for use in making a variety of polyethylene plastics. Another important reaction of acetylene is its combustion, which also is the basis of the acetylene welding technologies. Otherwise, its major applications involve its conversion to acrylic acid derivatives. Acetylides are formed with many metal ions by reactions of it with solutions of their salts, e.g., silver acetylide and copper acetylide, are powerful explosives. Copper acetylide is also formed by reacting acetylene with metallic copper or its alloys; these materials are therefore unsuitable for installations for handling acetylene.

Walter Reppe discovered that in the presence of metal catalysts, acetylene can react to give a wide range of industrially significant chemicals.

- With alcohols, hydrogen cyanide, hydrogen chloride, or carboxylic acids to give vinyl compounds.

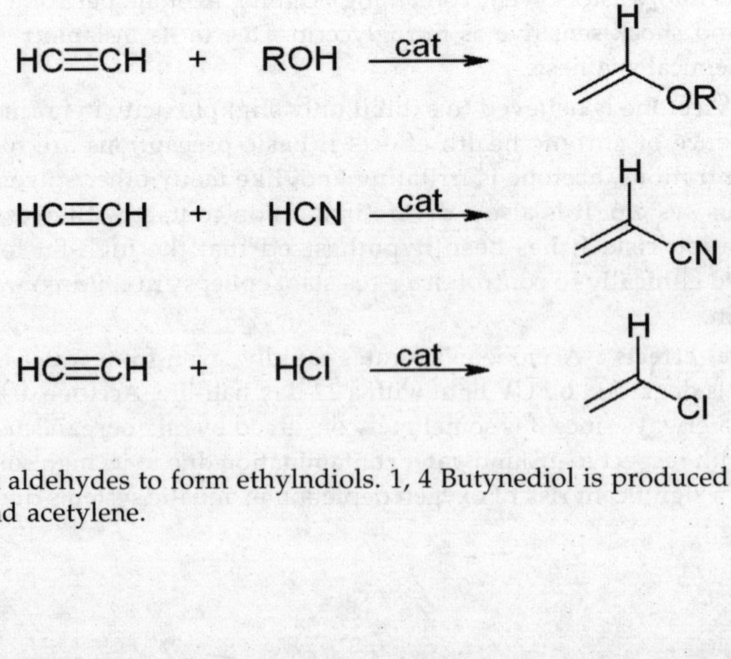

It reacts with aldehydes to form ethylndiols. 1, 4 Butynediol is produced industrially from formaldehyde and acetylene.

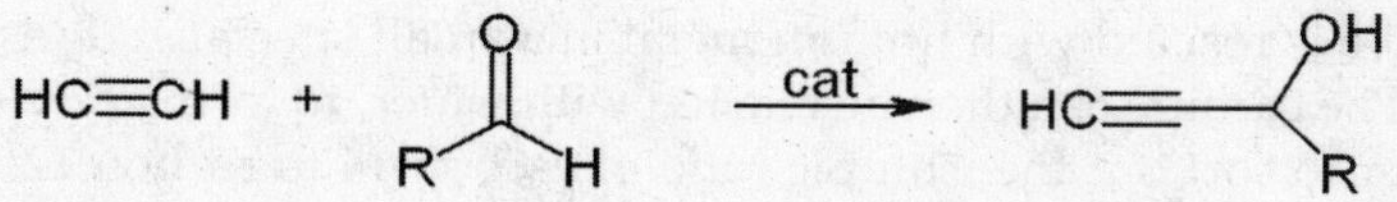

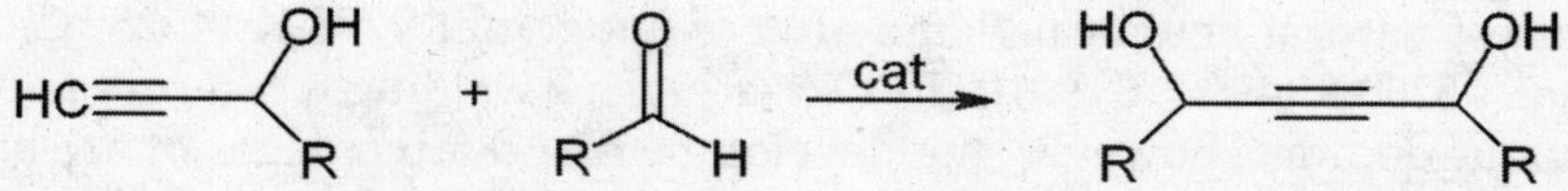

With carbon monoxide to give acrylic acid, or acrylic esters, which can be used to produce acrylic glass.

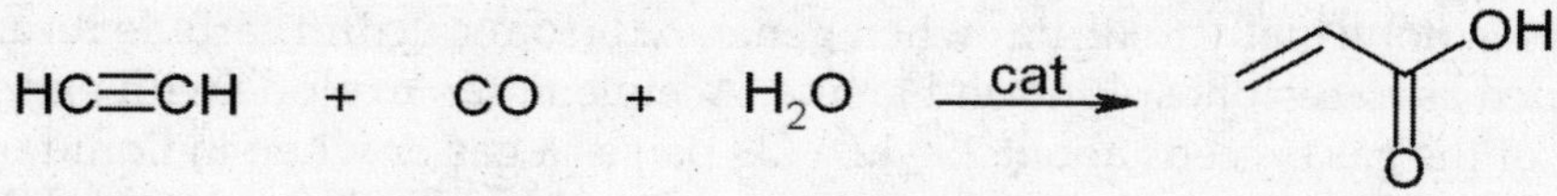

Cyclicization to give benzene and cyclooctatetraene:

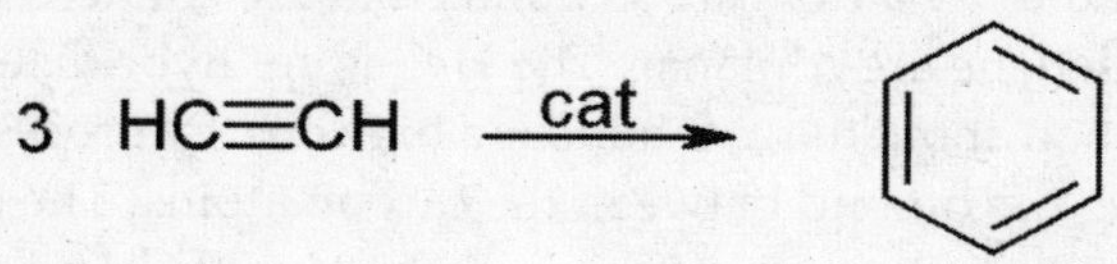

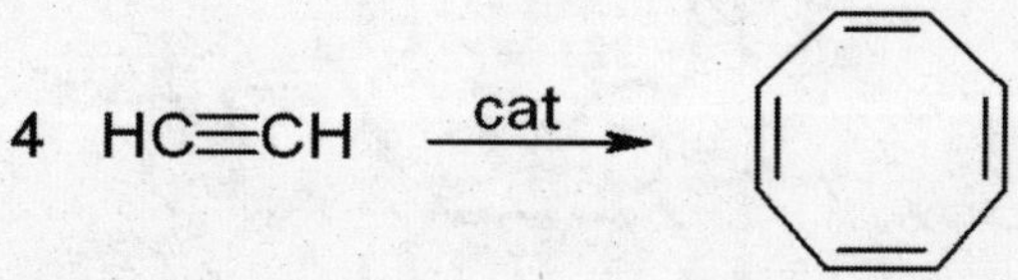

Mikhail Kucherov described the hydration of acetylene to acetaldehyde using mercury(II) bromide as catalyst.

The polymerization of acetylene with Ziegler-Natta catalysts produces polyacetylene films. Polyacetylene, a chain of CH centres with alternating single and double bonds, was the one of first discovered organic semiconductors. Its reaction with iodine produces a highly electrically conducting material.

Acetylene was widely used for illumination, including street lighting in some towns. Some early automobiles used carbide lamps before the adoption of electric headlights.

Acetylene is sometimes used for carburization (that is, hardening) of steel when the object is too large to fit into a furnace.

Acetylene is used to volatilize carbon in radiocarbon dating. The carbonaceous material in an archeological sample is reacted with lithium metal in a small specialized research furnace to form lithium carbide The carbide can then be reacted with water, to form acetylene gas to be fed into mass spectrometer to sort out the isotopic ratio of carbon 14 to carbon 12.

Acetylene is consumed for oxyacetylene gas welding and cutting due to the high temperature of the flame; combustion of acetylene with oxygen produces a flame of over 3600 K (3300 °C, 6000 °F), releasing 11.8 kJ/g. Oxyacetylene is the hottest burning common fuel gas. Acetylene is the third hottest natural chemical flame after cyanogen at 4798 K (4525 °C, 8180 °F) and dicyanoacetylene's 5260 K (4990 °C, 9010 °F). Oxy-acetylene welding was a very popular welding process in previous decades, however, the development and advantages of arc-based welding processes have made oxy-fuel welding nearly extinct. Acetylene usage for welding has dropped significantly. However, oxy-fuel cutting is still very popular and oxy-acetylene cutting is present in nearly every metal fabrication shop. For use in welding and cutting, the working pressures must be controlled by a regulator, or the gas will spontaneously combust.

Acetylene is not usually toxic but when generated from calcium carbide it can contain toxic impurities such as traces phosphine and arsine. Acetylene can explode with extreme violence if the pressure of the gas exceeds about 200 kPa (39 psi) as a gas or when in liquid or solid form. It is therefore shipped and stored dissolved in acetone or dimethylformamide (DMF), contained in a metal cylinder with a porous filling which renders it safe to transport and use.

8.7. ETHYLENE

Ethylene (IUPAC name: **ethene**) has formula C_2H_4. It is the simplest alkene. It contains carbon-carbon double bond, therefore, ethylene is called an *unsaturated hydrocarbon* or an *olefin*. It is extremely important in industry and also has a role in biology as a hormone. Ethylene has four hydrogen atoms bound to a pair of carbon atoms that are connected by a double bond. All six atoms that comprise ethylene are coplanar. The H-C-is sp^2 hybridized carbon. The molecule is also relatively rigid: rotation about the C-C bond is a high energy process that requires breaking the σ-bond, while retaining the σ-bond between the carbon atoms. The double bond is a region of high electron density, and most reactions occur at this double bond position.

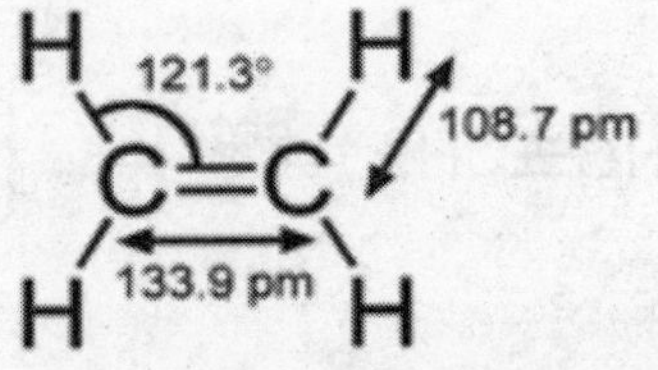

Ethylene is produced in the petrochemical industry by steam cracking. In this process, gaseous or light liquid hydrocarbons are heated to 750–950 °C, inducing numerous free radical reactions followed by immediate quench to freeze the reactions. This process converts large hydrocarbons into smaller ones and introduces unsaturation. Ethylene is separated from the resulting complex mixture by repeated compression and distillation. In a related process used in oil refineries, high molecular weight hydrocarbons are cracked over zeolite catalysts. Heavier feedstocks, such as naphtha and gas oils require at least two "quench towers" downstream of the cracking furnaces to recirculate pyrolysis-derived gasoline and process water. When cracking a mixture of ethane and propane, only one water quench tower is required.

The cracking process typically involves endothermic reaction such as

$$C_2H_6(g) \rightleftharpoons C_2H_4(g) + H_2(g) \qquad \Delta H = +138 \text{ kJ/mol}$$
$$C_3H_8(g) \rightleftharpoons C_2H_4(g) + CH_4(g) \qquad \Delta H = +81 \text{ kJ/mol}$$

To maximize the rate of cracking reactions, the temperatrue can be increased so that the gas particles move more quickly and collide more often. Secondly, the pressure can be increased which forces the gas particles closer together so that maximum collision occur.

Laboratory preparation : Laboratory scale preparation involves distilling of absolute ethanol with an excess of concentrated sulfuric acid and washing the distillate vapor stream in an aqueous solution of sodium hydroxide to remove the sulfur dioxide contaminant.

Properties of Ethylene : Ethylene is a colourless gas at room temperature and pressure. It has melting point of –169°C and boiling point –104°C. It has slightly sweet small and is non polar molecule soluble in non-polar solvents and insoluble in polar solvent like water. It is flammable in nature.

Chemical Reactions : Ethylene is an extremely important building block in the petrochemical industry. It can undergo many types of reactions which leads to a plethora of major chemical products. A list of some major types of reactions includes, 1) polymerization, 2) oxidation, 3) halogenation and hydrohalogenation, 4) alkylation, 5) hydration, 6) oligomerization, and 7) hydroformylation.

Ethylene is more reactive than alkanes because of two reasons :

- It has a double bond, a π-bond(pi) and one σ-bond (sigma), where the π-bond is weak and the σ-bond is strong. The presence of the σ-bond makes it a high energy molecule. Thus bromine water decolourises readily when it is added to ethylene.
- High electron density at the double bond makes it react readily. It is broken in an addition reaction to produce many useful products

Ethylene reacts with halogens to produce halogenated hydrocarbons 1,2-$C_2H_4X_2$. Major intermediates from the halogenation and hydrohalogenation of ethylene include: ethylene dichloride, ethyl chloride and ethylene dibromide. Some products in this group are: polyvinyl chloride, trichloroethylene, perchloroethylene, methyl chloroform, polyvinylidiene chloride and copolymers, and ethyl bromide. It also reacts with water to form ethanol in presence of catalyst like sulphuric or phosphoric acid. Under high pressure, and, in the presence of a catalytic metal (platinum, rhodium, nickel), it reacts with hydrogen to form ethane. Ethylene under goes alkylation reaction to form ethylbenzene, ethyl toluene, ethyl anilines, 1,4-hexadiene and aluminium alkyls. These products also act as an intermediates for polystyrene, unsaturated polyesters and ethylene-propylene terpolymers.

Ethylene is used as an intermediate in the manufacture of other chemicals and in the synthesis of monomers. Ethylene is chlorinated to produce 1,2-dichloroethane (ethylene dichloride), which can be converted to vinyl chloride (the monomer precursor to plastic polyvinyl chloride) or combined with benzene to produce ethylbenzene, which is used in the manufacture of polystyrene, another important plastic. Some of the reactions are given below :

Reactions of Ethylene :

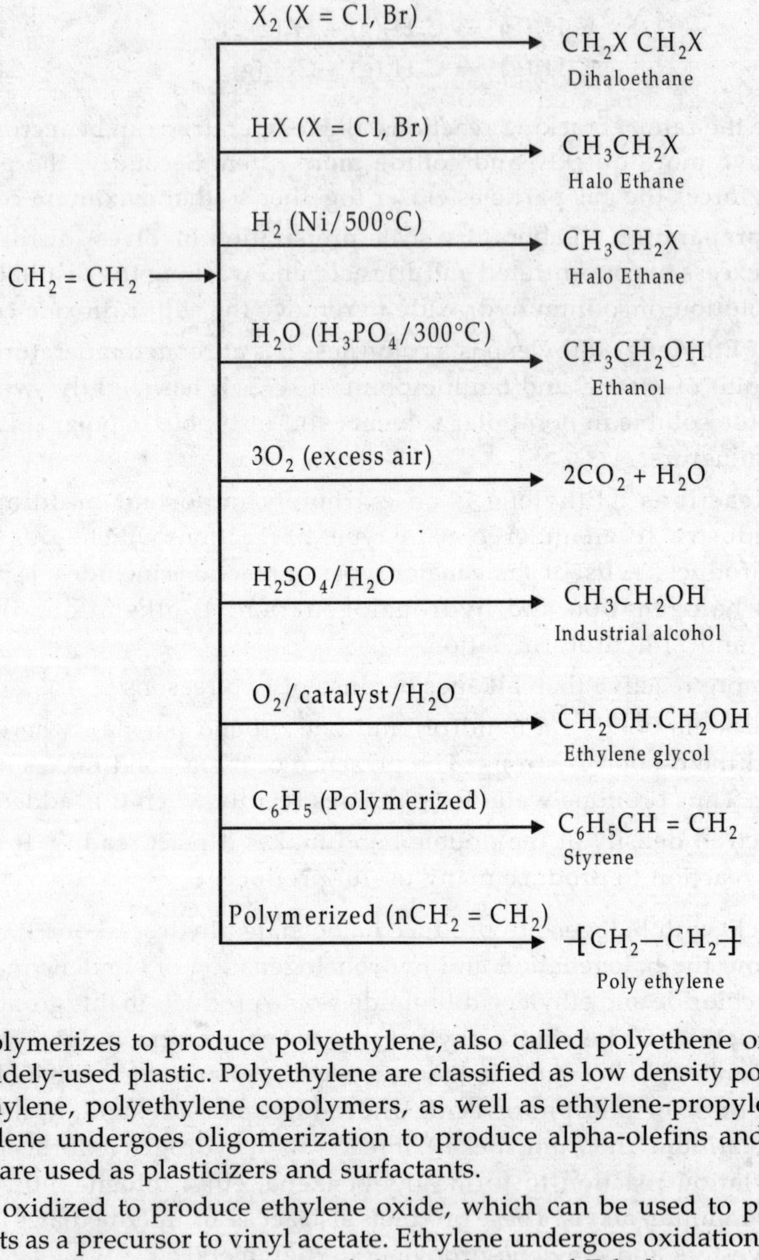

Ethylene polymerizes to produce polyethylene, also called polyethene or polythene, the world's most widely-used plastic. Polyethylene are classified as low density polyethylene, high density polyethylene, polyethylene copolymers, as well as ethylene-propylene co- and ter-polymers. Ethylene undergoes oligomerization to produce alpha-olefins and linear primary alcohols. These are used as plasticizers and surfactants.

Ethylene is oxidized to produce ethylene oxide, which can be used to produce ethylene glycol. It also acts as a precursor to vinyl acetate. Ethylene undergoes oxidation by palladium to give acetaldehyde. Major intermediates of the oxidation of Ethylene are ethylene oxide, acetaldehyde, vinyl acetate and ethylene glycol. Some important product derived from ethylene include - polyesters, polyurethane, morpholine, ethanolamines and glycol ethers.

The hydroformylation (Oxo-reaction) of ethylene results in propionaldehyde with its primary products of propionic acid and n-propyl alcohol

Uses : Ethylene is used to create ethylene oxide, ethylene dichloride, and polyethylene. In smaller quantities, ethylene is used as an anesthetic agent (in an 85% ethylene/15% oxygen ratio), to hasten fruit ripening, as well as a welding gas. Polyethylene of various density and melt flow account for more than 50% of world ethylene demand. The primary use of polyethylene is in film applications for packaging, carrier bags and trash liners. Other applications include injection moulding, pipe extrusion, wire and cable sheathing and insulation, as well as extrusion coating of paper and cardboard. Ethylene derivatives include: ethylene oxide, styrene monomer (via ethyl benzene) and linear higher olefins. Ethylene oxide is a key raw material in the production of surfactants and detergents. It is also used to manufacture ethylene glycol (widely used as automotive antifreeze), higher molecular weight glycols, and styrene monomer is used principally in polystyrene for packaging and insulation, as well as in styrene butadiene rubber for tires and footwear. Linear higher olefins are used as base materials for the manufacture of detergents, plasticizers, synthetic lubricants and additives, but also as co-monomers in the production of polyethylene. Ethylene is a key component in Levinstein sulfur mustard, a chemical weapon agent.

Ethylene acts physiologically as a hormone in plants. It exists as a gas and acts at trace levels throughout the life of the plant by stimulating or regulating the ripening of fruit, the opening of flowers, and the abscission (or shedding) of leaves. Its biosynthesis starts from methionine with 1- aminocyclopropane-1-carboxylic acid (ACC) as a key intermediate (Fig. 8.5).

Ethylene Biosynthesis in Plants

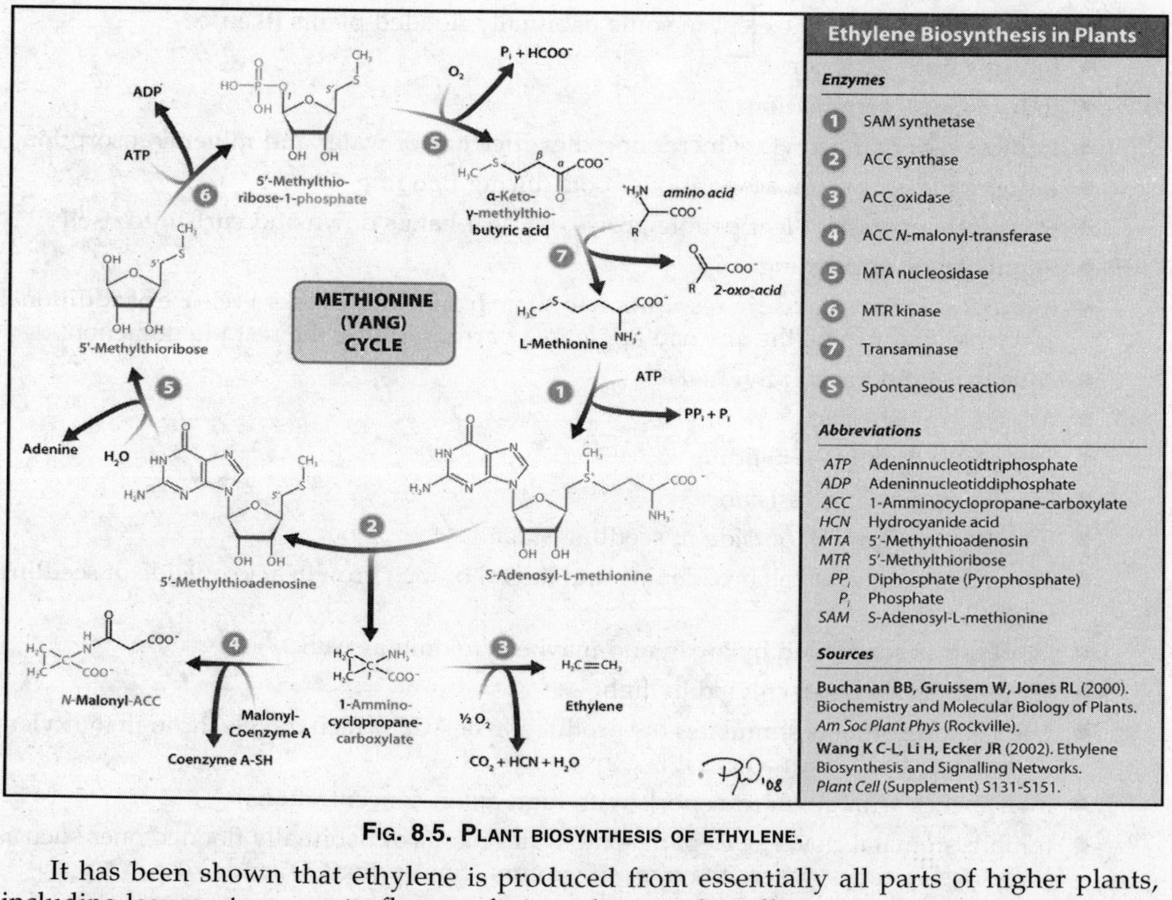

FIG. 8.5. PLANT BIOSYNTHESIS OF ETHYLENE.

It has been shown that ethylene is produced from essentially all parts of higher plants, including leaves, stems, roots, flowers, fruits, tubers, and seedlings.

The biosynsthesis of the hormone starts with conversion of the amino acid methionine to S-adenosyl-L-methionine (SAM, also called Adomet) by the enzyme Met Adenosyltransferase. SAM is then converted to 1-aminocyclopropane-1-carboxylic-acid (ACC) by the enzyme ACC synthase (ACS); the activity of ACS is the rate-limiting step in ethylene production, therefore regulation of this enzyme is key for the ethylene biosynthesis. The final step requires oxygen and involves the action of the enzyme ACC-oxidase (ACO), formerly known as the Ethylene Forming Enzyme (EFE). Ethylene biosynthesis can be induced by endogenous or exogenous ethylene. ACC synthesis increases with high levels of auxins, specially Indole Acetic Acid (IAA), and cytokinins. ACC synthase is inhibited by abscisic acid.

Like the other plant hormones, ethylene is considered to have pleiotropic effects. List of plant responses to ethylene

- Seedling triple response, thickening and shortening of hypocotyl with pronounced apical hook. This is thought to be a seedling's reaction to an obstacle in the soil such a stone, allowing it to push past the obstruction.
- In pollination, when the pollen reaches the stigma, the precursor of the ethylene, ACC, is secreted to the petal, the ACC releases ethylene with ACC oxidase.
- Stimulates leaf and flower senescence
- Stimulates senescence of mature xylem cells in preparation for plant use
- Inhibits shoot growth except in some habitually flooded plants like rice
- Induces leaf abscission
- Induces seed germination
- Induces root hair growth – increasing the efficiency of water and mineral absorption
- Induces the growth of adventitious roots during flooding
- Stimulates epinasty – leaf petiole grows out, leaf hangs down and curls into itself
- Stimulates fruit ripening
- Induces a climacteric rise in respiration in some fruit which causes a release of additional ethylene. This can be the one bad apple in a barrel spoiling the rest phenomenon.
- Affects neighboring individuals
- Affects gravistropism
- Stimulates nutational bending
- Disease/wounding resistance
- Inhibits stem growth outside of seedling stage
- Stimulates stem and cell broadening and lateral branch growth also outside of seedling stage
- Synthesis is stimulated by auxin and maybe cytokinin as well
- Ethylene levels are decreased by light
- The flooding of roots stimulates the production of ACC which travels through the xylem to the stem and leaves where it is converted to the gas
- Interference with auxin transport (with high auxin concentrations)
- Inhibits stomatal closing except in some water plants or habitually flooded ones such as some rice varieties, where the opposite occurs (conserving CO_2 and O_2)
- Where ethylene induces stomatal closing, it also induces stem elongation
- Induces flowering in pineapples

Depending on the concentration, ethylene gas can cause a pleasant odor, euphoria, nausea, hyperglycemia, a variety of psychological effects, blood pressure changes, hypoxia, loss of consciousness, or death. Ethylene has a pleasant sweet faint odor, and has a slightly sweet taste.

8.8. POLYETHYLENE

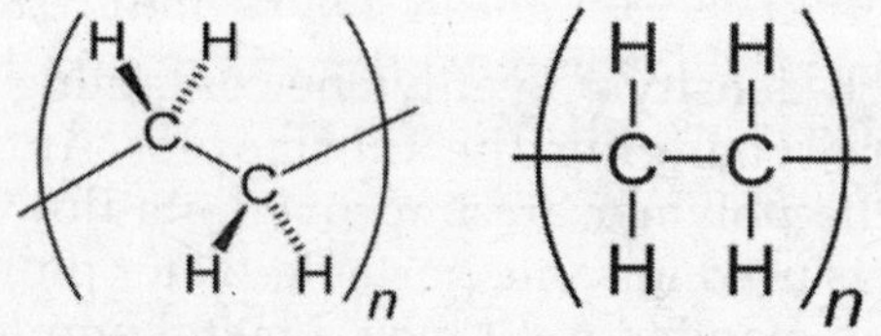

Polyethylene or **polythene** (IUPAC name **polyethene)**) is a thermoplastic commodity heavily used in consumer products.

Polyethylene is created through polymerization of ethene. It can be produced through radical polymerization, anionic addition polymerization, ion coordination polymerization or cationic addition polymerization. This is because ethene does not have any substituent groups that influence the stability of the propagation head of the polymer. Each of these methods results in a different type of polyethylene.

Polyethylene is classified into several different categories based mostly on its density and branching. The mechanical properties of polyethylene depend significantly on variables such as the extent and type of branching, the crystal structure and the molecular weight.

- Ultra high molecular weight polyethylene (UHMWPE)
- Ultra low molecular weight polyethylene (ULMWPE or PE-WAX)
- High molecular weight polyethylene (HMWPE)
- High density polyethylene (HDPE)
- High density cross-linked polyethylene (HDXLPE)
- Cross-linked polyethylene (PEX or XLPE)
- Medium density polyethylene (MDPE)
- Low density polyethylene (LDPE)
- Linear low density polyethylene (LLDPE)
- Very low density polyethylene (VLDPE)

UHMWPE is polyethylene with a molecular weight numbering in the millions, usually between 3.1 and 5.67 million. The high molecular weight makes it a very tough material, but results in less efficient packing of the chains into the crystal structure as evidenced by densities of less than high density polyethylene (for example, 0.930–0.935 g/cm^3). UHMWPE can be made using Ziegler catalysts .Since it has toughness and because of its cut, wear and excellent chemical resistance, UHMWPE is used in a diverse range of applications. These include bottle handling, machine parts, moving parts on weaving machines, bearings, gears, artificial joints, edge protection on ice rinks and butchers' chopping boards. It is commonly used for the construction of articular portions of implants used for hip and knee replacements.

HDPE is defined by a density of greater or equal to 0.941 g/cm^3. HDPE has a low degree of branching and thus stronger intermolecular forces and tensile strength. HDPE can be produced by chromium/silica catalysts, Ziegler-Natta catalysts or metallocene catalysts. The lack of branching is ensured by an appropriate choice of catalyst (for example, chromium catalysts or Ziegler-Natta catalysts) and reaction conditions. HDPE is used in products and packaging such as milk jugs, detergent bottles, margarine tubs, garbage containers and water pipes.

PEX is a medium- to high-density polyethylene containing cross-link bonds introduced into the polymer structure, changing the thermoplast into an elastomer. The high-temperature properties of the polymer are improved, its flow is reduced and its chemical resistance is enhanced. PEX is used in some portable-water plumbing systems because tubes made of the material can be expanded to fit over a metal nipple and it will slowly return to its original shape, forming a permanent, water-tight, connection.

MDPE is defined by a density range of 0.926–0.940 g/cm^3. MDPE can be produced by chromium/silica catalysts, Ziegler-Natta catalysts or metallocene catalysts. MDPE has good shock and drop resistance properties. It is less notch sensitive than HDPE, stress cracking resistance is better than HDPE. MDPE is typically used in gas pipes and fittings, sacks, shrink film, packaging film, carrier bags and screw closures.

LLDPE is defined by a density range of 0.915–0.925 g/cm^3. LLDPE is a substantially linear polymer with significant numbers of short branches, commonly made by copolymerization of ethylene with short-chain alpha-olefins (for example, 1-butene, 1-hexene and 1-octene). LLDPE has higher tensile strength than LDPE, it exhibits higher impact and puncture resistance than LDPE. Lower thickness (gauge) films can be blown, compared with LDPE, with better environmental stress cracking resistance but is not as easy to process. LLDPE is used in packaging, particularly film for bags and sheets. Lower thickness may be used compared to LDPE for cable covering, toys, lids, buckets, containers and pipe. While other applications are available, LLDPE is used predominantly in film applications due to its toughness, flexibility and relative transparency.

LDPE is defined by a density range of 0.910–0.940 g/cm^3. LDPE has a high degree of short and long chain branching, which means that the chains do not pack into the crystal structure as well. It has, therefore, less strong intermolecular forces as the instantaneous-dipole induced-dipole attraction is less. This results in a lower tensile strength and increased ductility. LDPE is created by free radical polymerization. The high degree of branching with long chains gives molten LDPE unique and desirable flow properties. LDPE is used for both rigid containers and plastic film applications such as plastic bags and film wrap.

VLDPE is defined by a density range of 0.880–0.915 g/cm^3. VLDPE is a substantially linear polymer with high levels of short-chain branches, commonly made by copolymerization of ethylene with short-chain alpha-olefins (for example, 1-butene, 1-hexene and 1-octene). VLDPE is most commonly produced using metallocene catalysts due to the greater co-monomer incorporation exhibited by these catalysts. VLDPEs are used for tubing, ice and frozen food bags, food packaging and stretch wrap as well as impact modifiers when blended with other polymers.

Recently much research activity has focused on the nature and distribution of long chain branches in polyethylene. In HDPE a relatively small number of these branches, perhaps 1 in 100 or 1,000 branches per backbone carbon, can significantly affect the rheological properties of the polymer.

Ethylene copolymers : In addition to copolymerization with alpha-olefins, ethylene can also be copolymerized with a wide range of other monomers and ionic composition that creates ionized free radicals. Common examples include vinyl acetate (the resulting product is ethylene-vinyl acetate copolymer, or EVA, widely used in athletic-shoe sole foams) and a variety of acrylates (applications include packaging and sporting goods).

Depending on the crystallinity and molecular weight, a melting point and glass transition may or may not be observable. The temperature at which these occur varies strongly with the type of polyethylene. For common commercial grades of medium- and high-density polyethylene, the melting point is typically in the range 120 to 130 °C (250 to 265 °F). The melting point for average, commercial, low-density polyethylene is typically 105 to 115 °C (220 to 240 °F).

Most LDPE, MDPE and HDPE grades have excellent chemical resistance and do not dissolve at room temperature because of their crystallinity. Polyethylene (other than cross-linked polyethylene) usually can be dissolved at elevated temperatures in aromatic hydrocarbons such as toluene or xylene, or in chlorinated solvents such as trichloroethane or trichlorobenzene.

The wide use of polyethylene makes it an important environmental issue. Though it can be recycled, most of the commercial polyethylene ends up in landfills and in the oceans. Polyethylene is not considered biodegradable, as it takes several centuries until it is efficiently degraded. Effort should be taken for proper disposal and effective utilization of polythene wastes.

8.9. ISOPROPYL ALCOHOL

Isopropyl alcohol (also **propan-2-ol, 2-propanol**) is a common name for **isopropanol.** It is a colorless, flammable chemical compound with a strong odor. It has the molecular formula C_3H_7OH and is the simplest example of a **secondary alcohol,** where the alcohol carbon is attached to two other carbons. It is an isomer of propanol.

Isopropyl alcohol is produced by combining water and propene. It occurs through indirect hydration via the sulfuric acid process and direct hydration. The former process, utilizes low-quality propylene, while the latter process, uses high-purity propylene. These processes give predominantly isopropyl alcohol rather than 1-propanol because the addition of water or sulfuric acid to propylene follows Markovnikov's rule.

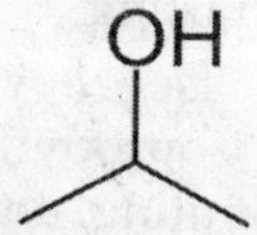

In the indirect process propylene reacts with sulfuric acid to form a mixture of sulfate esters. Subsequent hydrolysis of these esters produces isopropyl alcohol. In direct hydration propylene and water reacts, either in gas or liquid phases, at high pressures in the presence of solid or supported acidic catalysts. In both the processes isopropyl alcohol can be separated from water and other by-products by distillation. Isopropyl alcohol and water form an azeotrope and simple distillation gives a material which is 87.9% by weight isopropyl alcohol and 12.1% by weight water. Pure (anhydrous) isopropyl alcohol is made by azeotropic distillation of the "wet" isopropyl alcohol using either diisopropyl ether or cyclohexane as azeotroping agents.

The physical properties of isopropyl alcohol is listed below in Table 8.1

TABLE 8.1. Physical Properties of Isopropyl Alcohol.

Molecular Weight	60.09
Boiling Point	82.5 °C
Melting Point	–88.5 °C
Flash Point	11.7 oC (53 °F) (closed cup)
Vapor Pressure	44 mm Hg at 25 °C
Density/Specific Gravity	0.7851 at 20/4 °C (water = 1)
Log/Octanol Water Partition Coefficient	0.05
Dissociation Constants	pka: 17.1 at 25 °C
Autoignition Temp	455.6 °C (852 °F)
Henry's Law Constant	8.07 x 10-6 atm-m^3/mole
Vapor Density	2.08 (air = 1)
Conversion Factor	1 ppm = 2.45 mg/m^3

Isopropyl alcohol is oxidized or dehydrogenated to give acetone

$$(CH_3)_2CH\text{-}OH \longrightarrow (CH_3)_2CO + H_2$$

Isopropanol may be converted to 2-Bromopropane using phosphorus tribromide, or dehydrated to propylene by heating with sulfuric acid. Isopropanol is often used as a hydride source in the Meerwein-Ponndorf-Verley reduction. Like most alcohols, isopropyl alcohol reacts with active metals such as potassium to form alkoxides which can be called **isopropoxides.** The reaction with aluminium (initiated by a trace of mercury) is used to prepare the catalyst aluminium isopropoxide. Isopropanol has a maximum absorbance at 204 nm in an ultraviolet-visible spectrum.

Isopropyl alcohol vapor is more dense than air and is highly flammable with a very wide combustible range. It should be kept away from heat and open flame. When mixed with air or other oxidizers it can explode. Isopropyl alcohol has also been reported to form explosive peroxides. Like many organic solvents, long term application to the skin can cause defatting.

Isopropyl alcohol is oxidized by the liver into acetone by alcohol dehydrogenase. Symptoms of isopropyl alcohol poisoning include flushing, headache, dizziness, CNS depression, nausea, vomiting, anesthesia, and coma.

Uses : Isopropyl alcohol is cheaply available. Like acetone, it dissolves in a wide range of non polar compounds. It is relatively nontoxic and dries (evaporates) quickly. Thus it is used widely as a solvent and as a cleaning fluid, especially for dissolving lipophilic contaminants such as oil. Examples of this use include cleaning electronic devices such as contact pins (like those on ROM cartridges), magnetic tape and disk heads (such as those in audio and video tape recorders and floppy disk drives), the lenses of lasers in optical disc drives (e.g. CD, DVD) and removing thermal paste from IC packages (such as CPUs.) It is also used to clean LCD and glass computer monitor screens and used by many music shops to give second-hand or worn records newer-looking sheens. Isopropyl alcohol also works well at removing smudges, dirt, and fingerprints from cell phones. It is effective at removing residual glue from some sticky labels . It can also be used to remove stains from most fabrics, wood, cotton, etc. Isopropyl alcohol is also used to remove brake fluid traces from hydraulic disk brake systems, so that the brake fluid does not contaminate the brake pads, which would result in poor braking. In addition it can also be used to clean paintballs or other oil based products so that they may be reused, commonly known as "repainting". As a preservative (for biological specimens) isopropyl alcohol provides a cost-effective (when compared to pure ethanol) and comparatively non-toxic alternative to formaldehyde and other synthetic preservatives. When used for the preservation of specimens in solution, concentrations of 90–99% are optimal, though concentrations as low as 70% can be used. Sterilizing pads typically contain a 60–70% solution of isopropanol in water. Isopropanol is a major ingredient in "gas dryer" fuel additives. In significant quantities, water is a problem in fuel tanks, as it separates from the gasoline, and can freeze in the supply lines at cold temperatures. The isopropanol does not remove the water from the gasoline; rather, the isopropanol solubilizes the water in the gasoline. Once soluble, the water does not pose the same risk as insoluble water as it will no longer accumulate in the supply lines and freeze. Isopropanol is often sold in aerosol cans as a windscreen de-icer. Isopropanol is used as a water-drying aid for treating otitis externa, better known as swimmer's ear

Unlike ethanol or methanol, isopropanol can be separated from aqueous solutions by adding a salt such as sodium chloride, sodium sulfate, or any of several other inorganic salts, since the alcohol is much less soluble in saline solutions than in salt-free water The process is colloquially called salting out, and causes concentrated isopropanol to separate into a distinct layer.

8.10. PROPENE

Propene, also known as **propylene**, is an unsaturated organic compound having the chemical formula C_3H_6. It has one double bond, and is the second simplest member of the alkene class of hydrocarbons, and is abundant in occurence

Propene is obtained from petroleum or natural gas deposits (and coal to a lesser extent). It is a fossil fuel. Propene is extracted from these by fractional distillation during oil refining. It is also manufactured by cracking. The products from this process contain a mixture of products and the propene is separated from the other products by fractional distillation.

At room temperature, propene is a colourless, odourless gas, though when used as a fuel, it is mixed with minute quantities of foul-smelling sulfurous compounds (mercaptans) so that gas leaks can be readily detected. It has a slightly lower boiling point than propane and is thus more volatile. It lacks strongly polar bonds, yet the molecule has a small dipole moment due to its reduced symmetry.

Propene is the raw material for the production of polypropylene, a versatile polymer widely used in several different grades for packaging. Most propene is polymerized using Ziegler-Natta catalysis, which produces isotactic polypropylene. Along with benzene, propene is a key feedstock in the cumene process, a reaction carried out on industrial scales to produce acetone and phenol. Propene is also used during the production of many other chemical products such as isopropanol (propan-2-ol), acrylonitrile, and propylene oxide (epoxypropane). Propene resembles other alkenes in that it undergoes addition reactions relatively easily at room temperature. The relative weakness of its double bond (which is less strong than two single bonds) explains the reactivity of propene. Alkene reactions include:

- polymerization,
- oxidation,
- halogenation and hydrohalogenation,
- alkylation,
- hydration,
- oligomerization, and
- hydroformylation.

8.11. POLYPROPYLENE

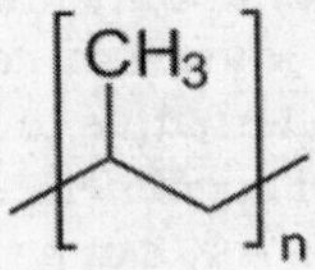

Polypropylene or **polypropene (PP)** is a thermoplastic polymer, used in a wide variety of applications, including packaging, textiles, stationary, plastic parts and reusable containers of various types, laboratory equipment, loudspeakers, automotive components, and polymer banknotes. It has unusual resistant to many chemical solvents, bases and acids.

Most commercial polypropylene is prepared by **co-ordination addition polymerization** Ziegler-Natta polymerization is a method of vinyl polymerization. It is used to prepare polymer of definite tacticity. It was discovered by Ziegler-Natta. It is especially useful, because it can make polymers such as linear unbranched polyethylene and isotactic polypropylene

Ziegler-Natta catalyst, is usually $TiCl_3$ or $TiCl_4$, along with an aluminum based co-catalyst, (Tri ethyl alumunium $Al(C_2H_5)_3$. The catalyst/co-catalyst pair are $TiCl_3$ and $Al(C_2H_5)_2Cl$, or $TiCl_4$ with $Al(C_2H_5)_3$.

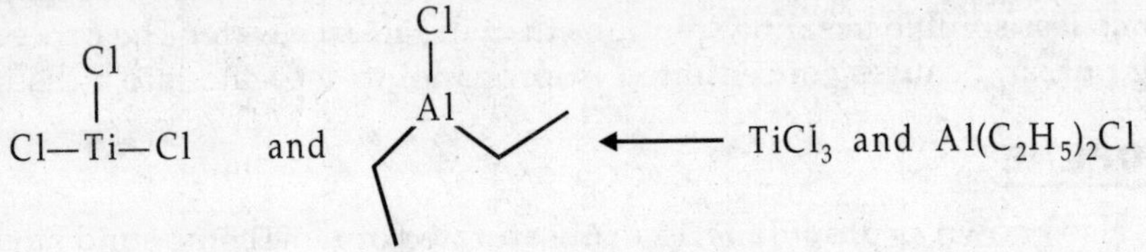

These are two sets of Zieglar-Natta catalyst/co-catalyst systems.
Either way, there are four chlorine atoms.

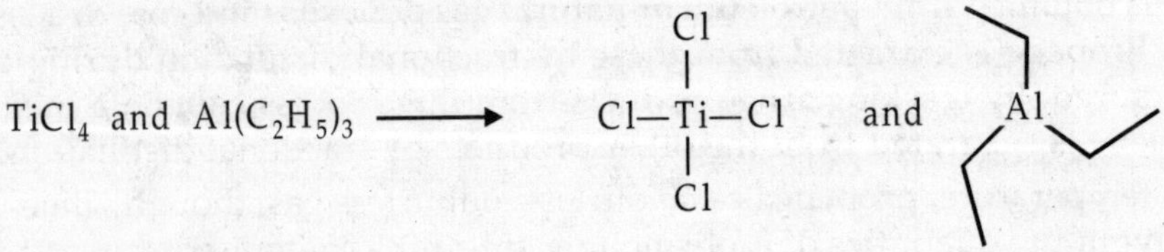

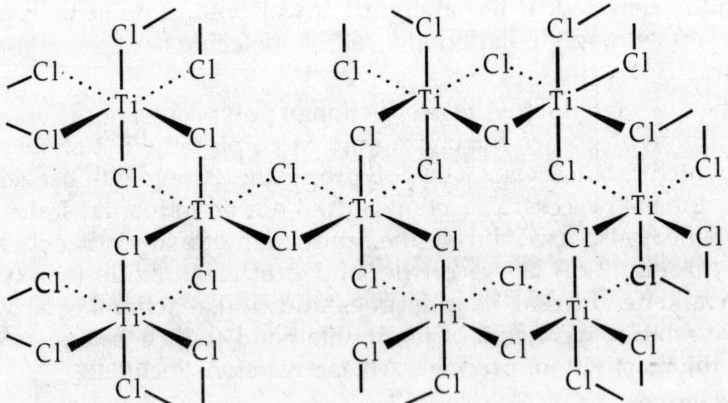

a crystal of α-TiCl₃

Each titanium atom is coordinated to six chlorine atoms, with octahedral geometry as shown above. In the interior of the crystal, each titanium is surrounded by six chlorines, but on the surface, a titanium atom is surrounded on one side by five chlorine atoms, but on the other side by empty space.

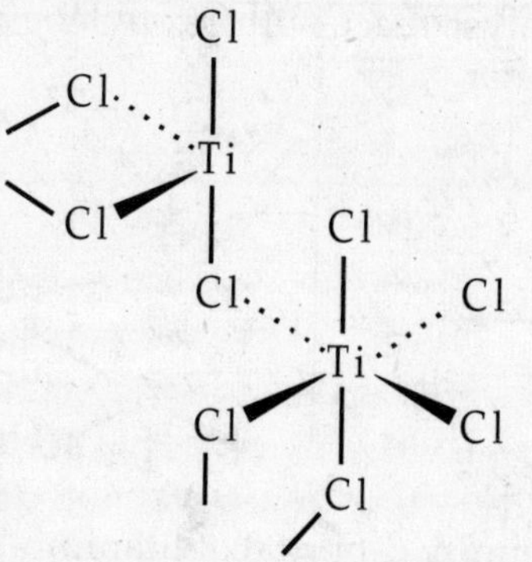

While the titanium on the interior has six chlorine neighbours, the surface titanium only has five.

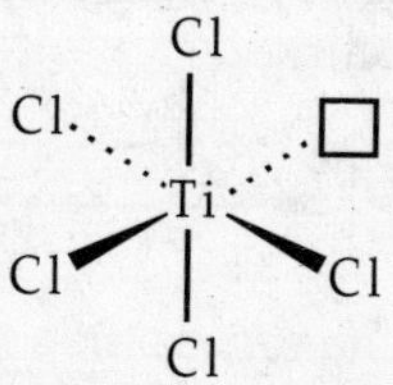

A surface titanium has an empty orbital.

$Al(C_2H_5)_2Cl$ donates one of its ethyl groups to the impoverished titanium, leaving behind one of the chlorines in the process. It still has an empty orbital

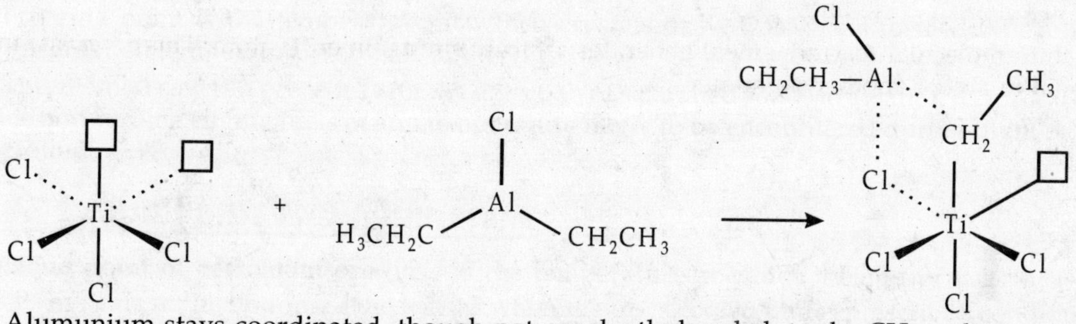

Alumunium stays coordinated, though not covalently bonded, to the CH_2 carbon atom of the ethyl group it just donated to the titanium. And it also coordinates itself to one of the chlorine atoms adjacent to the titanium. But titanium still has one empty orbital left to be filled. This is being filled by making it react with propylene monomer. There are two electrons in the p-system of a carbon-carbon double bond. Those electrons can be used to fill the empty orbital of the titanium. Propylene and the titanium form a complex

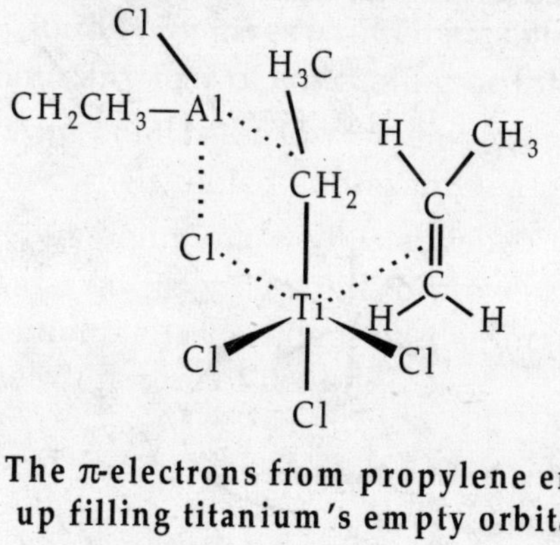

The π-electrons from propylene end up filling titanium's empty orbital

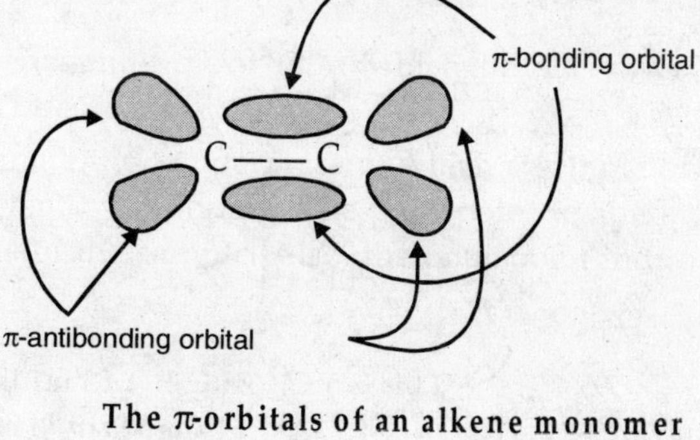

The π-orbitals of an alkene monomer

Isotactic Polymerization : Zeigler Natta catalyst react with propylene monomer and the electron reshuffling takes place in following manner

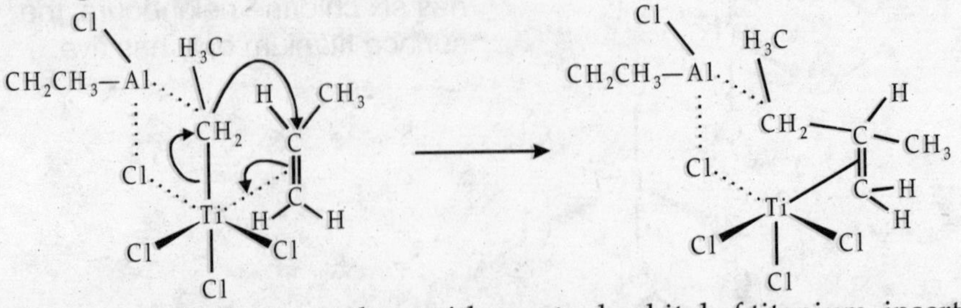

The π-electron cloud of alkene overlaps with empty d-orbital of titanium, insertion of alkene molecule between the Ti-C bonds occurs.

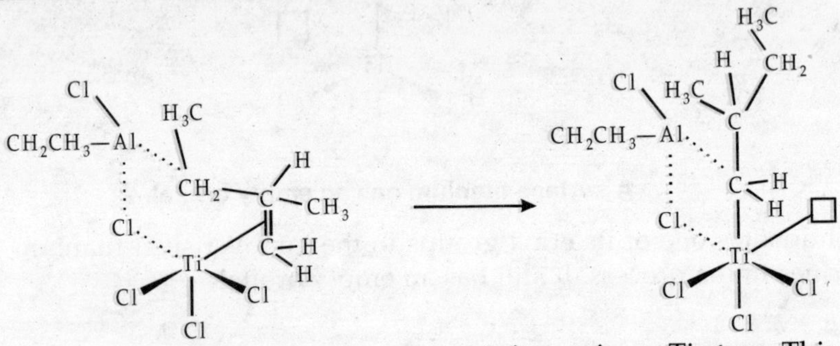

Intramolecular rearrangement generates a vacant site again on Ti atom. This gives an empty orbital to accept another monomer.

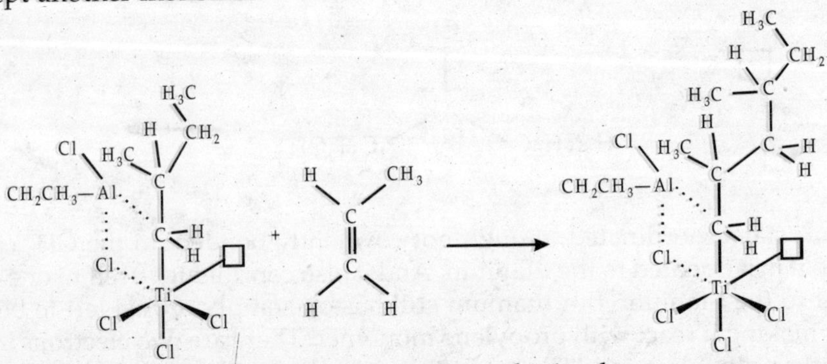

With this mechanism isotactic polypropylene are obtained.

Syndiotactic Polymerization : The catalyst used is based on vanadium rather than titanium. That system is $VCl_4/Al(C_2H_5)_2Cl$.

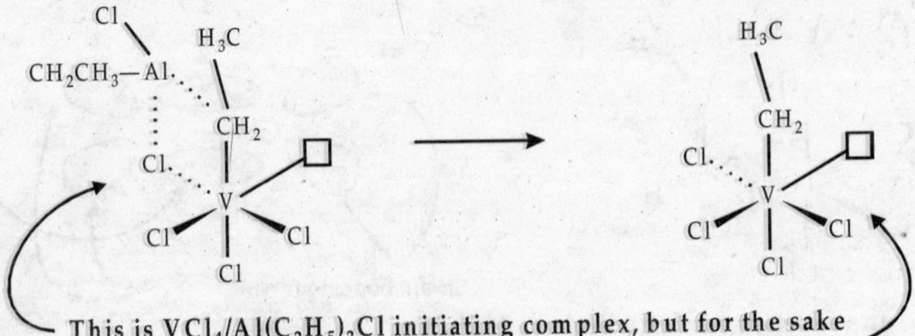

This is $VCl_4/Al(C_2H_5)_2Cl$ initiating complex, but for the sake of simplicity it is drawn like this.

It reacts in the same way as the titanium system does when a propylene molecule comes its way. First the propylene complexes with the vanadium, then the electrons shift just like before, and the propylene is inserted between the metal and the ethyl group.

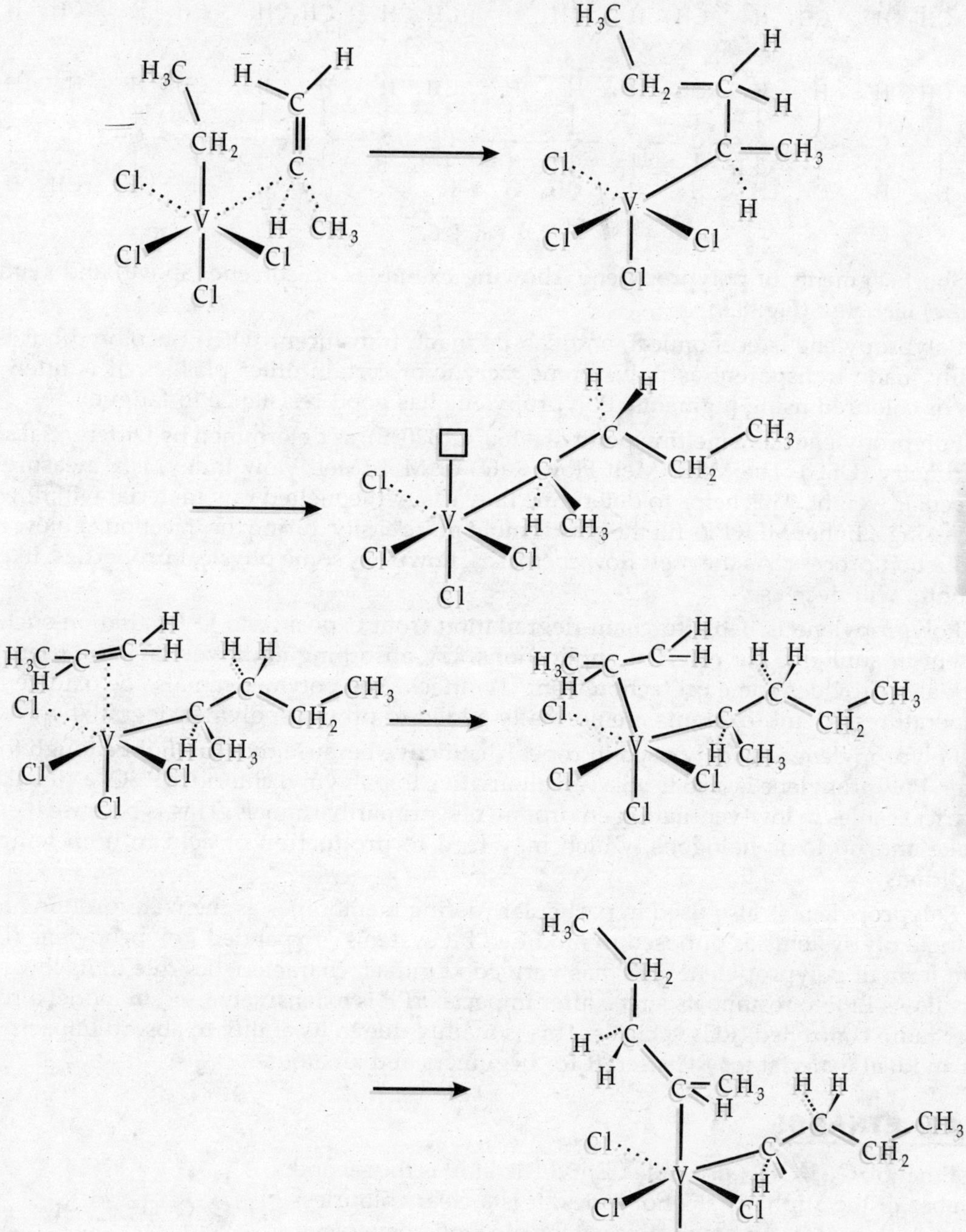

The second propylene adds to the chain, the chain changes position again and its back in the position where it started. Since the growing polymer chain switches positions with each propylene monomer added, the methyl groups end up on alternating sides of the chain, giving us a syndiotactic polymer.

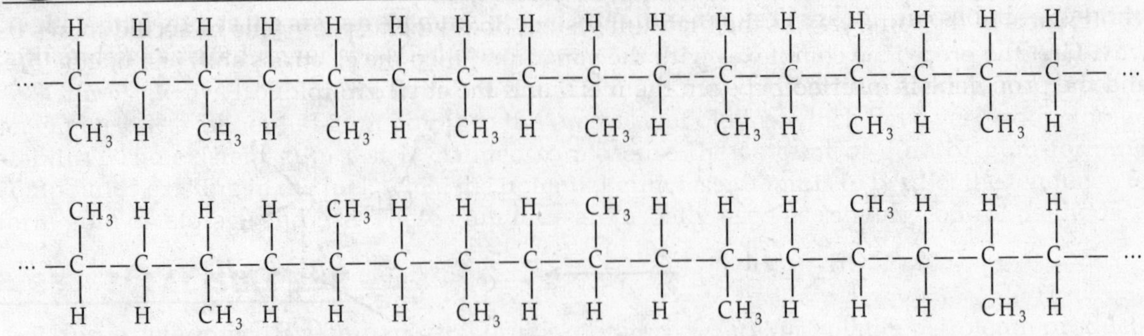

Fig. 8.6.

Short segments of polypropylene, showing examples of isotactic (above) and syndiotactic (below) tacticity. (Fig. 8.6)

Polypropylene is economical, and can be made translucent when uncolored but is not as readily made transparent as polystyrene, acrylic or certain other plastics. It is often opaque and/or coloured using pigments. Polypropylene has good resistance to fatigue.

Polypropylene has a melting point of ~160°C (320°F), as determined by Differential scanning calorimetry (DSC). The MFR (Melt Flow Rate) or MFI (Melt Flow Index) is a measure of PP's molecular weight. This helps to determine how easily the melted raw material will flow during processing. Higher MFR PPs fill the plastic mold more easily during the injection or blow molding production process. As the melt flow increases, however, some physical properties, like impact strength, will decrease.

Polypropylene is liable to chain degradation from exposure to UV radiation such as that present in sunlight. For external applications, UV-absorbing additives must be used. Carbon black also provides some protection from UV attack. The polymer can also be oxidized at high temperatures, so anti-oxidants are normally added to prevent polymer degradation.

Polypropylene is widely used in ropes, distinctive because they are light enough to float in water. Polypropylene is also used as an alternative to polyvinyl chloride (PVC) as insulation for electrical cables in low-ventilation environments, primarily tunnels. This is because it emits less smoke and no toxic halogens, which may lead to production of acid in high temperature conditions.

Polypropylene is also used in particular roofing membranes as the waterproofing top layer of single ply systems as opposed to modified bit systems . Expanded Polypropylene (EPP) is a foam form of polypropylene. EPP has very good impact characteristics due to its low stiffness, this allows EPP to resume its shape after impacts. EPP is extensively used in model aircraft and other radio controlled (RC) vehicles. This is mainly due to its ability to absorb impacts, making this an ideal material for RC aircraft for beginners and amateurs.

8.12. ETHANOL

Ethanol, C_2H_5OH, (also called Ethyl Alcohol) is the second member of the aliphatic alcohol series. It is a clear colourless liquid with a pleasant smell. Except for alcoholic beverages, nearly all the ethanol used industrially is a mixture of 95% ethanol and 5% water, known as 95% alcohol.

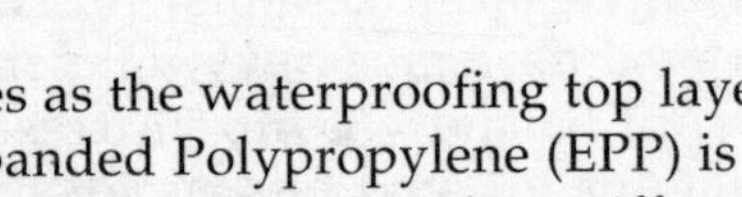

Ethanol

Ethanol is prepared as 95% alcohol (*i.e.* a 95% solution of ethanol in water) by distillation of the solution which results from the fermentation of sugars.

The industrial method for the production of ethanol are

Reaction of Ethene with Steam

$$\underset{\text{Ethene}}{C_2H_4} + \underset{\text{Steam}}{H_2O} \Longrightarrow \underset{\text{Ethanol}}{C_2H_5OH}$$

Alcoholic Fermentation

A solution of sucrose, to which yeast is added, is heated. An enzyme, invertase, which is present in yeast is added and this acts as a catalyst to convert the sucrose into glucose and fructose.

$$C_{12}H_{22}O_{11} \text{ (Sucrose)} + H_2O \Longrightarrow \underset{\text{Glucose}}{C_6H_{12}O_6} + \underset{\text{Fructose}}{C_6H_{12}O_6}$$

The glucose, $C_6H_{12}O_6$, and fructose, $C_6H_{12}O_6$, formed are then converted into ethanol and carbon dioxide by another enzyme, zymase, which is also present in yeast.

$$C_6H_{12}O_6 \xrightarrow{\text{zymase}} \underset{\text{Ethanol}}{2\ C_2H_5OH} + 2CO_2$$

The fermentation process takes three days and is carried out at a temperature between 250°C and 300°C. The ethanol is then obtained by fractional distillation.

Absolute Ethanol : Ethanol is initially obtained in admixture with water. The ethanol is then extracted from this solution by fractional distillation.

The boiling point of ethanol, (78.3°C) is significantly lower than the boiling point of water, (100°C) therefore these material cannot be separated completely by distillation. Instead, an azeotropic mixture (i.e. a mixture of 95% ethanol and 5% water) is obtained, and the boiling point of the azeotrope is 78.15°C. In a distillation, the most volatile material (i.e. the material that has the lowest boiling point) is the first material to distill from the distillation flask, and this material is the azeotrope of 95% ethanol which has the lowest boiling point. If an efficient fractionating column is used, there is obtained first 95% alcohol, then a small intermediate fraction of lower concentration, and then water. But no matter how efficient the fractionating column used, 95% alcohol cannot be further concentrated by distillation.

Ethanol is a colourless liquid with a pleasant smell. It is completely miscible with water and organic solvents and is very hydroscopic.

Melting Point	-1150C
Boiling Point	780C
Specific Gravity	0.79.

Combustion of Ethanol

Ethanol burns with a pale blue, non luminous flame to form carbon dioxide and steam.

$$\underset{\text{Ethanol}}{C_2H_5OH} + 3O_2 \Longrightarrow 2CO_2 + 3H_2O$$

Oxidation of Ethanol

Ethanol is oxidised to form ethanal, (*i.e.* acetaldehyde) by

- Acidified Potassium Dichromate, $K_2Cr_2O_7$, or

- Acidified Sodium Dichromate, $Na_2Cr_2O_7$, or
- Acidified potassium permanganate, $KMnO_4$,

$$C_2H_5OH \overset{[O]}{==>} CH_3CHO + H_2O$$

Ethanol → Ethanal

The ethanal is further oxidised to ethanoic acid (i.e. acetic acid) if the oxidising agent is in excess.

$$CH_3CHO \overset{[O]}{==>} CH_3COOH$$

Ethanal → Ethanoic Acid

The oxidising agent usually used for this reaction is a mixture of sodium dichromate or potassium dichromate and sulphuric acid which react together to provide oxygen atoms as follows.

$$Na_2Cr_2O_7 + 4\,H_2SO_4 ==> Na_2SO_4 + Cr_2(SO_4)_3 + 4H_2O + 3[O]$$

Dehydration of Ethanol

When ethanol is mixed with concentrated sulphuric acid with the acid in excess and heated to 170 °C, ethylene is formed. (One mole of ethanol loses one mole of water)

$$C_2H_5OH \overset{H_2SO_4}{==>} C_2H_4 + H_2O$$

When ethanol is mixed with concentrated sulphuric acid with the alcohol in excess and heated to 140°C, diethyl ether distils over (two moles of ethanol loses one mole of water) .

$$2\,C_2H_5OH \overset{H_2SO_4}{==>} C_2H_5OC_2H_5 + H_2O$$

Reaction of Ethanol with Sodium

Sodium reacts with ethanol at room temp to liberate hydrogen. The hydrogen atom of the hydroxyl group is replaced by a sodium atom, forming sodium ethoxide.

$$C_2H_5OH + Na ==> C_2H_5ONa + H_2$$

Apart from this reaction, ethanol and the other alcohols show no acidic properties.

Dehydrogenation of Ethanol

Ethanol can also be oxidised to ethanal (i.e. acetaldehyde) by passing its vapour over copper heated to 300 °C,. Two atoms of hydrogen are eliminated from each molecule to form hydrogen gas and hence this process is termed dehydrogenation.

$$C_2H_5OH ==> CH_3CHO + H_2$$

Ethanol → Ethanal

Esterification of Ethanol

Ethanol, C_2H_5OH, reacts with organic acids to form esters.

$$C_2H_5OH + CH_3COOH \overset{H(+)}{==>} CH_3COOC_2H_5 + H_2O$$

C_2H_5OH	+	CH_3COOH	==>	$CH_3COOC_2H_5$	+ H_2O
Ethanol Acid		Ethanoic Acetate		Ethyl	Water

Halogenation or Substitution of Ethanol with PCl_5

Ethanol reacts with phosphorus pentachloride at room temperature to form hydrogen chloride, ethyl chloride (i.e. chloroethane) and phosphoryl chloride.

$$C_2H_5OH + PCl_5 \Longrightarrow C_2H_5Cl + POCl_3 + HCl$$

Ethanol Phosphorus Ethyl Phosphorus Hydrogen

Pentachloride Chloride Pentachloride Chloride

Halogenation or Substitution of Ethanol with HCl

Ethanol reacts with hydrogen chloride to form ethyl chloride (i.e. chloroethane) and water. A dehydrating agent (e.g. zinc chloride) is used as a catalyst.

$$C_2H_5OH + HCl \xRightarrow{ZnCl_2} C_2H_5Cl + H_2O$$

Ethanol Ethyl

Chloride

Ethanol is used

- in the manufacture of alcoholic drinks, e.g. Vodka, etc.,
- as a widely used solvent for paint, varnish and drugs,
- in the manufacture of ethanal, (i.e. acetaldehyde), and ethanoic acid, (i.e. acetic acid),
- as a fuel (e.g. in Gasahol),
- as the fluid in thermometers, and
- in preserving biological specimens.

8.13. VINYL CHLORIDE

Vinyl chloride is the organic compound with the formula $CH_2{:}CHCl$. This colourless compound is an important industrial chemical chiefly used to produce the polymer polyvinyl chloride (PVC). At room temperature, vinyl chloride is a gas with a sickly sweet odor that is easily condensed. It is highly toxic.

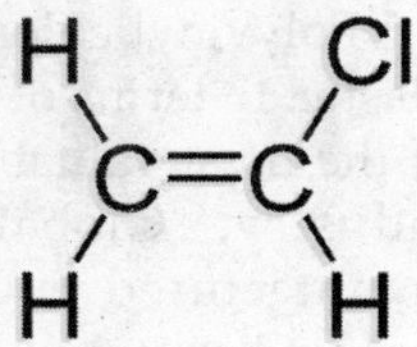

Vinyl chloride is produced by hydrochlorination of acetylene and the dehydrochlorination of ethylene dichloride (1,2-dichloroethane).

The production of vinyl chloride from ethylene dichloride (EDC) consists of a series of well-defined steps. Ethylene dichloride (EDC) is prepared by reacting ethylene and chlorine. In the presence of iron(III) chloride as a catalyst, these compounds react exothermically:

$$CH_2{=}CH_2 + Cl_2 \longrightarrow ClCH_2CH_2Cl$$

When heated to 500 °C at 15–30 atm (1.5 to 3 MPa) pressure, EDC decomposes to produce vinyl chloride and HCl.

$$ClCH_2CH_2Cl \longrightarrow CH_2{=}CHCl + HCl$$

A propylene refrigerant can be used to chill the outlet stream prior to a series of distillation towers. The last distillation tower has pure HCl going from the top and product vinyl chloride coming out of the bottom. The recycled HCl is used to produce more EDC. The recycling process involves a copper(II) chloride-catalyzed oxychlorination of ethylene. Oxychlorination entails the combined action of oxygen and hydrogen chloride to produce chlorine in situ:

$$CH_2{=}CH_2 + 2\,HCl + \tfrac{1}{2}\,O_2 \longrightarrow ClCH_2CH_2Cl + H_2O$$

Due to the economical advantages of this recycling as well as the low cost of ethylene, most vinyl chloride has been produced by this technique. By-products of the oxychlorination reaction, are recovered. One useful byproduct of the oxychlorination is ethyl chloride, a topical anesthetic

Acetylene, produced by the hydrolysis of calcium carbide, is treated with hydrogen chloride to give vinyl chloride:

$$C_2H_2 + HCl \longrightarrow CH_2{=}CHCl$$

The method is not widely practiced owing to the cost of the acetylene and the associated environmental impact of its production.

Vinyl chloride is a chemical intermediate, not a final product. Due to the hazardous nature of vinyl chloride to human health there are no end products that use vinyl chloride in its monomer form. Polyvinyl chloride is very stable, storable, and less acutely hazardous than the monomer.Vinyl chloride was briefly used as an inhalational anaesthetic, in a similar way to ethyl chloride, though its toxicity forced this practice to be abandoned.

Dioxins is produced as a byproduct of vinyl chloride manufacture and from incineration of waste PVC in domestic garbage. Dioxins are a global health threat because they persist in the environment and can travel long distances. Dioxins results in immune system suppression, reproductive disorders, a variety of cancers, and endometriosis. Oxychlorination reactions [used to make vinyl chloride and some chlorinated solvents] generate polychlorinated dibenzodioxins (PCDDs) and dibenzofurans (PCDFs). The reactions include all of the ingredients and conditions necessary to form PCDD/PCDFs.... which are harmful to human health. In other words, dioxins are an undesirable byproduct of polymerizing PVC and eliminating the production of dioxins while maintaining the polymerization reaction may be difficult. Dioxins created by vinyl chloride production are released by on-site incinerators, flares, boilers, wastewater treatment systems and even in trace quantities in vinyl resins.

8.14. POLYVINYL CHLORIDE

Polyvinyl chloride (PVC) is a widely-used plastic. In terms of revenue generated, it is one of the most valuable products of the chemical industry. Globally, over 50% of PVC manufactured is used in construction. As a building material PVC is cheap, and easy to assemble. In recent years, PVC has been replacing traditional building materials such as wood, concrete and clay in many areas. Despite appearing to be an ideal building material, PVC has high environmental and human health costs.

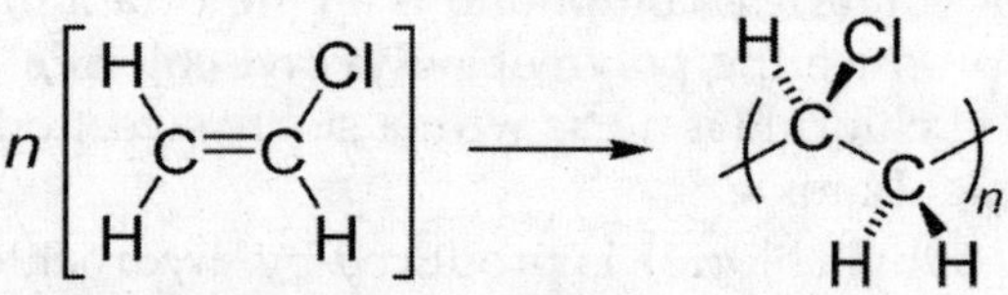

It is obtained by heating a water –emulsion of vinyl chloride in presence of small amount hydrogen peroxide under pressure.

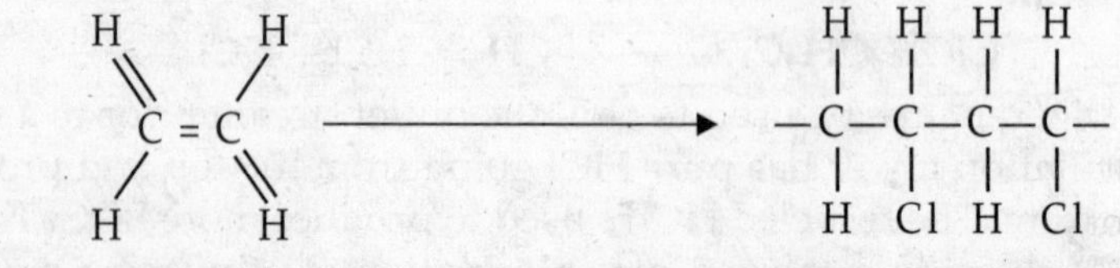

Vinyl Chloride Monomer   Polyvinyl Chloride Polymer

Polyvinyl chloride is produced by polymerization of the vinyl chloride monomer (VCM). Since about 57% of its mass is chlorine, creating a given mass of PVC requires less petroleum than many other polymers. By far the most widely used production process is suspension polymerization. In this process, VCM and water are introduced into the polymerization reactor and a polymerization initiator, along with other chemical additives, are added to initiate the polymerization reaction. The contents of the reaction vessel are continually mixed to maintain the suspension and ensure a uniform particle size of the PVC resin. The reaction is exothermic, and thus requires a cooling mechanism to maintain the reactor contents at the appropriate temperature. As the volumes also contract during the reaction (PVC is denser than VCM), water is continually added to the mixture to maintain the suspension.

Once the reaction has run its course, the resulting PVC slurry is degassed and stripped to remove excess VCM (which is recycled into the next batch) then passed though a centrifuge to remove most of the excess water. The slurry is then dried further in a hot air bed and the resulting powder is sieved before storage or pelletization. In normal operations, the resulting PVC has a VCM content of less than 1 part per million.

Other production processes, such as micro-suspension polymerization and emulsion polymerization, produce PVC with smaller particle sizes (10ìm vs 120-150ìm for suspension PVC) with slightly different properties and with somewhat different sets of applications.

The product of the polymerization process is unmodified PVC. Before PVC can be made into finished products, it almost always requires conversion into a compound by the incorporation of additives such as heat stabilizers, UV stabilizers, lubricants, plasticizers, processing aids, impact modifiers, thermal modifiers, fillers, flame retardants, biocides, blowing agents and smoke suppressors, and pigments.

Poly vinyl chlorides are

- Odourless
- Inflammable
- Chemically inert
- Resistant to light,atmospheric oxygen
- It has greater stiffness compared to polyethylene, but is brittle

The properties of PVC are listed below in Table 8.2

TABLE 8.2. The Properties of PVC.

Polyvinyl chloride	
Density	1390 kg/m^3[1]
Young's modulus (E)	2900-3300 MPa
Tensile strength (π_t)	50-80 MPa
Elongation at break	20-40%
Notch test	2-5 kJ/m^2
Glass temperature	82 °C[1]
Melting point	100–260 °C[1]
Vicat B	85 °C[2]
Heat transfer coefficient (λ)	0.16 W/(m·K)
Effective heat of combustion	17.95 MJ/kg
Linear expansion coefficient (α)	8 10^{-5}/K
Specific heat (c)	0.9 kJ/(kg·K)
Water absorption (ASTM)	0.04-0.4

PVC has side chains incorporating chlorine atoms, which form strong bonds. PVC in its normal form is stiff, strong, heat and weather resistant, and is now used for making plumbing, gutters, house siding, enclosures for computers and other electronics gear, and compact-disk media. PVC can also be softened with chemical processing, and in this form it is now used for food packaging, and raingear. Most vinyl products are believed to be generally harmless when used properly. However, some of the additives and softeners leach out of certain vinyl products. Even though soft PVC toys have been made for babies for years, studies find that these additives leach out of soft toys into the mouths of the children chewing on them. In Europe, phthalate additives in PVC toys for children under the age of three have been banned and in the USA, most companies have voluntarily stopped manufacturing PVC toys for this age group or have eliminated the phthalates.

The other uses include

Rigid PVC or unplasticized PVC : It is used for making

- Sheets
- Light fittings
- Safety helmets
- Tyres
- Cycle and motorcycle mudguards

Plasticized PVC is used for

- Packing rain coats, tablecloths, curtains etc.
- Injection moulding of toys, tool handles, radio components etc.
- Making chemical containers, conveyor belts etc.

REVIEW QUESTIONS

1. What are primary petrochemicals ?
2. Discuss briefly about the various intermediates and derivatives of petrochemicals and the major end product obtained.
3. What is cumene process for preparing acetone ? Give some of the uses of acetone.
4. Give the following chemical reactions of acetylene.

   (*i*) $CH{\equiv}CH + ROH \longrightarrow$

   (*ii*) $CH{\equiv}CH + HCl \longrightarrow$

   (*iii*) $CH{\equiv}CH + RCHO \longrightarrow$

   (*iv*) $CH{\equiv}CH \xrightarrow{\text{Cyclization}}$
5. Give the chemical reaction and uses of ethylene.
6. Explain the mechanism to obtain sydicotactic and Isotactic polypropylene.
7. What are HDPE, PEX, MDPE, LLDPE, VLDPE and LDPE ?
8. Explain the industrial method for preparation of ethanol.
9. Give the chemical properties of ethanol.
10. Give the preparation, properties and uses of PVC.

❑❑❑

BIBLIOGRAPHY

- G.D. Hobson Editor, *Modern Petroleum Technology* (5th ed. ed.), Wiley, Chichester (1984) Part II.
- G.N.Sarkar, Advanced Petrochemicals, First Edition, Khanna Publishers, 2008
- Ram Prasad, Petroleum Refining Technology, First Edition, Khanna Publishers, 2008
- G. Pipenger, Making 'premium' diesel fuel. *Hydrocarbon Processing* 76 2 (1997), pp. 63–78.
- M.E. LePera, Fuel quality vs engine types. *Hydrocarbon Processing* 61 1 (1982), pp. 139–142.
- SEA News Circular. The Solvent Extractors Association of India, Vol. X, 1996.
- Anjana Srivastava and Ram Prasad, Triglycerides-based diesel fuel., *Renewable and Sustainable Energy Review,*Vol.4, 111-133, 2000
- K.S. Marckley. *Fatty Acids* (2nd ed. ed.), Interscience, New York (1960).
- Goering CE, Schwab AW, Daugherty MJ, Pryde EH, Heakin AJ, Fuel properties of eleven vegetable oils, Trans. ASAE 1982;1472–1483.
- J.P. Kapur, S.D. Bhasin and K.C. Mathur, Oilseeds for food and industrial applications. *Chemical Age of India* 33 9 (1982), pp. 475–482.
- Y. Ali, M.A. Hanna and S.L. Cuppett, Fuel properties of tallow and soybean oil esters. *J. Am. Oil Chem. Soc.* 72 12 (1995), pp. 1557–15649. P.S. Rao and K.V. Gopalakrishnan, Vegetable oils and their methylesters as fuels for diesel engines. *Indian Journal of Technology* 29 6 (1991), pp. 292–297.
- Ziejewski M, Kaufman KR, Pratt GL. Vegetable oil as diesel fuel. Seminar II, Northern Regional Research Center, Peoria, Illinois, 19–20 October, 1983.
- A.W. Schwab, G.J. Dykstra, E. Selke, S.C. Sorenson and E.H. Pryde, Diesel fuel from thermal decomposition of soybean oil. *J. Am. Oil Chem. Soc.* 65 11 (1988), pp. 1781–1785.
- Bagby MO. Vegetable oils for diesel fuel: opportunities for development. International Winter Meeting of the ASAE, Hyatt Regency Chicago, 15–18 December, 1987.
- A.W. Schwab, M.O. Bagby and B. Freedman, Preparation and properties of diesel fuels from vegetable oils. *Fuel* 66 (1987), pp. 1372–1378.
- J. Otera, Transesterification. *Chem. Rev.* 93 4 (1993), pp. 1449–1470.
- T. Krawczyk, Biodiesel-alternative fuel makes inroads but hurdles remain. *INFORM* 7 8 (1996), pp. 800–815.
- B. Freedman, R.O. Butterfield and E.H. Pryde, Transesterification kinetics of soybean oil. *J. Am. Oil Chem. Soc.* 63 10 (1986), pp. 1375–1380.
- H. Noureddini and D. Zhu, Kinetics of transesterification of soybean oil. *J. Am. Oil Chem. Soc.* 74 11 (1997), pp. 1457–1463.
- R.O. Feuge and A.T. Gros, Modification of vegetable oils. VII. Alkali catalyzed interesterification of peanut oil with ethanol. *J. Am. Oil Chem. Soc.* 26 3 (1949), pp. 97–102.
- R.O. Dunn and M.O. Bagby, Low-temperature properties of triglyceride-based diesel fuels: transesterified methyl esters and petroleum middle distillate/ester blends. *J. Am. Oil Chem. Soc.* 72 8 (1995), pp. 895–904.
- D.Y.Z. Chang, J.H. Van Gerpen, I. Lee, L.A. Johnson, E.G. Hammond and S.J. Marley, Fuel properties and emissions of soybean oil esters as diesel fuel. *J. Am. Oil Chem. Soc.* 73 11 (1996), pp. 1549–1555.

- Hui YH, editors. Bailey's Industrial Oils Fats: Industrial and Consumer Non Edible Products From Oils and Fats, 5th ed. Wiley-Interscience New York, 1996.
- B. Freedman, E.H. Pryde and T.L. Mounts, Variables affecting the yields of fatty esters from transesterified vegetable oils. *J. Am. Oil Chem. Soc.* 61 10 (1984), pp. 1638–1643.
- R. Fillieres, B. Benjelloun-Mlayah and M. Delmas, Ethanolysis of rapeseed oil: quantitation of ethyl esters, mono-, di-, and triglycerides and glycerol by high performance size-exclusion chromatography. *J. Am. Oil Chem. Soc.* 72 4 (1995), pp. 427–432.
- K. Krisnangkura and R. Simamaharnnop, Continuous transmethylation of palm oil in an organic solvent. *J. Am. Oil Chem. Soc.* 69 2 (1992), pp. 166–169.
- M.W. Formo, Ester reactions of fatty materials. *J. Am. Oil Chem. Soc.* 31 11 (1954), pp. 548–559.
- M. Mittelbach, Lipase catalysed alcoholysis of sunflower oil. *J. Am. Oil Chem. Soc.* 67 3 (1990), pp. 168–170.
- R. Varese and M. Varese, Methyl ester biodiesel: opportunity or necessity. *INFORM* 7 8 (1996), pp. 816–824.
- S.J. Clark, L. Wagner, M.D. Schrock and P.G. Piennaar, Methyl and ethyl soybean esters as renewable fuels for diesel engines. *J. Am. Oil Chem. Soc.* 61 10 (1984), pp. 1632–1638.
- Lee, L.A. Johnson and E.G. Hammond, Use of branched-chain esters to reduce the crystallization temperature of biodiesel. *J. Am. Oil Chem. Soc.* 72 10 (1995), pp. 1155–1160.
- Lee, L.A. Johnson and E.G. Hammond, Reducing the crystallization temperature of biodiesel by winterizing methyl soyate. *J. Am. Oil Chem. Soc.* 73 5 (1996), pp. 631–636
- R.O. Dunn, M.W. Shockley and M.O. Bagby, Improving the low temperature properties of alternative diesel fuels: vegetable oil-derived methyl esters. *J. Am. Oil Chem. Soc.* 73 10 (1996), pp. 1719–1728.
- P. Bondioli, A. Gasparoli, A. Lanzani, E. Fedeli, S. Veronese and M. Sala, Storage stability of biodiesel. *J. Am. Oil Chem. Soc.* 72 6 (1995), pp. 699–702.
- G. Ustun, Separation of fatty acid methyl esters from tall oil by selective adsorption. *J. Am. Oil Chem. Soc.* 73 2 (1996), pp. 203–210.
- T. Murayama, Evaluating vegetable oils as a diesel fuel. *INFORM* 5 10 (1994), pp. 1138–1145.
- N.C. Kothari, H.K.Mulchandani, C.S.Sharma and J.Prakash, Dearomatising of naptha-Technoeconomic aspects, CEW, Vol.XXIII, No.3, pp 56-60 (1988)
- Sunita Rattan, Text Book on Engineering Chemistry, 5th Edition , Kataria Publisher (2009)
- K.D. Pathak and BC Subba Rao studies in synthetic Ester type lubricants, Part-I-Synthetic Lubricants from oliphatic dibasic acids, Indian Journal of Technology vol. 1, No. 1 pp 30-35 (1963).
- C.V. Agarwal, V.K. Verma and D.S. Sarma, E. Valuation of Certain Organic Additives as extreme pressure lubricants, Indian Journal Technology, Vol. 14, No. 12 pp. 647-650 (1978).
- A.K. Biswas, Role of cutting fluids in metal cutting operations, CEW XX No. 1 pp. 73-75 (1985).
- S. Shelley, Gearning up for synthetic lubricants, Chem. Engg. pp. 67-69 (July 1993).
- C.C. Bucher G.D. Growes, Petroleum Waxex, Industrial and Engg. Chemistry, Vol. 19 pp. 718 (1927).
- R.E. Dymond, World Markets for petroleum coke, hydrocarbon processing, vol. 70, No. 9, p.162 (1991).
- B.G.S. Marzana, Electric desalting in oil refineries, The Institute of Petroleum, April-June, pp. 113-117 (1989).
- Anjana Srivastava "Thesis–Studies on Triglyceride based diesel fuel" (2000).

INDEX

A

B

C

D

E

F

G

H

I

K

L

M

N

O

P

Q

R

❑❑❑

Win Prizes

ATTENTION READERS

Win Prizes

We request for your frank assessment regarding various aspects of the book as given below :

PETROCHEMICALS

By : Dr. ANJANA SRIVASTAVA

Please Fill up the Sheet in Neat and Clean Writing

(i) What topics of your syllabus, important for your examination point of view, are not covered in this book ?

....

....

....

....

(ii) What are the chapters wherein the treatment of the subject matter is not systematic or organised or updated ?

....

....

....

....

(iii) Have you come across misprints/mistakes/factual inaccuracies in the book ? Please specify the chapters and the page numbers.

....

....

....

....

(iv) What other books on the same subject have you found/heard better than the present book ? Please specify in terms of price and quality.

....

....

....

....

(v) Further suggestions and comments for the improvement of this book.

....

....

....

....

(PS : If need be please attach a separate sheet to write your views)

OTHER DETAILS

(i) Who recommended you this book ? (Please tick (✓) in the box near the option relevant to you.

☐ Teacher ☐ Friends ☐ Bookseller

....

....

....

....

(ii) Name of the recommending teacher, his designation and address:

....

....

....

....

(iii) Name and address of the bookseller you purchased the book from :

....

....

....

....

(iv) Name and address of your institution (Please mention the University or Board as the case may be)

....

....

....

....

(v) Your name and complete address :

....

....

....

....

(vi) Course you are enrolled in :

....

....

....

....

The best assessment will be awarded each month : The award will be in the form of our publications as decided by the Editorial Board.

Please mail the filled up coupon at your earliest to :

S.K. KATARIA & SONS®
4760-61/23, Ansari Road,
Daryaganj, New Delhi-110002
☎ 23243489, 23269324, 22733315,
Fax : 91-11-23243489
email : katariabooks@yahoo.com